To Anne and Frances

who had to put up with it all

U0795687

This is volume 140 in the PURE AND APPLIED MATHEMATICS series
Founding Editors: Paul A. Smith and Samuel Eilenberg

SOBOLEV SPACES

Second Edition

Robert A. Adams and John J. F. Fournier

Department of Mathematics
The University of British Columbia
Vancouver, Canada

ACADEMIC PRESS
An imprint of Elsevier Science

Amsterdam Boston Heidelberg London New York Oxford
Paris San Diego San Francisco Singapore Sydney Tokyo

SOBOLEV SPACES, second edition
Robert A. Adams and John J. F. Fournier
ISBN: 9780120441433
Copyright © 2003 Elsevier Ltd. All rights reserved.
Authorized Chinese translation published by Beijing World Publishing Corporation.

《索伯列夫空间》（第 2 版）
ISBN: 9787523218600
Copyright ©Elsevier Ltd. and Beijing World Publishing Corporation. All rights reserved.
No part of this publication may be reproduced or transmitted in any form or by any means, electronic or mechanical, including photocopying, recording, or any information storage and retrieval system, without permission in writing from Elsevier (Singapore) Pte Ltd. Details on how to seek permission, further information about the Elsevier' s permissions policies and arrangements with organizations such as the Copyright Clearance Center and the Copyright Licensing Agency, can be found at our website: www.elsevier.com/permissions.
This book and the individual contributions contained in it are protected under copyright by Elsevier Ltd. and Beijing World Publishing Corporation (other than as may be noted herein).
Online resources are not available with this reprint.

This edition of *SOBOLEV SPACES* is published by Beijing World Publishing Corporation under arrangement with ELSEVIER LTD.
This edition is authorized for sale in China only, excluding Hong Kong, Macau and Taiwan. Unauthorized export of this edition is a violation of the Copyright Act. Violation of this Law is subject to Civil and Criminal Penalties.
本版由 ELSEVIER LTD. 授权世界图书出版有限公司北京分公司在中国大陆地区（不包括香港、澳门以及台湾地区）出版发行。
本版仅限在中国大陆地区（不包括香港、澳门以及台湾地区）出版及标价销售。未经许可之出口，视为违反著作权法，将受民事及刑事法律之制裁。
本书封底贴有 Elsevier 防伪标签，无标签者不得销售。

Notice

Knowledge and best practice in this field are constantly changing. As new research and experience broaden our understanding, changes in research methods, professional practices, or medical treatment may become necessary. Practitioners and researchers must always rely on their own experience and knowledge in evaluating and using any information, methods, compounds or experiments described herein. Because of rapid advances in the medical sciences, in particular, independent verification of diagnoses and drug dosages should be made. To the fullest extent of the law, no responsibility is assumed by Elsevier, authors, editors or contributors in relation to the adaptation or for any injury and/or damage to persons or property as a matter of products liability, negligence or otherwise, or from any use or operation of any methods, products, instructions, or ideas contained in the material herein.

CONTENTS

PREFACE

This monograph presents an introductory study of of the properties of certain Banach spaces of weakly differentiable functions of several real variables that arise in connection with numerous problems in the theory of partial differential equations, approximation theory, and many other areas of pure and applied mathematics. These spaces have become associated with the name of the late Russian mathematician S. L. Sobolev, although their origins predate his major contributions to their development in the late 1930s.

Even by 1975 when the first edition of this monograph was published, there was a great deal of material on these spaces and their close relatives, though most of it was available only in research papers published in a wide variety of journals. The monograph was written to fill a perceived need for a single source where graduate students and researchers in a wide variety of disciplines could learn the essential features of Sobolev spaces that they needed for their particular applications. No attempt was made even at that time for complete coverage. To quote from the Preface of the first edition:

> The existing mathematical literature on Sobolev spaces and their generalizations is vast, and it would be neither easy nor particularly desirable to include everything that was known about such spaces between the covers of one book. An attempt has been made in this monograph to present all the core material in sufficient generality to cover most applications, to give the reader an overview of the subject that is difficult to obtain by reading research papers, and finally . . . to provide a ready reference for someone requiring a result about Sobolev spaces for use in some application.

This remains as the purpose and focus of this second edition. During the intervening twenty-seven years the research literature has grown exponentially, and there are now several other books in English that deal in whole or in part with Sobolev spaces. (For example, see [Ad2], [Bu1], [Mz1], [Tr1], [Tr3], and [Tr4].) However, there is still a need for students in other disciplines than mathematics, and in other areas of mathematics than just analysis to have available a book that describes these spaces and their core properties based only a background in mathematical analysis at the senior undergraduate level. We have tried to make this such a book.

The organization of this book is similar but not identical to that of the first edition:

Chapter 1 remains a potpourri of standard topics from real and functional analysis, included, mainly without proofs, because they provide a necessary background for what follows.

Chapter 2 on the Lebesgue Spaces $L^p(\Omega)$ is much expanded and reworked from the previous edition. It provides, in addition to standard results about these spaces, a brief treatment of mixed-norm L^p spaces, weak-L^p spaces, and the Marcinkiewicz interpolation theorem, all of which will be used in a new treatment of the Sobolev Imbedding Theorem in Chapter 4. For the most part, complete proofs are given, as they are for much of the rest of the book.

Chapter 3 provides the basic definitions and properties of the Sobolev spaces $W^{m,p}(\Omega)$ and $W_0^{m,p}(\Omega)$. There are minor changes from the first edition.

Chapter 4 is now completely concerned with the imbedding properties of Sobolev Spaces. The first half gives a more streamlined presentation and proof of the various imbeddings of Sobolev spaces into L^p spaces, including traces on subspaces of lower dimension, and spaces of continuous and uniformly continuous functions. Because the approach to the Sobolev Imbedding Theorem has changed, the roles of Chapters 4 and 5 have switched from the first edition. The latter part of Chapter 4 deals with situations where the regularity conditions on the domain Ω that are necessary for the full Sobolev Imbedding Theorem do not apply, but some weaker imbedding results are still possible.

Chapter 5 now deals with interpolation, extension, and approximation results for Sobolev spaces. Part of it is expanded from material in Chapter 4 of the first edition with newer results and methods of proof.

Chapter 6 deals with establishing compactness of Sobolev imbeddings. It is only slightly changed from the first edition.

Chapter 7 is concerned with defining and developing properties of scales of spaces with fractional orders of smoothness, rather than the integer orders of the Sobolev spaces themselves. It is completely rewritten and bears little resemblance to the corresponding chapter in the first edition. Much emphasis is placed on real interpolation methods. The J-method and K-method are fully presented and used to develop the theory of Lorentz spaces and Besov spaces and their imbeddings, but both families of spaces are also provided with intrinsic characterizations. A key theorem identifies lower dimensional traces of functions in Sobolev spaces as constituting certain Besov spaces. Complex interpolation is used to introduce Sobolev spaces of fractional order (also called spaces of Bessel potentials) and Fourier transform methods are used to characterize and generalize these spaces to yield the Triebel Lizorkin spaces and illuminate their relationship with the Besov spaces.

Chapter 8 is very similar to its first edition counterpart. It deals with Orlicz and Orlicz-Sobolev spaces which generalize L^p and $W^{m,p}$ spaces by allowing the role of the function t^p to be assumed by a more general convex function $A(t)$. An important result identifies a certain Orlicz space as a target for an imbedding of $W^{m,p}(\Omega)$ in a limiting case where there is an imbedding into $L^p(\Omega)$ for $1 \le p < \infty$ but not into $L^\infty(\Omega)$.

This monograph was typeset by the authors using TeX on a PC running Linux-Mandrake 8.2. The figures were generated using the mathematical graphics software package *MG* developed by R. B. Israel and R. A. Adams.

RAA & JJFF
Vancouver, August 2002

List of Spaces and Norms

Space	Norm	Paragraph
$B^{s;p,q}(\Omega)$	$\|\cdot\,; B^{s;p,q}(\Omega)\|$	7.32
$B^{s;p,q}(\mathbb{R}^n)$	$\|\cdot\,; B^{s;p,q}(\mathbb{R}^n)\|$	7.67
$\dot{B}^{s;p,q}(\mathbb{R}^n)$		7.68
$C^m(\Omega),\ C^\infty(\Omega)$		1.26
$C_0(\Omega),\ C_0^\infty(\Omega)$		1.26
$C^m(\overline{\Omega})$	$\|\cdot\,; C^m(\overline{\Omega})\|$	1.28
$C^{m,\lambda}(\overline{\Omega})$	$\|\cdot\,; C^{m,\lambda}(\overline{\Omega})\|$	1.29
$C_B^m(\Omega)$	$\|\cdot\,; C_B^m(\Omega)\|$	1.27, 4.2
$C^j(\overline{\Omega})$	$\|\cdot\,; C^j(\overline{\Omega})\|$	4.2
$C^{j,\lambda}(\overline{\Omega})$	$\|\cdot\,; C^{j,\lambda}(\overline{\Omega})\|$	4.2
$C^{j,\lambda,q}(\overline{\Omega})$	$\|\cdot\,; C^{j,\lambda,q}(\overline{\Omega})\|$	7.35
$\mathscr{D}(\Omega)$		1.56
$\mathscr{D}'(\Omega)$		1.57
$E_A(\Omega)$	$\|\cdot\|_A = \|\cdot\|_{A,\Omega}$	8.14

1

PRELIMINARIES

1.1 (Introduction) Sobolev spaces are vector spaces whose elements are functions defined on domains in n-dimensional Euclidean space $\mathbb{R}^n$ and whose partial derivatives satisfy certain integrability conditions. In order to develop and elucidate the properties of these spaces and mappings between them we require some of the machinery of general topology and real and functional analysis. We assume that readers are familiar with the concept of a vector space over the real or complex scalar field, and with the related notions of dimension, subspace, linear transformation, and convex set. We also expect the reader will have some familiarity with the concept of topology on a set, at least to the extent of understanding the concepts of an open set and continuity of a function.

In this chapter we outline, mainly without any proofs, those aspects of the theories of topological vector spaces, continuity, the Lebesgue measure and integral, and Schwartz distributions that will be needed in the rest of the book. For a reader familiar with the basics of these subjects, a superficial reading to settle notations and review the main results will likely suffice.

Notation

1.2 Throughout this monograph the term *domain* and the symbol Ω will be reserved for a nonempty open set in n-dimensional real Euclidean space $\mathbb{R}^n$. We shall be concerned with the differentiability and integrability of functions defined on Ω; these functions are allowed to be complex-valued unless the contrary is explicitly stated. The complex field is denoted by $\mathbb{C}$. For $c \in \mathbb{C}$ and two functions u and v, the scalar multiple cu, the sum $u + v$, and the product uv are always defined pointwise:

$$(cu)(x) = cu(x),$$
$$(u + v)(x) = u(x) + v(x),$$
$$(uv)(x) = u(x)v(x)$$

at all points x where the right sides make sense.

A typical point in $\mathbb{R}^n$ is denoted by $x = (x_1, \ldots, x_n)$; its norm is given by $|x| = (\sum_{j=1}^n x_j^2)^{1/2}$. The inner product of two points x and y in $\mathbb{R}^n$ is $x \cdot y = \sum_{j=1}^n x_j y_j$.

If $\alpha = (\alpha_1, \ldots, \alpha_n)$ is an n-tuple of nonnegative integers α_j, we call α a *multi-index* and denote by x^α the monomial $x_1^{\alpha_1} \cdots x_n^{\alpha_n}$, which has degree $|\alpha| = \sum_{j=1}^n \alpha_j$. Similarly, if $D_j = \partial/\partial x_j$, then

$$D^\alpha = D_1^{\alpha_1} \cdots D_n^{\alpha_n}$$

denotes a differential operator of order $|\alpha|$. Note that $D^{(0,\ldots,0)}u = u$.

If α and β are two multi-indices, we say that $\beta \le \alpha$ provided $\beta_j \le \alpha_j$ for $1 \le j \le n$. In this case $\alpha - \beta$ is also a multi-index, and $|\alpha - \beta| + |\beta| = |\alpha|$. We also denote

$$\alpha! = \alpha_1! \cdots \alpha_n!$$

and if $\beta \le \alpha$,

$$\binom{\alpha}{\beta} = \frac{\alpha!}{\beta!(\alpha - \beta)!} = \binom{\alpha_1}{\beta_1} \cdots \binom{\alpha_n}{\beta_n}.$$

The reader may wish to verify the Leibniz formula

$$D^\alpha(uv)(x) = \sum_{\beta \le \alpha} \binom{\alpha}{\beta} D^\beta u(x) D^{\alpha-\beta} v(x)$$

valid for functions u and v that are $|\alpha|$ times continuously differentiable near x.

1.3 If $G \subset \mathbb{R}^n$ is nonempty, we denote by $\overline{G}$ the closure of G in $\mathbb{R}^n$. We shall write $G \Subset \Omega$ if $\overline{G} \subset \Omega$ and $\overline{G}$ is a compact (that is, closed and bounded) subset of $\mathbb{R}^n$. If u is a function defined on G, we define the *support* of u to be the set

$$\operatorname{supp}(u) = \overline{\{x \in G : u(x) \ne 0\}}.$$

We say that u has *compact support* in Ω if $\operatorname{supp}(u) \Subset \Omega$. We denote by "bdry G" the boundary of G in $\mathbb{R}^n$, that is, the set $\overline{G} \cap \overline{G^c}$, where G^c is the complement of G in $\mathbb{R}^n$; $G^c = \mathbb{R}^n - G = \{x \in \mathbb{R}^n : x \notin G\}$.

If $x \in \mathbb{R}^n$ and $G \subset \mathbb{R}^n$, we denote by "dist(x, G)" the distance from x to G, that is, the number $\inf_{y \in G} |x - y|$. Similarly, if $F,\ G \subset \mathbb{R}^n$ are both nonempty,

$$\operatorname{dist}(F, G) = \inf_{y \in F} \operatorname{dist}(y, G) = \inf_{\substack{x \in G \\ y \in F}} |y - x|.$$

Topological Vector Spaces

1.4 (Topological Spaces) If X is any set, a *topology* on X is a collection $\mathcal{O}$ of subsets of X which contains

(i) the whole set X and the empty set $\emptyset$,

(ii) the union of any collection of its elements, and

(iii) the intersection of any finite collection of its elements.

The pair $(X, \mathcal{O})$ is called a *topological space* and the elements of $\mathcal{O}$ are the *open sets* of that space. An open set containing a point x in X is called a *neighbourhood* of x. The complement $X - U = \{x \in X : x \notin U\}$ of any open set U is called a *closed* set. The closure $\overline{S}$ of any subset $S \subset X$ is the smallest closed subset of X that contains S.

Let $\mathcal{O}_1$ and $\mathcal{O}_2$ be two topologies on the same set X. If $\mathcal{O}_1 \subset \mathcal{O}_2$, we say that $\mathcal{O}_2$ is *stronger* than $\mathcal{O}_1$, or that $\mathcal{O}_1$ is *weaker* than $\mathcal{O}_2$.

A topological space $(X, \mathcal{O})$ is called a *Hausdorff space* if every pair of distinct points x and y in X have disjoint neighbourhoods.

The *topological product* of two topological spaces $(X, \mathcal{O}_X)$ and $(Y, \mathcal{O}_Y)$ is the topological space $(X \times Y, \mathcal{O})$, where $X \times Y = \{(x, y) : x \in X,\ y \in Y\}$ is the Cartesian product of the sets X and Y, and $\mathcal{O}$ consists of arbitrary unions of sets of the form $\{O_X \times O_Y : O_X \in \mathcal{O}_X,\ O_Y \in \mathcal{O}_Y\}$.

Let $(X, \mathcal{O}_X)$ and $(Y, \mathcal{O}_Y)$ be two topological spaces. A function f from X into Y is said to be *continuous* if the preimage $f^{-1}(O) = \{x \in X : f(x) \in O\}$ belongs to $\mathcal{O}_X$ for every $O \in \mathcal{O}_Y$. Evidently the stronger the topology on X or the weaker the topology on Y, the more such continuous functions f there will be.

1.5 (Topological Vector Spaces) We assume throughout this monograph that all vectors spaces referred to are taken over the complex field unless the contrary is explicitly stated.

A *topological vector space*, hereafter abbreviated TVS, is a Hausdorff topological space that is also a vector space for which the vector space operations of addition and scalar multiplication are continuous. That is, if X is a TVS, then the mappings

$$(x, y) \to x + y \qquad \text{and} \qquad (c, x) \to cx$$

from the topological product spaces $X \times X$ and $\mathbb{C} \times X$, respectively, into X are continuous. (Here $\mathbb{C}$ has its usual topology induced by the Euclidean metric.)

X is a *locally convex* TVS if each neighbourhood of the origin in X contains a convex neighbourhood of the origin.

We outline below those aspects of the theory of topological and normed vector spaces that play a significant role in the study of Sobolev spaces. For a more thorough discussion of these topics the reader is referred to standard textbooks on functional analysis, for example [Ru1] or [Y].

1.6 (Functionals) A scalar-valued function defined on a vector space X is called a *functional*. The functional f is linear provided

$$f(ax + by) = af(x) + bf(y), \qquad x, y \in X, \quad a, b \in \mathbb{C}.$$

If X is a TVS, a functional on X is continuous if it is continuous from X into $\mathbb{C}$ where $\mathbb{C}$ has its usual topology induced by the Euclidean metric.

The set of all continuous, linear functionals on a TVS X is called the *dual* of X and is denoted by X'. Under pointwise addition and scalar multiplication X' is itself a vector space:

$$(f + g)(x) = f(x) + g(x), \quad (cf)(x) = cf(x), \qquad f, g \in X', \ x \in X, \ c \in \mathbb{C}.$$

X' will be a TVS provided a suitable topology is specified for it. One such topology is the *weak-star topology*, the weakest topology with respect to which the functional F_x, defined on X' by $F_x(f) = f(x)$ for each $f \in X'$, is continuous for each $x \in X$. This topology is used, for instance, in the space of Schwartz distributions introduced in Paragraph 1.57. The dual of a normed vector space can be given a stronger topology with respect to which it is itself a normed space. (See Paragraph 1.11.)

Normed Spaces

1.7 (Norms) A *norm* on a vector space X is a real-valued function f on X satisfying the following conditions:

(i) $f(x) \geq 0$ for all $x \in X$ and $f(x) = 0$ if and only if $x = 0$,

(ii) $f(cx) = |c| f(x)$ for every $x \in X$ and $c \in \mathbb{C}$,

(iii) $f(x + y) \leq f(x) + f(y)$ for every $x, y \in X$.

A *normed space* is a vector space X provided with a norm. The norm will be denoted $\|\cdot\,; X\|$ except where other notations are introduced.

If $r > 0$, the set

$$B_r(x) = \{y \in X : \|y - x\,; X\| < r\}$$

is called the *open ball* of radius r with center at $x \in X$. Any subset $A \subset X$ is called *open* if for every $x \in A$ there exists $r > 0$ such that $B_r(x) \subset A$. The open sets thus defined constitute a topology for X with respect to which X is a TVS. This topology is the *norm topology* on X. The closure of $B_r(x)$ in this topology is

$$\overline{B_r(x)} = \{y \in X : \|y - x\,; X\| \le r\}.$$

A TVS X is *normable* if its topology coincides with the topology induced by some norm on X. Two different norms on a vector space X are equivalent if they induce the same topology on X. This is the case if and only if there exist two positive constants a and b such that,

$$a\,\|x\|_1 \le \|x\|_2 \le b\,\|x\|_1$$

for all $x \in X$, where $\|x\|_1$ and $\|x\|_2$ are the two norms.

Let X and Y be two normed spaces. If there exists a one-to-one linear operator L mapping X onto Y having the property $\|L(x)\,; Y\| = \|x\,; X\|$ for every $x \in X$, then we call L an *isometric isomorphism* between X and Y, and we say that X and Y are *isometrically isomorphic*. Such spaces are often identified since they have identical structures and only differ in the nature of their elements.

1.8 A sequence $\{x_n\}$ in a normed space X is *convergent* to the limit x_0 if and only if $\lim_{n\to\infty} \|x_n - x_0\,; X\| = 0$ in $\mathbb{R}$. The norm topology of X is completely determined by the sequences it renders convergent.

A subset S of a normed space X is said to be *dense* in X if each $x \in X$ is the limit of a sequence of elements of S. The normed space X is called *separable* if it has a countable dense subset.

1.9 (Banach Spaces) A sequence $\{x_n\}$ in a normed space X is called a *Cauchy sequence* if and only if for every $\epsilon > 0$ there exists an integer N such that $\|x_m - x_n\,; X\| < \epsilon$ holds whenever $m, n > N$. We say that X is *complete* and a *Banach space* if every Cauchy sequence in X converges to a limit in X. Every normed space X is either a Banach space or a dense subset of a Banach space Y called the *completion* of X whose norm satisfies

$$\|x\,; Y\| = \|x\,; X\| \qquad \text{for every } x \in X.$$

1.10 (Inner Product Spaces and Hilbert Spaces) If X is a vector space, a functional $(\cdot\,,\,\cdot)_X$ defined on $X \times X$ is called an *inner product* on X provided that for every $x, y \in X$ and $a, b \in \mathbb{C}$

(i) $(x, y)_X = \overline{(y, x)_X}$, (where $\overline{c}$ denotes the complex conjugate of $c \in \mathbb{C}$)

(ii) $(ax + by, z)_X = a(x, z)_X + b(y, z)_X$,

(iii) $(x, x)_X = 0$ if and only if $x = 0$,

Equipped with such a functional, X is called an *inner product space,* and the functional

$$\|x\,;X\| = \sqrt{(x, x)_X} \tag{1}$$

is, in fact, a norm on X If X is complete (i.e. a Banach space) under this norm, it is called a *Hilbert space.* Whenever the norm on a vector space X is obtained from an inner product via (1), it satisfies the *parallelogram law*

$$\|x + y\,; X\|^2 + \|x - y\,; X\|^2 = 2\,\|x\,; X\|^2 + 2\,\|y\,; X\|^2\,. \tag{2}$$

Conversely, if the norm on X satisfies (2) then it comes from an inner product as in (1).

1.11 (The Normed Dual) A norm on the dual X' of a normed space X can be defined by setting

$$\left\|x'\,; X'\right\| = \sup\left\{|x'(x)| \;:\; \|x\,; X\| \le 1\right\},$$

for each $x' \in X'$. Since $\mathbb{C}$ is complete, with the topology induced by this norm X' is a Banach space (whether or not X is) and it is called the *normed dual* of X. If X is infinite dimensional, the norm topology of X' is stronger (has more open sets) than the weak-star topology defined in Paragraph 1.6.
The following theorem shows that if X is a Hilbert space, it can be identified with its normed dual.

1.12 THEOREM (The Riesz Representation Theorem) Let X be a Hilbert space. A linear functional x' on X belongs to X' if and only if there exists $x \in X$ such that for every $y \in X$ we have

$$x'(y) = (y, x)_X,$$

and in this case $\left\|x'\,; X'\right\| = \|x\,; X\|$. Moreover, x is uniquely determined by $x' \in X'$. ∎

A vector subspace M of a normed space X is itself a normed space under the norm of X, and so normed is called a *subspace* of X. A closed subspace of a Banach space is itself a Banach space.

1.13 THEOREM (The Hahn-Banach Extension Theorem) Let M be a subspace of the normed space X. If $m' \in M'$, then there exists $x' \in X'$ such that $\left\|x'\,; X'\right\| = \left\|m'\,; M'\right\|$ and $x'(m) = m'(m)$ for every $m \in M$. ∎

1.14 (Reflexive Spaces) A natural linear injection of a normed space X into its second dual space $X'' = (X')'$ is provided by the mapping J whose value Jx at $x \in X$ is given by

$$Jx(x') = x'(x), \qquad x' \in X'.$$

Since $|Jx(x')| \le \|x' ; X'\| \, \|x ; X\|$, we have

$$\|Jx ; X''\| \le \|x ; X\| .$$

However, the Hahn-Banach Extension Theorem assures us that for any $x \in X$ we can find $x' \in X'$ such that $\|x' ; X'\| = 1$ and $x'(x) = \|x ; X\|$. Therefore J is an isometric isomorphism of X into X''.

If the range of the isomorphism J is the entire space X'', we say that the normed space X is *reflexive*. A reflexive space must be complete, and hence a Banach space.

1.15 THEOREM Let X be a normed space. X is reflexive if and only if X' is reflexive. X is separable if X' is separable. Hence if X is separable and reflexive, so is X'. ∎

1.16 (Weak Topologies and Weak Convergence) The *weak topology* on a normed space X is the weakest topology on X that still renders continuous each x' in the normed dual X' of X. Unless X is finite dimensional, the weak topology is weaker than the norm topology on X. It is a consequence of the Hahn-Banach Theorem that a closed, convex set in a normed space is also closed in the weak topology of that space.

A sequence convergent with respect to the weak topology on X is said to *converge weakly*. Thus x_n converges weakly to x in X provided $x'(x_n) \to x'(x)$ in $\mathbb{C}$ for every $x' \in X'$. We denote norm convergence of a sequence $\{x_n\}$ to x in X by $x_n \to x$, and we denote weak convergence by $x_n \rightharpoonup x$. Since we have $|x'(x_n - x)| \le \|x' ; X'\| \, \|x_n - x ; X\|$, we see that $x_n \to x$ implies $x_n \rightharpoonup x$. The converse is generally not true (unless X is finite dimensional).

1.17 (Compact Sets) A subset A of a normed space X is called *compact* if every sequence of points in A has a subsequence converging in X to an element of A. (This definition is equivalent in normed spaces to the definition of compactness in a general topological space; A is compact if whenever A is a subset of the union of a collection of open sets, it is a subset of the union of a finite subcollection of those sets.) Compact sets are closed and bounded, but closed and bounded sets need not be compact unless X is finite dimensional. A is called *precompact* in X if its closure $\overline{A}$ in the norm topology of X is compact. A is called *weakly sequentially compact* if every sequence in A has a subsequence converging weakly in X to a point in A. The reflexivity of a Banach space can be characterized in terms of this property.

1.18 THEOREM A Banach space is reflexive if and only if its closed unit ball $\overline{B_1(0)} = \{x \in X : \|x ; X\| \le 1\}$ is weakly sequentially compact. ∎

1.19 THEOREM A set A is precompact in a Banach space X if and only if for every positive number ϵ there is a finite subset N_ϵ of points of X such that

$$A \subset \bigcup_{y \in N_\epsilon} B_\epsilon(y).$$

A set N_ϵ with this property is called a *finite ϵ-net* for A. ∎

1.20 (Uniform Convexity) Any normed space is locally convex with respect to its norm topology. The norm on X is called *uniformly convex* if for every number ϵ satisfying $0 < \epsilon \le 2$, there exists a number $\delta(\epsilon) > 0$ such that if $x, y \in X$ satisfy $\|x\,; X\| = \|y\,; X\| = 1$ and $\|x - y\,; X\| \ge \epsilon$, then $\|(x+y)/2\,; X\| \le 1 - \delta(\epsilon)$. The normed space X itself is called "uniformly convex" in this case. It should be noted, however, that uniform convexity is a property of the norm—X may have another equivalent norm that is not uniformly convex. Any normable space is called *uniformly convex* if it possesses a uniformly convex norm. The parallelogram law (2) shows that a Hilbert space is uniformly convex.

1.21 THEOREM A uniformly convex Banach space is reflexive. ∎

The following two theorems will be used to establish the separability, reflexivity, and uniform convexity of the Sobolev spaces introduced in Chapter 3.

1.22 THEOREM Let X be a Banach space and M a subspace of X closed with respect to the norm topology of X. Then M is also a Banach space under the norm inherited from X. Furthermore

(i) M is separable if X is separable,
(ii) M is reflexive if X is reflexive,
(iii) M is uniformly convex if X is uniformly convex. ∎

The completeness, separability, and uniform convexity of M follow easily from the corresponding properties of X. The reflexivity of M is a consequence of Theorem 1.18 and the fact that M, being closed and convex, is closed in the weak topology of X.

1.23 THEOREM For $j = 1, 2, \ldots, n$ let X_j be a Banach space with norm $\|\cdot\|_j$. The Cartesian product $X = \prod_{j=1}^n X_j$, consisting of points $(x_1, \ldots, x_n)$ with $x_j \in X_j$, is a vector space under the definitions

$$x + y = (x_1 + y_1, \ldots, x_n + y_n), \qquad cx = (cx_1, \ldots, cx_n),$$

and is a Banach space with respect to any of the equivalent norms

$$\|x\|_{(p)} = \left(\sum_{j=1}^n \|x_j\|_j^p \right)^{1/p}, \qquad 1 \le p < \infty,$$

$$\|x\|_{(\infty)} = \max_{1 \le j \le n} \|x_j\|_j.$$

Furthermore,

(i) if X_j is separable for $1 \le j \le n$, then X is separable,

(ii) if X_j is reflexive for $1 \le j \le n$, then X is reflexive,

(iii) if X_j is uniformly convex for $1 \le j \le n$, then X is uniformly convex. More precisely, $\|\cdot\|_{(p)}$ is a uniformly convex norm on X provided $1 < p < \infty$. ∎

The functionals $\|\cdot\|_{(p)}$, $1 \le p \le \infty$, are norms on X, and X is complete with respect to each of them. Equivalence of these norms follows from the inequalities

$$\|x\|_{(\infty)} \le \|x\|_{(p)} \le \|x\|_{(1)} \le n\,\|x\|_{(\infty)}\,.$$

The separability and uniform convexity of X are readily deduced from the corresponding properties of the spaces X_j. The reflexivity of X follows from that of X_j, $1 \le j \le n$, via Theorem 1.18 or via the natural isomorphism between X' and $\prod_{j=1}^{n} X'_j$.

1.24 (Operators) Since the topology of a normed space X is determined by the sequences it renders convergent, an operator f defined on X into a topological space Y is continuous if and only if $f(x_n) \to f(x)$ in Y whenever $x_n \to x$ in X. Such is also the case for any topological space X whose topology is determined by the sequences it renders convergent. (These are called *first countable spaces*.) Let X, Y be normed spaces and f an operator from X into Y. We say that f is *compact* if $f(A)$ is precompact in Y whenever A is bounded in X. (A bounded set in a normed space is one which is contained in the ball $B_R(0)$ for some R.) If f is continuous and compact, we say that f is *completely continuous.* We say that f is *bounded* if $f(A)$ is bounded in Y whenever A is bounded in X.

Every compact operator is bounded. Every bounded linear operator is continuous. Therefore, every compact linear operator is completely continuous. The norm of a linear operator f is $\sup\{\|f(x)\,;Y\| : \|x\,;X\| \le 1\}$.

1.25 (Imbeddings) We say the normed space X is *imbedded* in the normed space Y, and we write $X \to Y$ to designate this imbedding, provided that

(i) X is a vector subspace of Y, and

(ii) the identity operator I defined on X into Y by $Ix = x$ for all $x \in X$ is continuous.

Since I is linear, (ii) is equivalent to the existence of a constant M such that

$$\|Ix\,;Y\| \le M\,\|x\,;X\|, \qquad x \in X.$$

Sometimes the requirement that X be a subspace of Y and I be the identity map is weakened to allow as imbeddings certain canonical transformations of X into Y. Examples are trace imbeddings of Sobolev spaces as well as imbeddings of Sobolev spaces into spaces of continuous functions. See Chapter 5.

We say that X is *compactly imbedded* in Y if the imbedding operator I is compact.

Spaces of Continuous Functions

1.26 Let Ω be a domain in $\mathbb{R}^n$. For any nonnegative integer m let $C^m(\Omega)$ denote the vector space consisting of all functions ϕ which, together with all their partial derivatives $D^\alpha\phi$ of orders $|\alpha| \le m$, are continuous on Ω. We abbreviate $C^0(\Omega) \equiv C(\Omega)$. Let $C^\infty(\Omega) = \bigcap_{m=0}^{\infty} C^m(\Omega)$.

The subspaces $C_0(\Omega)$ and $C_0^\infty(\Omega)$ consist of all those functions in $C(\Omega)$ and $C^\infty(\Omega)$, respectively, that have compact support in Ω.

1.27 (Spaces of Bounded, Continuous Functions) Since Ω is open, functions in $C^m(\Omega)$ need not be bounded on Ω. We define $C_B^m(\Omega)$ to consist of those functions $\phi \in C^m(\Omega)$ for which $D^\alpha u$ is bounded on Ω for $0 \le |\alpha| \le m$. $C_B^m(\Omega)$ is a Banach space with norm given by

$$\left\|\phi\,;C_B^m(\Omega)\right\| = \max_{0\le\alpha\le m}\sup_{x\in\Omega}|D^\alpha\phi(x)|.$$

1.28 (Spaces of Bounded, Uniformly Continuous Functions) If $\phi \in C(\Omega)$ is bounded and uniformly continuous on Ω, then it possesses a unique, bounded, continuous extension to the closure $\overline{\Omega}$ of Ω. We define the vector space $C^m(\overline{\Omega})$ to consist of all those functions $\phi \in C^m(\Omega)$ for which $D^\alpha\phi$ is bounded and uniformly continuous on Ω for $0 \le |\alpha| \le m$. (This convenient abuse of notation leads to ambiguities if Ω is unbounded; e.g., $C^m(\overline{\mathbb{R}^n}) \ne C^m(\mathbb{R}^n)$ even though $\overline{\mathbb{R}^n} = \mathbb{R}^n$.) $C^m(\overline{\Omega})$ is a closed subspace of $C_B^m(\Omega)$, and therefore also a Banach space with the same norm

$$\left\|\phi\,;C^m(\overline{\Omega})\right\| = \max_{0\le\alpha\le m}\sup_{x\in\Omega}|D^\alpha\phi(x)|.$$

1.29 (Spaces of Hölder Continuous Functions) If $0 < \lambda \le 1$, we define $C^{m,\lambda}(\overline{\Omega})$ to be the subspace of $C^m(\overline{\Omega})$ consisting of those functions ϕ for which, for $0 \le \alpha \le m$, $D^\alpha\phi$ satisfies in Ω a Hölder condition of exponent λ, that is, there exists a constant K such that

$$|D^\alpha\phi(x) - D^\alpha\phi(y)| \le K|x-y|^\lambda, \qquad x, y \in \Omega.$$

$C^{m,\lambda}(\overline{\Omega})$ is a Banach space with norm given by

$$\left\|\phi\,;C^{m,\lambda}(\overline{\Omega})\right\| = \left\|\phi\,;C^m(\overline{\Omega})\right\| + \max_{0\le|\alpha|\le m}\sup_{\substack{x,y\in\Omega\\x\ne y}}\frac{|D^\alpha\phi(x) - D^\alpha\phi(y)|}{|x-y|^\lambda}.$$

It should be noted that for $0 < \nu < \lambda \le 1$,

$$C^{m,\lambda}(\overline{\Omega}) \subsetneq C^{m,\nu}(\overline{\Omega}) \subsetneq C^m(\overline{\Omega}).$$

Since Lipschitz continuity (that is, Hölder continuity of exponent 1) does not imply everywhere differentiability, it is clear that $C^{m,1}(\overline{\Omega}) \not\subset C^{m+1}(\overline{\Omega})$. In general, $C^{m+1}(\overline{\Omega}) \not\subset C^{m,1}(\overline{\Omega})$ either, but the inclusion is possible for many domains Ω, for instance convex ones as can be seen by using the Mean-Value Theorem. (See Theorem 1.34.)

1.30 If Ω is bounded, the following two well-known theorems provide useful criteria for the denseness and compactness of subsets of $C(\overline{\Omega})$. If $\phi \in C(\overline{\Omega})$, we may regard ϕ as defined on $\overline{\Omega}$, that is, we identify ϕ with its unique continuous extension to the closure of Ω.

1.31 THEOREM (The Stone-Weierstrass Theorem) Let Ω be a bounded domain in $\mathbb{R}^n$. A subset $\mathscr{A}$ of $C(\overline{\Omega})$ is dense in $C(\overline{\Omega})$ if it has the following four properties:

(i) If $\phi, \psi \in \mathscr{A}$ and $c \in \mathbb{C}$, then $\phi + \psi$, $\phi\psi$, and $c\phi$ all belong to $\mathscr{A}$.

(ii) If $\phi \in \mathscr{A}$, then $\overline{\phi} \in \mathscr{A}$, where $\overline{\phi}$ is the complex conjugate of ϕ.

(iii) If $x, y \in \overline{\Omega}$ and $x \neq y$, there exists $\phi \in \mathscr{A}$ such that $\phi(x) \neq \phi(y)$.

(iv) If $x \in \overline{\Omega}$, there exists $\phi \in \mathscr{A}$ such that $\phi(x) \neq 0$. ∎

1.32 COROLLARY If Ω is bounded in $\mathbb{R}^n$, then the set P of all polynomials in $x = (x_1, \dots, x_n)$ having rational-complex coefficients is dense in $C(\overline{\Omega})$. (A *rational-complex* number is a number of the form $c_1 + ic_2$ where c_1 and c_2 are rational numbers.) Hence $C(\overline{\Omega})$ is separable.

Proof. The set of all polynomials in x is dense in $C(\overline{\Omega})$ by the Stone-Weierstrass Theorem. Any polynomial can be uniformly approximated on the compact set $\overline{\Omega}$ by elements of the countable set P, which is therefore also dense in $C(\overline{\Omega})$. ∎

1.33 THEOREM (The Ascoli-Arzela Theorem) Let Ω be a bounded domain in $\mathbb{R}^n$. A subset K of $C(\overline{\Omega})$ is precompact in $C(\overline{\Omega})$ if the following two conditions hold:

(i) There exists a constant M such that $|\phi(x)| \leq M$ holds for every $\phi \in K$ and $x \in \Omega$.

(ii) For every $\epsilon > 0$ there exists $\delta > 0$ such that if $\phi \in K$, $x, y \in \Omega$, and $|x - y| < \delta$, then $|\phi(x) - \phi(y)| < \epsilon$. ∎

The following is a straightforward imbedding theorem for the various continuous function spaces introduced above. It is a preview of the main attraction, the Sobolev imbedding theorem of Chapter 5.

1.34 THEOREM Let m be a nonnegative integer and let $0 < \nu < \lambda \leq 1$. Then the following imbeddings exist:

$$C^{m+1}(\overline{\Omega}) \to C^m(\overline{\Omega}), \tag{3}$$

$$C^{m,\nu}(\overline{\Omega}) \to C^m(\overline{\Omega}), \tag{4}$$
$$C^{m,\lambda}(\overline{\Omega}) \to C^{m,\nu}(\overline{\Omega}). \tag{5}$$

If Ω is bounded, then imbeddings (4) and (5) are compact. If Ω is convex, we have the further imbeddings

$$C^{m+1}(\overline{\Omega}) \to C^{m,1}(\overline{\Omega}), \tag{6}$$
$$C^{m+1}(\overline{\Omega}) \to C^{m,\lambda}(\overline{\Omega}). \tag{7}$$

If Ω is convex and bounded, then imbeddings (3) is compact, and so is (7) if $\lambda < 1$.

Proof. The existence of imbeddings (3) and (4) follows from the obvious inequalities

$$\left\|\phi\,;C^m(\overline{\Omega})\right\| \le \left\|\phi\,;C^{m+1}(\overline{\Omega})\right\|,$$
$$\left\|\phi\,;C^m(\overline{\Omega})\right\| \le \left\|\phi\,;C^{m,\lambda}(\overline{\Omega})\right\|.$$

To establish (5) we note that for $|\alpha| \le m$,

$$\sup_{\substack{x,y\in\Omega\\ 0<|x-y|<1}} \frac{|D^\alpha\phi(x)-D^\alpha\phi(y)|}{|x-y|^\nu} \le \sup_{x,y\in\Omega} \frac{|D^\alpha\phi(x)-D^\alpha\phi(y)|}{|x-y|^\lambda}$$

and

$$\sup_{\substack{x,y\in\Omega\\ |x-y|\ge 1}} \frac{|D^\alpha\phi(x)-D^\alpha\phi(y)|}{|x-y|^\nu} \le 2\sup_{x\in\Omega}|D^\alpha\phi(x)|,$$

from which we conclude that

$$\left\|\phi\,;C^{m,\nu}(\overline{\Omega})\right\| \le 2\left\|\phi\,;C^{m,\lambda}(\overline{\Omega})\right\|.$$

If Ω is convex and $x, y \in \Omega$, then by the Mean-Value Theorem there is a point $z \in \Omega$ on the line segment joining x and y such that $D^\alpha\phi(x) - D^\alpha\phi(y)$ is given by $(x-y)\cdot\nabla D^\alpha\phi(z)$, where $\nabla u = (D_1u, \ldots, D_nu)$. Thus

$$|D^\alpha\phi(x)-D^\alpha\phi(y)| \le n|x-y|\left\|\phi\,;C^{m+1}(\overline{\Omega})\right\|, \tag{8}$$

and so

$$\left\|\phi\,;C^{m,1}(\overline{\Omega})\right\| \le n\left\|\phi\,;C^{m+1}(\overline{\Omega})\right\|.$$

Thus (6) is proved, and (7) follows from (5) and (6).

Now suppose that Ω is bounded. If A is a bounded set in $C^{0,\lambda}(\overline{\Omega})$, then there exists M such that $\left\|\phi\,;C^{0,\lambda}(\overline{\Omega})\right\| \le M$ for all $\phi \in A$. But then $|\phi(x)-\phi(y)| \le M|x-y|^\lambda$ for all $\phi \in A$ and all $x, y \in \Omega$, whence A is precompact in $C(\overline{\Omega})$ by the Ascoli-Arzela Theorem 1.33. This proves the compactness of (4) for $m = 0$. If $m \ge 1$ and

A is bounded in $C^{m,\lambda}(\overline{\Omega})$, then A is bounded in $C^{0,\lambda}(\overline{\Omega})$ and there is a sequence $\{\phi_j\} \subset A$ such that $\phi_j \to \phi$ in $C(\overline{\Omega})$. But $\{D_1\phi_j\}$ is also bounded in $C^{0,\lambda}(\overline{\Omega})$ so there exists a subsequence of $\{\phi_j\}$ which we again denote by $\{\phi_j\}$ such that $D_1\phi_j \to \psi_1$ in $C(\overline{\Omega})$. Convergence in $C(\overline{\Omega})$ being uniform convergence on Ω, we have $\psi_1 = D_1\phi$. We may continue to extract subsequences in this manner until we obtain one for which $D^\alpha\phi_j \to D^\alpha\phi$ in $C(\overline{\Omega})$ for each α satisfying $0 \le |\alpha| \le m$. This proves the compactness of (4). For (5) we argue as follows:

$$\begin{aligned} \frac{|D^\alpha\phi(x) - D^\alpha\phi(y)|}{|x-y|^\nu} &= \left(\frac{|D^\alpha\phi(x) - D^\alpha\phi(y)|}{|x-y|^\lambda}\right)^{\nu/\lambda} |D^\alpha\phi(x) - D^\alpha\phi(y)|^{1-\nu/\lambda} \\ &\le \text{const}\, |D^\alpha\phi(x) - D^\alpha\phi(y)|^{1-\nu/\lambda} \end{aligned} \tag{9}$$

for all ϕ in a bounded subset of $C^{m,\lambda}(\overline{\Omega})$. Since (9) shows that any sequence bounded in $C^{m,\lambda}(\overline{\Omega})$ and converging in $C^m(\overline{\Omega})$ is Cauchy and so converges in $C^{m,\nu}(\overline{\Omega})$, the compactness of (5) follows from that of (4).

Finally, if Ω is both convex and bounded, the compactness of (3) and (7) follows from composing the continuous imbedding (6) with the compact imbeddings (4) and (5) for the case $\lambda = 1$. ∎

1.35 The existence of imbeddings (6) and (7), as well as the compactness of (3) and (7), can be obtained under less restrictive hypotheses than the convexity of Ω. For instance, if every pair of points $x, y \in \Omega$ can be joined by a rectifiable arc in Ω having length not exceeding some fixed multiple of $|x - y|$, then we can obtain an inequality similar to (8) and carry out the proof. We leave it to the reader to show that (6) is not compact.

The Lebesgue Measure in $\mathbb{R}^n$

1.36 Many of the vector spaces considered in this monograph consist of functions integrable in the Lebesgue sense over domains in $\mathbb{R}^n$. While we assume that most readers are familiar with Lebesgue measure and integration, we nevertheless include here a brief discussion of that theory, especially those aspects of it relevant to the study of the L^p spaces and Sobolev spaces considered hereafter. All proofs are omitted. For a more complete and systematic discussion of the Lebesgue theory, as well as more general measures and integrals, we refer the reader to any of the books [Fo], [Ro], [Ru2], and [Sx].

1.37 (Sigma Algebras) A collection Σ of subsets of $\mathbb{R}^n$ is called a *σ-algebra* if the following conditions hold:

(i) $\mathbb{R}^n \in \Sigma$.

(ii) If $A \in \Sigma$, then its complement $A^c \in \Sigma$.

(iii) If $A_j \in \Sigma$, $j = 1, 2, \ldots$, then $\bigcup_{j=1}^{\infty} \in \Sigma$.

It follows from (i)–(iii) that:

(iv) The empty set $\emptyset \in \Sigma$.

(v) If $A_j \in \Sigma$, $j = 1, 2, \ldots$, then $\bigcap_{j=1}^{\infty} \in \Sigma$.

(vi) If $A, B \in \Sigma$, then $A - B = A \cap B^c \in \Sigma$.

1.38 (Measures) By a *measure* μ on a σ-algebra Σ we mean a function on Σ taking values in either $\mathbb{R} \cup \{+\infty\}$ (a *positive measure*) or $\mathbb{C}$ (a *complex measure*) which is *countably additive* in the sense that

$$\mu\left(\bigcup_{j=1}^{\infty} A_j\right) = \sum_{j=1}^{\infty} \mu(A_j)$$

whenever $A_j \in \Sigma$, $j = 1, 2, \ldots$ and the sets A_j are pairwise disjoint, that is, $A_j \cap A_k = \emptyset$ for $j \neq k$. For a complex measure the series on the right must converge to the same sum for all permutations of the indices in the sequence $\{A_j\}$, and so must be absolutely convergent. If μ is a positive measure and if $A, B \in \Sigma$ and $A \subset B$, then $\mu(A) \le \mu(B)$. Also, if $A_j \in \Sigma$, $j = 1, 2, \ldots$ and $A_1 \subset A_2 \subset \cdots$, then $\mu\left(\bigcup_{j=1}^{\infty} A_j\right) = \lim_{j \to \infty} \mu(A_j)$.

1.39 THEOREM (Existence of Lebesgue Measure) There exists a σ-algebra Σ of subsets of $\mathbb{R}^n$ and a positive measure μ on Σ having the following properties:

(i) Every open set in $\mathbb{R}^n$ belongs to Σ.

(ii) If $A \subset B$, $B \in \Sigma$, and $\mu(B) = 0$, then $A \in \Sigma$ and $\mu(A) = 0$.

(iii) If $A = \{x \in \mathbb{R}^n : a_j \le x_j \le b_j,\ j = 1, 2, \ldots, n\}$, then $A \in \Sigma$ and $\mu(A) = \prod_{j=1}^{n}(b_j - a_j)$.

(iv) μ is translation invariant. This means that if $x \in \mathbb{R}^n$ and $A \in \Sigma$, then $x + A = \{x + y : y \in A\} \in \Sigma$, and $\mu(x + A) = \mu(A)$. ∎

The elements of Σ are called *(Lebesgue) measurable subsets* of $\mathbb{R}^n$, and μ is called the *(Lebesgue) measure* in $\mathbb{R}^n$. (We normally suppress the word "Lebesgue" in these terms as it is the measure on $\mathbb{R}^n$ we mainly use.) For $A \in \Sigma$ we call $\mu(A)$ the *measure of* A or the *volume of* A, since Lebesgue measure is the natural extension of volume in $\mathbb{R}^3$. While we make no formal distinction between "measure" and "volume" for sets that are easily visualized geometrically, such as balls, cubes, and domains, and we write $\operatorname{vol}(A)$ in place of $\mu(A)$ in these cases. Of course the terms *length* and *area* are more appropriate in $\mathbb{R}^1$ and $\mathbb{R}^2$.

The reader may wonder whether in fact all subsets of $\mathbb{R}^n$ are Lebesgue measurable. The answer depends on the axioms of one's set theory. Under the most common axioms the answer is no; it is possible using the Axiom of Choice to construct a nonmeasurable set. There is a version of set theory where every subset of $\mathbb{R}^n$ is measurable, but the Hahn-Banach theorem 1.13 becomes false in that version.

1.40 (Almost Everywhere) If $B \subset A \subset \mathbb{R}^n$ and $\mu(B) = 0$, then any condition that holds on the set $A - B$ is said to hold *almost everywhere* (abbreviated a.e.) in A. It is easily seen that any countable set in $\mathbb{R}^n$ has measure zero. The converse is, however, not true.

1.41 (Measurable Functions) A function f defined on a measurable set and having values in $\mathbb{R} \cup \{-\infty, +\infty\}$ is itself called *measurable* if the set

$$\{x : f(x) > a\}$$

is measurable for every real a. Some of the more important aspects of this definition are listed in the following theorem.

1.42 THEOREM (a) If f is measurable, so is $|f|$.

(b) If f and g are measurable and real-valued, so are $f + g$ and fg.

(c) If $\{f_j\}$ is a sequence of measurable functions, then $\sup_j f_j$, $\inf_j f_j$, $\limsup_{j\to\infty} f_j$, and $\liminf_{j\to\infty} f_j$ are measurable.

(d) If f is continuous and defined on a measurable set, then f is measurable.

(e) If f is continuous on $\mathbb{R}$ into $\mathbb{R}$ and g is measurable and real-valued, then the composition $f \circ g$ defined by $f \circ g(x) = f\big((g(x)\big)$ is measurable.

(f) **(Lusin's Theorem)** If f is measurable and $f(x) = 0$ for $x \in A^c$ where $\mu(A) < \infty$, and if $\epsilon > 0$, then there exists a function $g \in C_0(\mathbb{R}^n)$ such that $\sup_{x\in\mathbb{R}^n} g(x) \le \sup_{x\in\mathbb{R}^n} f(x)$ and $\mu(\{x \in \mathbb{R}^n : f(x) \ne g(x)\}) < \epsilon$. ▮

1.43 (Characteristic and Simple Functions) Let $A \subset \mathbb{R}^n$. The function χ_A defined by

$$\chi_A(x) = \begin{cases} 1 & \text{if } x \in A \\ 0 & \text{if } x \notin A \end{cases}$$

is called the *characteristic function* of A. A real-valued function s on $\mathbb{R}^n$ is called a *simple function* if its range is a finite set of real numbers. If for every x, we have $s(x) \in \{a_1, \dots, a_n\}$, then $s = \sum_{j=1}^{m} \chi_{A_j}(x)$, where $A_j = \{x \in \mathbb{R}^n : s(x) = a_j\}$, and s is measurable if and only if $A_1, A_2, \dots, A_m$ are all measurable. Because of the following approximation theorem, simple functions are a very useful tool in integration theory.

1.44 THEOREM Given a real-valued function f with domain $A \subset \mathbb{R}^n$ there is a sequence $\{s_j\}$ of simple functions converging pointwise to f on A. If f is bounded, $\{s_j\}$ may be chosen so that the convergence is uniform. If f is measurable, each s_j may be chosen measurable. If f is nonnegative-valued, the sequence $\{s_j\}$ may be chosen to be monotonically increasing at each point. ▮

The Lebesgue Integral

1.45 We are now in a position to define the *(Lebesgue) integral* of a measurable, real-valued function defined on a measurable subset $A \subset \mathbb{R}^n$. For a simple function $s = \sum_{j=1}^{m} a_j \chi_{A_j}$, where $A_j \subset A$, A_j measurable, we define

$$\int_A s(x)\,dx = \sum_{j=1}^{m} a_j \mu(A_j). \tag{10}$$

If f is measurable and nonnegative-valued on A, we define

$$\int_A f(x)\,dx = \sup \int_A s(x)\,dx, \tag{11}$$

where the supremum is taken over measurable, simple functions s vanishing outside A and satisfying $0 \le s(x) \le f(x)$ in A. If f is a nonnegative simple function, then the two definitions of $\int_A f(x)\,dx$ given by (10) and (11) coincide. Note that the integral of a nonnegative function may be $+\infty$.

If f is measurable and real-valued, we set $f = f^+ - f^-$, where $f^+ = \max(f, 0)$ and $f^- = -\min(f, 0)$ are both measurable and nonnegative. We define

$$\int_A f(x)\,dx = \int_A f^+(x)\,dx - \int_A f^-(x)\,dx$$

provided at least one of the integrals on the right is finite. If *both integrals* are finite, we say that f is (Lebesgue) *integrable* on A. The class of integrable functions on A is denoted $L^1(A)$.

1.46 THEOREM Assume all of the functions and sets appearing below are measurable.

(a) If f is bounded on A and $\mu(A) < \infty$, then $f \in L^1(A)$.

(b) If $a \le f(x) \le b$ for all $x \in A$ and if $\mu(A) < \infty$, then

$$a\,\mu(A) \le \int_A f(x)\,dx \le b\,\mu(A).$$

(c) If $f(x) \le g(x)$ for all $x \in A$, and if both integrals exist, then

$$\int_A f(x)\,dx \le \int_A g(x)\,dx.$$

(d) If $f, g \in L^1(A)$, then $f + g \in L^1(A)$ and

$$\int_A (f+g)(x)\,dx = \int_A f(x)\,dx + \int_A g(x)\,dx.$$

(e) If $f \in L^1(A)$ and $c \in \mathbb{R}$, then $cf \in L^1(A)$ and

$$\int_a (cf)(x)\,dx = c\int_A f(x)\,dx.$$

(f) If $f \in L^1(A)$, then $|f| \in L^1(A)$ and

$$\left|\int_A f(x)\,dx\right| \le \int_A |f(x)|\,dx.$$

(g) If $f \in L^1(A)$ and $B \subset A$, then $f \in L^1(B)$. If, in addition, $f(x) \ge 0$ for all $x \in A$, then

$$\int_B f(x)\,dx \le \int_A f(x)\,dx.$$

(h) If $\mu(A) = 0$, then $\int_A f(x)\,dx = 0$.

(i) If $f \in L^1(A)$ and $\int_B f(x) = 0$ for every $B \subset A$, then $f(x) = 0$ a.e. on A. ∎

One consequence of part (i) and the additivity of the integral is that sets of measure zero may be ignored for purposes of integration. That is, if f and g are measurable on A and if $f(x) = g(x)$ a.e. on A, then $\int_A f(x)\,dx = \int_A g(x)\,dx$. Accordingly, two elements of $L^1(A)$ are considered identical if they are equal almost everywhere. Thus the elements of $L_1(A)$ are actually not functions but equivalence classes of functions; two functions belong to the same element of $L_1(A)$ if they are equal a.e. on A. Nevertheless, we will continue to refer (loosely) to the elements of $L_1(A)$ as functions on A.

1.47 THEOREM If f is either an element of $L^1(\mathbb{R}^n)$ or measurable and nonnegative on $\mathbb{R}^n$, then the set function λ defined by

$$\lambda(A) = \int_A f(x)\,dx$$

is countably additive, and hence a measure on the σ-algebra of Lebesgue measurable subsets of $\mathbb{R}^n$. ∎

The following three theorems are concerned with the interchange of integration and limit processes.

1.48 THEOREM (The Monotone Convergence Theorem) Let $A \subset \mathbb{R}^n$ be measurable and let $\{f_j\}$ be a sequence of measurable functions satisfying $0 \le f_1(x) \le f_2(x) \le \cdots$ for every $x \in A$. Then

$$\lim_{j\to\infty}\int_A f_j(x)\,dx = \int_A \left(\lim_{j\to\infty} f_j(x)\right)dx. \;∎$$

1.49 THEOREM (Fatou's Lemma) Let $A \subset \mathbb{R}^n$ be measurable and let $\{f_j\}$ be a sequence of nonnegative measurable functions. Then

$$\int_A \left(\liminf_{j\to\infty} \right) dx \le \liminf_{j\to\infty} \int_A f_j(x)\, dx. \blacksquare$$

1.50 THEOREM (The Dominated Convergence Theorem) Let $A \subset \mathbb{R}^n$ be measurable and let $\{f_j\}$ be a sequence of measurable functions converging to a limit pointwise on A. If there exists a function $g \in L^1(A)$ such that $|f_j(x)| \le g(x)$ for every j and all $x \in A$, then

$$\lim_{j\to\infty} \int_A f_j(x)\, dx = \int_A \left(\lim_{j\to\infty} f_j(x) \right) dx. \blacksquare$$

1.51 (Integrals of Complex-Valued Functions) The integral of a complex-valued function over a measurable set $A \subset \mathbb{R}^n$ is defined as follows. Set $f = i+i v$, where u and v are real-valued and call f measurable if and only if u and v are measurable. We say f is integrable over A, and write $f \in L^1(A)$, provided $|f| = (u^2+v^2)^{1/2}$ belongs to $L^1(A)$ in the sense described in Paragraph 1.45. For $f \in L^1(A)$, and only for such f, the integral is defined by

$$\int_A f(x)\, dx = \int_A u(x)\, dx + i \int_A v(x)\, dx.$$

It is easily checked that $f \in L^1(A)$ if and only if $u, v \in L^1(A)$. Theorem 1.42(a,b,d–f), Theorem 1.46(a,d–i), Theorem 1.47 (assuming $f \in L^1(\mathbb{R}^n)$), and Theorem 1.50 all extend to cover the case of complex f.

The following theorem enables us to express certain complex measures in terms of Lebesgue measure μ. It is the converse of Theorem 1.47.

1.52 THEOREM (The Radon-Nikodym Theorem) Let λ be a complex measure defined on the σ-algebra Σ of Lebesgue measurable subsets of $\mathbb{R}^n$. Suppose that $\lambda(A) = 0$ for every $A \in \Sigma$ for which $\mu(A) = 0$. Then there exists $f \in L^1(\mathbb{R}^n)$ such that for every $A \in \Sigma$

$$\lambda(A) = \int_A f(x)\, dx.$$

The function f is uniquely determined by λ up to sets of measure zero. $\blacksquare$

1.53 If f is a function defined on a subset A of $\mathbb{R}^{n+m}$, we may regard f as depending on the pair of variables (x, y) with $x \in \mathbb{R}^n$ and $y \in \mathbb{R}^m$. The integral of f over A is then denoted by

$$\int_A f(x, y)\, dx\, dy$$

or, if it is desired to have the integral extend over all of $\mathbb{R}^{n+m}$,

$$\int_{\mathbb{R}^{n+m}} f(x, y)\chi_A(x, y)\, dx\, dy,$$

where χ_A is the characteristic function of A. In particular, if $A \subset \mathbb{R}^n$, we may write

$$\int_A f(x)\, dx = \int_A f(x_1, \ldots, x_n)\, dx_1 \cdots dx_n.$$

1.54 THEOREM (Fubini's Theorem) Let f be a measurable function on $\mathbb{R}^{m+n}$ and suppose that at least one of the integrals

$$\begin{aligned} I_1 &= \int_{\mathbb{R}^{n+m}} |f(x, y)|\, dx, dy, \\ I_2 &= \int_{\mathbb{R}^m} \left(\int_{\mathbb{R}^n} |f(x, y)|\, dx \right) dy, \\ I_3 &= \int_{\mathbb{R}^n} \left(\int_{\mathbb{R}^m} |f(x, y)|\, dy \right) dx \end{aligned} \qquad (12)$$

exists and is finite. For I_2, we mean by this that there is an integrable function g on $\mathbb{R}^n$ such that $g(y)$ is equal to the inner integral for almost all y, and similarly for I_3. Then

(a) $f(\cdot, y) \in L^1(\mathbb{R}^n)$ for almost all $y \in \mathbb{R}^m$.

(b) $f(x, \cdot) \in L^1(\mathbb{R}^m)$ for almost all $x \in \mathbb{R}^n$.

(c) $\int_{R^m} f(\cdot, y)\, dy \in L^1(\mathbb{R}^n)$.

(d) $\int_{R^n} f(x, \cdot)\, dx \in L^1(\mathbb{R}^m)$.

(e) $I_1 = I_2 = I_3$.

Distributions and Weak Derivatives

1.55 We require in subsequent chapters some of the basic concepts and techniques of the Schwartz theory of distributions [Sch], and we present here a brief description of those aspects of the theory that are relevant for our purposes. Of special importance is the notion of weak or distributional derivative of an integrable function. One of the standard definitions of Sobolev spaces is phrased in terms of such derivatives. (See Paragraph 3.2.) Besides [Sch], the reader is referred to [Ru1] and [Y] for more complete treatments of the spaces $\mathscr{D}(\Omega)$ and $\mathscr{D}'(\Omega)$ introduced below, as well as useful generalizations of these spaces.

1.56 (Test Functions) Let Ω be a domain in $\mathbb{R}^n$. A sequence $\{\phi_j\}$ of functions belonging to $C_0^\infty(\Omega)$ is said to *converge in the sense of the space* $\mathscr{D}(\Omega)$ to the function $\phi \in C_0^\infty(\Omega)$ provided the following conditions are satisfied:

(i) there exists $K \Subset \Omega$ such that $\operatorname{supp}(\phi_j - \phi) \subset K$ for every j, and

(ii) $\lim_{j\to\infty} D^\alpha \phi_j(x) = D^\alpha \phi(x)$ uniformly on K for each multi-index α.

There is a locally convex topology on the vector space $C_0^\infty(\Omega)$ which respect to which a linear functional T is continuous if and only if $T(\phi_j) \to T(\phi)$ in $\mathbb{C}$ whenever $\phi_j \to \phi$ in the sense of the space $\mathscr{D}(\Omega)$. Equipped with this topology, $C_0^\infty(\Omega)$ becomes a TVS called $\mathscr{D}(\Omega)$ whose elements are called *test functions*. $\mathscr{D}(\Omega)$ is not a normable space. (We ignore the question of uniqueness of the topology asserted above. It uniquely determines the dual of $\mathscr{D}(\Omega)$ which is sufficient for our purposes.)

1.57 (Schwartz Distributions) The dual space $\mathscr{D}'(\Omega)$ of $\mathscr{D}(\Omega)$ is called the *space of (Schwartz) distributions* on Ω. $\mathscr{D}'(\Omega)$ is given the weak-star topology as the dual of $\mathscr{D}(\Omega)$, and is a locally convex TVS with that topology. We summarize the vector space and convergence operations in $\mathscr{D}'(\Omega)$ as follows: if S, T, T_j belong to $\mathscr{D}'(\Omega)$ and $c \in \mathbb{C}$, then

$$\begin{aligned} (S+T)(\phi) &= S(\phi) + T(\phi), & \phi &\in \mathscr{D}(\Omega), \\ (cT)(\phi) &= c\,T(\phi), & \phi &\in \mathscr{D}(\Omega), \end{aligned}$$

$T_j \to T$ in $\mathscr{D}'(\Omega)$ if and only if $T_j(\phi) \to T(\phi)$ in $\mathbb{C}$ for every $\phi \in \mathscr{D}(\Omega)$.

1.58 (Locally Integrable Functions) A function u defined almost everywhere on Ω is said to be *locally integrable* on Ω provided $u \in L^1(U)$ for every open $U \Subset \Omega$. In this case we write $u \in L^1_{\mathrm{loc}}(\Omega)$. Corresponding to every $u \in L^1_{\mathrm{loc}}(\Omega)$ there is a distribution $T_u \in \mathscr{D}'(\Omega)$ defined by

$$T_u(\phi) = \int_\Omega u(x)\phi(x)\,dx, \qquad \phi \in \mathscr{D}(\Omega). \tag{13}$$

Evidently T_u, thus defined, is a linear functional on $\mathscr{D}(\Omega)$. To see that it is continuous, suppose that $\phi_j \to \phi$ in $\mathscr{D}(\Omega)$. Then there exists $K \Subset \Omega$ such that $\operatorname{supp}(\phi_j - \phi) \subset K$ for all j. Thus

$$|T_u(\phi_j) - T_u(\phi)| \le \sup_{x\in K} |\phi_j(x) - \phi(x)| \int_K |u(x)|\,dx.$$

The right side of the above inequality tends to zero as $j \to \infty$ since $\phi_j \to \phi$ uniformly on K.

1.59 Not every distribution $T \in \mathscr{D}'(\Omega)$ is of the form T_u defined by (13) for some $u \in L^1_{\mathrm{loc}}(\Omega)$. Indeed, if $0 \in \Omega$, there can be no locally integrable function δ on Ω such that for every $\phi \in \mathscr{D}(\Omega)$

$$\int_\Omega \delta(x)\phi(x)\,dx = \phi(0).$$

However, the linear functional δ defined on $\mathscr{D}(\Omega)$ by

$$\delta(\phi) = \phi(0) \tag{14}$$

is easily seen to be continuous and hence a distribution on Ω. It is called a *Dirac distribution.*

1.60 (Derivatives of Distributions) Let $u \in C^1(\Omega)$ and $\phi \in \mathscr{D}(\Omega)$. Since ϕ vanishes outside some compact subset of Ω, we obtain by integration by parts in the variable x_j

$$\int_\Omega \left(\frac{\partial}{\partial x_j} u(x)\right) \phi(x)\, dx = -\int_\Omega u(x) \left(\frac{\partial}{\partial x_j} \phi(x)\right) dx.$$

Similarly, if $u \in C^{|\alpha|}(\Omega)$, then integration by parts $|\alpha|$ times leads to

$$\int_\Omega (D^\alpha u(x))\phi(x)\, dx = (-1)^{|\alpha|} \int_\Omega u(x) D^\alpha \phi(x)\, dx.$$

This motivates the following definition of the derivative $D^\alpha T$ of a distribution $T \in \mathscr{D}'(\Omega)$:

$$(D^\alpha T)(\phi) = (-1)^{|\alpha|} T(D^\alpha \phi). \tag{15}$$

Since $D^\alpha \phi \in \mathscr{D}(\Omega)$ whenever $\phi \in \mathscr{D}(\Omega)$, $D^\alpha T$ is a functional on $\mathscr{D}(\Omega)$, and it is clearly linear. We show that it is continuous, and hence a distribution on Ω. To this end suppose $\phi_j \to \phi$ in $\mathscr{D}(\Omega)$. Then

$$\operatorname{supp}\left(D^\alpha(\phi_j - \phi)\right) \subset \operatorname{supp}\left(\phi_j - \phi\right) \subset K$$

for some $K \Subset \Omega$. Moreover,

$$D^\beta\left(D^\alpha(\phi_j - \phi)\right) = D^{\beta+\alpha}(\phi_j - \phi)$$

converges to zero uniformly on K as $j \to \infty$ for each multi-index β. Hence $D^\alpha \phi_j \to D^\alpha \phi$ in $\mathscr{D}(\Omega)$. Since $T \in \mathscr{D}'(\Omega)$ it follows that

$$D^\alpha T(\phi_j) = (-1)^{|\alpha|} T(D^\alpha \phi_j) \to (-1)^{|\alpha|} T(D^\alpha \phi) = D^\alpha T(\phi)$$

in $\mathbb{C}$. Thus $D^\alpha T \in \mathscr{D}'(\Omega)$.

We have shown that every distribution in $\mathscr{D}'(\Omega)$ possesses derivatives of all orders in $\mathscr{D}'(\Omega)$ in the sense of definition (15). Furthermore, the mapping D^α from $\mathscr{D}'(\Omega)$ into $\mathscr{D}'(\Omega)$ is continuous; if $T_j \to T$ in $\mathscr{D}'(\Omega)$ and $\phi \in \mathscr{D}(\Omega)$, then

$$D^\alpha T_j(\phi) = (-1)^{|\alpha|} T_j(D^\alpha \phi) \to (-1)^{|\alpha|} T(D^\alpha \phi) = D^\alpha T(\phi).$$

1.61 EXAMPLES

1. If $0 \in \Omega$ and $\delta \in \mathscr{D}'(\Omega)$ is the Dirac distribution defined by (14), then $D^\alpha \delta$ is given by
$$D^\alpha \delta(\phi) = (-1)^{|\alpha|} D^\alpha \phi(0).$$

2. If $\Omega = \mathbb{R}$ (i.e., $n = 1$) and $H \in L^1_{\mathrm{loc}}(\mathbb{R})$ is the Heaviside function defined by
$$H(x) = \begin{cases} 1 & \text{if } x \geq 0 \\ 0 & \text{if } x < 0, \end{cases}$$
then the derivative $(T_H)'$ of the corresponding distribution T_H is δ. To see this, let $\phi \in \mathscr{D}(\mathbb{R})$ have support in the interval $[-a, a]$. Then
$$(T_H)'(\phi) = -T_H(\phi') = -\int_0^a \phi'(x)\,dx = \phi(0) = \delta(\phi).$$

1.62 (Weak Derivatives) We now define the concept of a function being the weak derivative of another function. Let $u \in L^1_{\mathrm{loc}}(\Omega)$. There may or may not exist a function $v_\alpha \in L^1_{\mathrm{loc}}(\Omega)$ such that $T_{v_\alpha} = D^\alpha T_u$ in $\mathscr{D}'(\Omega)$. If such a v_α exists, it is unique up to sets of measure zero and is called the *weak* or *distributional* partial derivative of u, and is denoted by $D^\alpha u$. Thus $D^\alpha u = v_\alpha$ in the weak (or distributional) sense provided $v_\alpha \in L^1_{\mathrm{loc}}(\Omega)$ satisfies
$$\int_\Omega u(x) D^\alpha \phi(x)\,dx = (-1)^{|\alpha|} \int_\Omega v_\alpha(x)\phi(x)\,dx$$
for every $\phi \in \mathscr{D}(\Omega)$.

If u is sufficiently smooth to have a continuous partial derivative $D^\alpha u$ in the usual (classical) sense, then $D^\alpha u$ is also a weak partial derivative of u. Of course, $D^\alpha u$ may exist in the weak sense without existing in the classical sense. We shall show in Theorem 3.17 that certain functions having weak derivatives (those in Sobolev spaces) can be suitably approximated by smooth functions.

1.63 Let us note in conclusion that distributions in Ω can be multiplied by smooth functions. If $T \in \mathscr{D}'(\Omega)$ and $\omega \in C^\infty(\Omega)$, the product $\omega T \in \mathscr{D}'(\Omega)$ is defined by
$$(\omega T)(\phi) = T(\omega\phi), \qquad \phi \in \mathscr{D}(\Omega).$$
If $T = T_u$ for some $u \in L^1_{\mathrm{loc}}(\Omega)$, then $\omega T = T_{\omega u}$. The Leibniz rule (see Paragraph 1.2) is easily checked to hold for $D^\alpha(\omega T)$.

2

THE LEBESGUE SPACES $L^p(\Omega)$

Definition and Basic Properties

2.1 (The Space $L^p(\Omega)$) Let Ω be a domain in $\mathbb{R}^n$ and let p be a positive real number. We denote by $L^p(\Omega)$ the class of all measurable functions u defined on Ω for which

$$\int_\Omega |u(x)|^p \, dx < \infty. \tag{1}$$

We identify in $L^p(\Omega)$ functions that are equal almost everywhere in Ω; the elements of $L^p(\Omega)$ are thus equivalence classes of measurable functions satisfying (1), two functions being equivalent if they are equal a.e. in Ω. For convenience, we ignore this distinction, and write $u \in L^p(\Omega)$ if u satisfies (1), and $u = 0$ in $L^p(\Omega)$ if $u(x) = 0$ a.e. in Ω. Evidently $cu \in L^p(\Omega)$ if $u \in L^p(\Omega)$ and $c \in \mathbb{C}$. To confirm that $L^p(\Omega)$ is a vector space we must show that if $u, v \in L^p(\Omega)$, then $u + v \in L^p(\Omega)$. This is an immediate consequence of the following inequality, which will also prove useful later on.

2.2 LEMMA If $1 \le p < \infty$ and $a, b \ge 0$, then

$$(a + b)^p \le 2^{p-1}(a^p + b^p). \tag{2}$$

Proof. If $p = 1$, then (2) is an obvious equality. For $p > 1$, the function t^p is convex on $[0, \infty)$; that is, its graph lies below the chord line joining the points

(a, a^p) and (b, b^p). Thus

$$\left(\frac{a+b}{2}\right)^p \le \frac{a^p + b^p}{2},$$

from which (2) follows at once. ∎

If $u, v \in L^p(\Omega)$, then integrating

$$|u(x) + v(x)|^p \le \big(|u(x)| + |v(x)|\big)^p \le 2^{p-1}\big(|u(x)|^p + |v(x)|^p\big)$$

over Ω confirms that $u + v \in L^p(\Omega)$.

2.3 (The L_p Norm) We shall verify presently that the functional $\|\cdot\|_p$ defined by

$$\|u\|_p = \left(\int_\Omega |u(x)^p\, dx\right)^{1/p}$$

is a norm on $L^p(\Omega)$ provided $1 \le p < \infty$. (It is not a norm if $0 < p < 1$.) In arguments where confusion of domains may occur, we use $\|\cdot\|_{p,\Omega}$ in place of $\|\cdot\|_p$. It is clear that $\|u\|_p \ge 0$ and $\|u\|_p = 0$ if and only if $u = 0$ in $L^p(\Omega)$. Moreover,

$$\|cu\|_p = |c|\, \|u\|_p, \qquad c \in \mathbb{C}.$$

Thus we will have shown that $\|\cdot\|_p$ is a norm on $L^p(\Omega)$ once we have verified the triangle inequality

$$\|u + v\|_p \le \|u\|_p + \|v\|_p,$$

which is known as *Minkowski's inequality*. We verify it in Paragraph 2.8 below, for which we first require Hölder's inequality.

2.4 THEOREM (Hölder's Inequality) Let $1 < p < \infty$ and let p' denote the *conjugate exponent* defined by

$$p' = \frac{p}{p-1}, \qquad \text{that is} \qquad \frac{1}{p} + \frac{1}{p'} = 1$$

which also satisfies $1 < p' < 1$. If $u \in L^p(\Omega)$ and $v \in L^{p'}(\Omega)$, then $uv \in L^1(\Omega)$, and

$$\int_\Omega |u(x)v(x)|\, dx \le \|u\|_p\, \|v\|_{p'}. \tag{3}$$

Equality holds if and only if $|u(x)|^p$ and $|v(x)|^{p'}$ are proportional a.e. in Ω.

Proof. Let $a, b > 0$ and let $A = \ln(a^p)$ and $B = \ln(b^{p'})$. Since the exponential function is strictly convex, $\exp\big((A/p) + (B/p')\big) \le (1/p)\exp A + (1/p')\exp B$, with equality only if $A = B$. Hence

$$ab \le (a^p/p) + (b^{p'}/p'),$$

with equality occurring if and only if $a^p = b^{p'}$. If either $\|u\|_p = 0$ or $\|v\|_{p'} = 0$, then $u(x)v(x) = 0$ a.e. in Ω, and (3) is satisfied. Otherwise we can substitute $a = |u(x)|/\|u\|_p$ and $b = |v(x)|/\|v\|_{p'}$ in the above inequality and integrate over Ω to obtain (3). ∎

2.5 COROLLARY If $p > 0, q > 0$ and $r > 0$ satisfy $(1/p) + (1/q) = 1/r$, and if $u \in L^p(\Omega)$ and $v \in L^q(\Omega)$, then $uv \in L^r(\Omega)$ and $\|uv\|_r \le \|u\|_p \|v\|_q$. To see this, we can apply Hölder's inequality to $|u|^r|v|^r$ with exponents p/r and $q/r = (p/r)'$. ∎

2.6 COROLLARY Hölder's inequality can be extended to products of more than two functions. Suppose $u = \prod_{j=1}^N u_j$ where $u_j \in L^{p_j}(\Omega)$, $1 \le j \le N$, where $p_j > 0$. If $\sum_{j=1}^N (1/p_j) = 1/q$, then $u \in L^q(\Omega)$ and $\|u\|_q \le \prod_{j=1}^N \|u_j\|_{p_j}$. This follows from the previous corollary by induction on N. ∎

2.7 LEMMA (A Converse of Hölder's Inequality) A measurable function u belongs to $L^p(\Omega)$ if and only if

$$\sup\left\{\int_\Omega |u(x)|v(x)\,dx \;:\; v(x) \ge 0 \text{ on } \Omega,\ \|v\|_{p'} \le 1\right\} \tag{4}$$

is finite, and then that supremum equals $\|u\|_p$.

Proof. This is obvious if $\|u\|_p = 0$. If $0 < \|u\|_p < \infty$, then for nonnegative v with $\|v\|_{p'} \le 1$ we have, by Hölder's inequality,

$$\int_\Omega |u(x)|v(x)\,dx \le \|u\|_p \|v\|_{p'} \le \|u\|_p,$$

and equality holds if $v = \left(|u|/\|u\|_p\right)^{p/p'}$, for which $\|v\|_{p'} = 1$.

Conversely, if $\|u\|_p = \infty$ we can find an increasing sequence s_j of nontrivial simple functions satisfying $0 \le s_j(x) \le |u(x)|$ on Ω for which $\|s_j\|_p \to \infty$. If $v_j = \left(|s_j|/\|s_j\|_p\right)^{p/p'}$, then

$$\int_\Omega |u(x)|v_j(x)\,dx \ge \int_\Omega s_j(x)v_j(x)\,dx = \|s_j\|_p$$

so the supremum (4) must be infinite. ∎

2.8 THEOREM (Minkowski's Inequality) If $1 \le p < \infty$, then

$$\|u+v\|_p \le \|u\|_p + \|v\|_p. \tag{5}$$

Proof. Inequality (5) certainly holds if $p = 1$ since

$$\int_\Omega |u(x) + v(x)|\, dx \le \int_\Omega |u(x)|\, dx + \int_\Omega |v(x)|\, dx.$$

For $1 < p < \infty$ observe that for $w \ge 0$, $\|w\|_{p'} \le 1$ we have, by Hölder's inequality,

$$\begin{aligned}\int_\Omega \big(|u(x)| + |v(x)|\big) w(x)\, dx &\le \int_\Omega |u(x)| w(x)\, dx + \int_\Omega |v(x)| w(x)\, dx \\ &\le \|u\|_p + \|v\|_p,\end{aligned}$$

whence $\|u + v\|_p \le \|u\|_p + \|v\|_p$ follows by Lemma 2.7. ∎

2.9 THEOREM (Minkowski's Inequality for Integrals) Let $1 \le p < \infty$. Suppose that f is measurable on $\mathbb{R}^m \times \mathbb{R}^n$, that $f(\cdot, y) \in L^p(\mathbb{R}^m)$ for almost all $y \in \mathbb{R}^n$, and that the function $y \to \|f(\cdot, y)\|_{p,\mathbb{R}^m}$ belongs to $L^1(\mathbb{R}^n)$. Then the function $x \to \int_{\mathbb{R}^n} f(x, y)\, dy$ belongs to $L^p(\mathbb{R}^m)$ and

$$\left(\int_{\mathbb{R}^m} \left|\int_{\mathbb{R}^n} f(x, y)\, dy\right|^p dx\right)^{1/p} \le \int_{\mathbb{R}^n} \left(\int_{\mathbb{R}^m} |f(x, y)|^p\, dx\right)^{1/p} dy.$$

That is,

$$\left\|\int_{\mathbb{R}^n} f(\cdot, y)\, dy\right\|_{p,\mathbb{R}^m} \le \int_{\mathbb{R}^n} \|f(\cdot, y)\|_{p,\mathbb{R}^m}\, dy.$$

Proof. Suppose initially that $f \ge 0$. When $p = 1$, the inequalities above become equalities given in Fubini's theorem. When $p > 1$, use a nonnegative function $\|w\|$ in the unit ball of $L^p(\Omega)$ as in Theorem 2.8. By Fubini's theorem and Hölder's inequality,

$$\begin{aligned}\int_{\mathbb{R}^m} \int_{\mathbb{R}^n} f(x, y)\, dy\, w(x)\, dx &= \int_{\mathbb{R}^{m+n}} f(x, y) w(x)\, dx\, dy \\ &\le \int_{\mathbb{R}^n} \|w\|_{p',\mathbb{R}^m} \|f(\cdot, y)\|_{p,\mathbb{R}^m}\, dy \\ &\le \int_{\mathbb{R}^n} \|f(\cdot, y)\|_{p,\mathbb{R}^m}\, dy.\end{aligned}$$

This case now follows by Lemma 2.7. For a general function f as above, split f into real and imaginary parts and split these as differences of nonnegative functions satisfying the hypotheses. It follows that the function mapping x to $\int_{R^n} f(x, y)\, dy$ belongs to $L^p(R^m)$. To get the norm estimate, replace f by $|f|$. ∎

2.10 (The Space $L^\infty(\Omega)$) A function u that is measurable on Ω is said to be *essentially bounded* on Ω if there is a constant K such that $|u(x)| \le K$ a.e. on Ω. The greatest lower bound of such constants K is called the essential supremum of $|u|$ on Ω, and is denoted by $\operatorname{ess\,sup}_{x\in\Omega}|u(x)|$. We denote by $L^\infty(\Omega)$ the vector space of all functions u that are essentially bounded on Ω, functions being once again identified if they are equal a.e. on Ω. It is easily checked that the functional $\|\cdot\|_\infty$ defined by

$$\|u\|_\infty = \operatorname*{ess\,sup}_{x\in\Omega} |u(x)|$$

is a norm on $L^\infty(\Omega)$. Moreover, Hölder's inequality (3) and its corollaries extend to cover the two cases $p = 1$, $p' = \infty$ and $p = \infty$, $p' = 1$.

2.11 THEOREM (An Interpolation Inequality) Let $1 \le p < q < r$, so that

$$\frac{1}{q} = \frac{\theta}{p} + \frac{1-\theta}{r}$$

for some θ satisfying $0 < \theta < 1$. If $u \in L^p(\Omega) \cap L^r(\Omega)$, then $u \in L^q(\Omega)$ and

$$\|u\|_q \le \|u\|_p^\theta\, \|u\|_r^{1-\theta}.$$

Proof. Let $s = p/(\theta q)$. Then $s \ge 1$ and $s' = s/(s-1) = r/((1-\theta)q)$ if $r < \infty$. In this case, by Hölder's inequality

$$\begin{aligned}\|u\|_q^q &= \int_O |u(x)|^{\theta q}|u(x)|^{(1-\theta)q}\,dx \\ &\le \left(\int_\Omega |u(x)|^{\theta q s}\,dx\right)^{1/s}\left(\int_\Omega |u(x)|^{(1-\theta)qs'}\,dx\right)^{1/s'} = \|u\|_p^{\theta q}\,\|u\|_r^{(1-\theta)q}\end{aligned}$$

and the result follows at once. The proof if $r = \infty$ is similar. ∎

The following two theorems establish reverse forms of Hölder's and Minkowski's inequalities for the case $0 < p < 1$. The latter inequality, which indicates that $\|\cdot\|_p$ is not a norm in this case, will be used to prove the Clarkson inequalities in Theorem 2.38.

2.12 THEOREM (A Reverse Hölder Inequality) Let $0 < p < 1$, so that $p' = p/(p-1) < 0$. If $f \in L^p(\Omega)$ and

$$0 < \int_\Omega |g(x)|^{p'}\,dx < \infty,$$

then

$$\int_\Omega |f(x)g(x)|\,dx \ge \left(\int_\Omega |f(x)|^p\,dx\right)^{1/p}\left(\int_\Omega |g(x)|^{p'}\,dx\right)^{1/p'}. \qquad (6)$$

Proof. We can assume $fg \in L^1(\Omega)$; otherwise the left side of (6) is infinite. Let $\phi = |g|^{-p}$ and $\psi = |fg|^p$ so that $\phi\psi = |f|^p$. Then $\psi \in L^q(\Omega)$, where $q = 1/p > 1$, and since $p' = -pq'$ where $q' = q/(q-1)$, we have $\phi \in L^{q'}(\Omega)$. By the direct form of Hölder's inequality (3) we have

$$\int_\Omega |f(x)|^p \, dx = \int_\Omega \phi(x)\psi(x)\, dx \le \|\psi\|_q \, \|\phi\|_{q'}$$
$$= \left(\int_\Omega |f(x)g(x)|\, dx\right)^p \left(\int_\Omega |g(x)|^{p'}\, dx\right)^{1-p}.$$

Taking pth roots and dividing by the last factor on the right side we obtain (6). ∎

2.13 THEOREM (A Reverse Minkowski Inequality) Let $0 < p < 1$. If $u, v \in L^p(\Omega)$, then

$$\||u| + |v|\|_p \ge \|u\|_p + \|v\|_p . \tag{7}$$

Proof. In $u = v = 0$ in $L^p(\Omega)$, then the right side of (7) is zero. Otherwise, the left side is greater than zero and we can apply the reverse Hölder inequality (6) to obtain

$$\||u| + |v|\|_p^p = \int_\Omega \big(|u(x)| + |v(x)|\big)^{p-1}\big(|u(x)| + |v(x)|\big)\, dx$$
$$\ge \left(\int_\Omega \big(|u(x)| + |v(x)|\big)^p \, dx\right)^{1/p'} \big(\|u\|_p + \|v\|_p\big)$$
$$= \||u| + |v|\|_p^{p/p'} \big(\|u\|_p + \|v\|_p\big)$$

and (7) follows by cancellation. ∎

Here is a useful imbedding theorem for L^p spaces over domains with finite volume.

2.14 THEOREM (An Imbedding Theorem for L^p Spaces) Suppose that $\mathrm{vol}(\Omega) = \int_\Omega 1\, dx < \infty$ and $1 \le p \le q \le \infty$. If $u \in L^q(\Omega)$, then $u \in L^p(\Omega)$ and

$$\|u\|_p \le \big(\mathrm{vol}(\Omega)\big)^{(1/p)-(1/q)} \|u\|_q . \tag{8}$$

Hence

$$L^q(\Omega) \to L^p(\Omega). \tag{9}$$

If $u \in L^\infty(\Omega)$, then

$$\lim_{p\to\infty} \|u\|_p = \|u\|_\infty . \tag{10}$$

Finally, if $u \in L^p(\Omega)$ for $1 \le p < \infty$ and if there exists a constant K such that for all such p

$$\|u\|_p \le K, \tag{11}$$

then $u \in L^\infty(\Omega)$ and

$$\|u\|_\infty \le K. \tag{12}$$

Proof. If $p = q$ or $q = \infty$, (8) and (9) are trivial. If $1 \le p < q < \infty$ and $u \in L^q(\Omega)$, Hölder's inequality gives

$$\int_\Omega |u(x)|^p\,dx \le \left(\int_\Omega |u(x)|^q\,dx\right)^{p/q} \left(\int_\Omega 1\,dx\right)^{1-(p/q)}$$

from which (8) and (9) follow immediately. If $u \in L^\infty(\Omega)$, we obtain from (8)

$$\limsup_{p\to\infty} \|u\|_p \le \|u\|_\infty\,. \tag{13}$$

On the other hand, for any $\epsilon > 0$ there exists a set $A \subset \Omega$ having positive measure $\mu(A)$ such that

$$|u(x)| \ge \|u\|_\infty - \epsilon \qquad \text{if } x \in A.$$

Hence

$$\int_\Omega |u(x)^p\,dx \ge \int_A |u(x)|^p\,dx \ge \mu(A)(\|u\|_\infty - \epsilon)^p.$$

It follows that $\|u\|_p \ge \big(\mu(A)\big)^{1/p}(\|u\|_\infty - \epsilon)$, whence

$$\liminf_{p\to\infty} \|u\|_p \ge \|u\|_\infty\,. \tag{14}$$

Equation (10) now follows from (13) and (14).

Now suppose (11) holds for $1 \le p < \infty$. If $u \notin L^\infty(\Omega)$ or else if (12) does not hold, then we can find a constant $K_1 > K$ and a set $A \subset \Omega$ with $\mu(A) > 0$ such that for $x \in A$, $|u(x)| \ge K_1$. The same argument used to obtain (14) now shows that

$$\liminf_{p\to\infty} \|u\|_p \ge K_1$$

which contradicts (11). ∎

2.15 COROLLARY $L^p(\Omega) \subset L^1_{\mathrm{loc}}(\Omega)$ for $1 \le p \le \infty$ and any domain Ω.

Completeness of $L^p(\Omega)$

2.16 THEOREM $L^p(\Omega)$ is a Banach space if $1 \le p \le \infty$.

Proof. First assume $1 \le p < \infty$ and let $\{u_n\}$ be a Cauchy sequence in $L^p(\Omega)$. There is a subsequence $\{u_{n_j}\}$ of $\{u_n\}$ such that

$$\left\|u_{n_{j+1}} - u_{n_j}\right\|_p \le \frac{1}{2^j}, \qquad j = 1, 2, \ldots.$$

Let $v_m(x) = \sum_{j=1}^{m} |u_{n_{j+1}}(x) - u_{n_j}(x)|$. Then

$$\|v_m\|_p \le \sum_{j=1}^{m} \frac{1}{2^j} < 1, \qquad m = 1, 2, \ldots.$$

Putting $v(x) = \lim_{m\to\infty} v_m(x)$, which may be infinite for some x, we obtain by the Monotone Convergence Theorem 1.48

$$\int_\Omega |v(x)|^p \, dx = \lim_{m\to\infty} \int_\Omega |v_m(x)|^p \, dx \le 1.$$

Hence $v(x) < \infty$ a.e. on Ω and the series

$$u_{n_1}(x) + \sum_{j=1}^{\infty} \left(u_{n_{j+1}}(x) - u_{n_j}(x) \right) \tag{15}$$

converges to a limit $u(x)$ a.e. on Ω by Theorem 1.50. Let $u(x) = 0$ wherever it is undefined by (15). Since (15) telescopes, we have

$$\lim_{m\to\infty} u_{n_m}(x) = u(x) \qquad \text{a.e. in } \Omega.$$

For any $\epsilon > 0$ there exists N such that if $m, n \ge N$, then $\|u_m - u_n\|_p < \epsilon$. Hence, by Fatou's lemma 1.49

$$\begin{aligned} \int_\Omega |u(x) - u_n(x)|^p \, dx &= \int_\Omega \lim_{j\to\infty} |u_{n_j}(x) - u_n(x)|^p \, dx \\ &\le \liminf_{j\to\infty} \int_\Omega |u_{n_j}(x) - u_n(x)|^p \, dx \le \epsilon^p \end{aligned}$$

if $n \ge N$. Thus $u = (u - u_n) + u_n \in L^p(\Omega)$ and $\|u - u_n\|_p \to 0$ as $n \to \infty$. Therefore $L^p(\Omega)$ is complete and so is a Banach space.

Finally, if $\{u_n\}$ is a Cauchy sequence in $L^\infty(\Omega)$, then there exists a set $A \subset \Omega$ having measure zero such that if $x \notin A$, then for every $n, m = 1, 2, \ldots$

$$|u_n(x)| \le \|u_n\|_\infty, \qquad |u_n(x) - u_m(x)| \le \|u_n - u_m\|_\infty.$$

Therefore, $\{u_n\}$ converges uniformly on $\Omega - A$ to a bounded function u. Setting $u = 0$ for $x \in A$, we have $u \in L^\infty(\Omega)$ and $\|u_n - u\|_\infty \to 0$ as $n \to \infty$. Thus $L^\infty(\Omega)$ is also complete and a Banach space. ∎

2.17 COROLLARY If $1 \le p \le \infty$, each Cauchy sequence in $L^p(\Omega)$ has a subsequence converging pointwise almost everywhere on Ω. ∎

2.18 COROLLARY $L^2(\Omega)$ is a Hilbert space with respect to the inner product

$$(u, v) = \int_\Omega u(x)\overline{v(x)}\,dx.$$

Hölder's inequality for $L^2(\Omega)$ is just the well-known Schwarz inequality

$$|(u, v)| \le \|u\|_2 \|v\|_2 . \blacksquare$$

Approximation by Continuous Functions

2.19 THEOREM $C_0(\Omega)$ is dense in $L^p(\Omega)$ if $1 \le p < \infty$.

Proof. Any $u \in L^p(\Omega)$ can be written in the form $u = u_1 - u_2 + i(u_3 - u_4)$ where, for $1 \le j \le 4$, $u_j \in L^p(\Omega)$ is real-valued and nonnegative. Thus it is sufficient to prove that if $\epsilon > 0$ and $u \in L^p(\Omega)$ is real-valued and nonnegative then there exists $\phi \in C_0(\Omega)$ such that $\|\phi - u\|_p < \epsilon$. By Theorem 1.44 for such a function u there exists a monotonically increasing sequence $\{s_n\}$ of nonnegative simple functions converging pointwise to u on Ω. Since $0 \le s_n(x) \le u(x)$, we have $s_n \in L^p(\Omega)$ and since $\big(u(x) - s_n(x)\big)^p \le \big(u(x)\big)^p$, we have $s_n \to u$ in $L^p(\Omega)$ by the Dominated Convergence Theorem 1.50. Thus there exists an $s \in \{s_n\}$ such that $\|u - s\|_p < \epsilon/2$. Since s is simple and $p < \infty$ the support of s has finite volume. We can also assume that $s(x) = 0$ if $x \in \Omega^c$. By Lusin's Theorem 1.42(f) there exists $\phi \in C_0(\mathbb{R}^n)$ such that

$$|\phi(x)| \le \|s\|_\infty \quad \text{for all } x \in \mathbb{R}^n,$$

and

$$\operatorname{vol}(\{x \in \mathbb{R}^n : \phi(x) \ne s(x)\}) < \left(\frac{\epsilon}{4\,\|s\|_\infty}\right)^p .$$

By Theorem 2.14

$$\begin{aligned}\|s - \phi\|_p &\le \|s - \phi\|_\infty \big(\operatorname{vol}(\{x \in \mathbb{R}^n : \phi(x) \ne s(x)\})\big)^{1/p} \\ &< 2\,\|s\|_\infty \left(\frac{\epsilon}{4\,\|s\|_\infty}\right) = \frac{\epsilon}{2}.\end{aligned}$$

It follows that $\|u - \phi\|_p < \epsilon$. $\blacksquare$

2.20 The above proof shows that the set of simple functions in $L^p(\Omega)$ is dense in $L^p(\Omega)$ for $1 \le p < \infty$. That this is also true for $L^\infty(\Omega)$ is a direct consequence of Theorem 1.44.

2.21 THEOREM $L^p(\Omega)$ is separable if $1 \le p < \infty$.

Proof. For $m = 1, 2, \dots$ let

$$\Omega_m = \big\{x \in \Omega : |x| \le m \quad \text{and} \quad \operatorname{dist}\big(x, \operatorname{bdry}(\Omega)\big) \ge 1/m\big\}.$$

Then Ω_m is a compact subset of Ω. Let P be the set of all polynomials on $\mathbb{R}^n$ having rational-complex coefficients, and let $P_m = \{\chi_m f : f \in P\}$ where χ_m is the characteristic function of Ω_m. As shown in Paragraph 1.32, P_m is dense in $C(\Omega_m)$. Moreover, $\bigcup_{m=1}^{\infty} P_m$ is countable.

If $u \in L^p(\Omega)$ and $\epsilon > 0$, there exists $\phi \in C_0(\Omega)$ such that $\|u - \phi\|_p < \epsilon/2$. If $1/m < \operatorname{dist}\big(\operatorname{supp}(\phi), \operatorname{bdry}(\Omega)\big)$, then there exists f in the set P_m such that $\|\phi - f\|_\infty < (\epsilon/2)\big(\operatorname{vol}(\Omega_m)\big)^{-1/p}$. It follows that

$$\|\phi - f\|_p \le \|\phi - f\|_\infty \big(\operatorname{vol}(\Omega_m)\big)^{1/p} < \epsilon/2$$

and so $\|u - f\|_p < \epsilon$. Thus the countable set $\bigcup_{m=1}^{\infty} P_m$ is dense in $L^p(\Omega)$ and $L^p(\Omega)$ is separable. ∎

2.22 $C_B^0(\Omega)$ is a proper closed subset of $L^\infty(\Omega)$ and so is not dense in that space. Therefore, neither are $C_0(\Omega)$ or $C_0^\infty(\Omega)$. In fact, $L^\infty(\Omega)$ is not separable.

Convolutions and Young's Theorem

2.23 (The Convolution Product) It is often useful to form a non-pointwise product of two functions that smooth out irregularities of each of them to produce a function better behaved locally than either factor alone. One such product is the *convolution* $u * v$ of two functions u and v defined by

$$u * v(x) = \int_{\mathbb{R}^n} u(x - y)v(y)\, dy \tag{16}$$

when the integral exists. For instance, if $u \in L^p(\mathbb{R}^n)$ and $v \in L^{p'}(\mathbb{R}^n)$, then the integral (16) converges absolutely by Hölder's inequality, and we have $|u*v(x)| \le \|u\|_p \|v\|_{p'}$ for all values of x. Moreover, $u*v$ is uniformly continuous in these cases. To see this, observe first that if $u \in L^p(\mathbb{R}^n)$ and $v \in C_0(\mathbb{R}^n)$, then applying Hölder's inequality to the convolution of u with differences between v and translates of v shows that $u * v$ is uniformly continuous. When $1 \le p' < \infty$ a general function v in $L^{p'}(\mathbb{R}^n)$ is the $L^{p'}$-norm limit of a sequence, $\{v_j\}$ say, of functions in $C_0(\mathbb{R}^n)$; then $u * v$ is the L^∞-norm limit of the sequence $\{u * v_j\}$, and so is still uniformly continuous. In any event, the change of variable $y = x - z$ shows that $u*v = v*u$. Thus $u*v$ is also uniformly continuous when $u \in L^1(\mathbb{R}^n)$ and $v \in L^\infty(\mathbb{R}^n)$.

2.24 THEOREM (Young's Theorem) Let $p, q, r \geq 1$ and suppose that $(1/p) + (1/q) + (1/r) = 2$. Then

$$\left|\int_{\mathbb{R}^n} (u * v)(x)w(x)\,dx\right| \leq \|u\|_p \,\|v\|_q \,\|w\|_r \tag{17}$$

holds for all $u \in L^p(\mathbb{R}^n)$, $v \in L^q(\mathbb{R}^n)$, $w \in L^r(\mathbb{R}^n)$.

Proof. For now, we prove this estimate when $u \in C_0(\mathbb{R}^n)$, and we explain in the proof of the Corollary below how to deal with more general functions u. This special case is the one we use in applications of convolution. The function mapping (x, y) to $u(x-y)$ is then jointly continuous on $\mathbb{R}^n \times \mathbb{R}^n$, and hence is a measurable function on on $\mathbb{R}^n \times \mathbb{R}^n$. This justifies the use of Fubini's theorem below. First observe that

$$\frac{1}{p'} + \frac{1}{q'} + \frac{1}{r'} = 3 - \frac{1}{p} - \frac{1}{q} - \frac{1}{r} = 1,$$

so the functions

$$\begin{aligned}
U(x, y) &= |v(y))|^{q/p'}|w(x)|^{r/p'}\\
V(x, y) &= |u(x-y))|^{p/q'}|w(x)|^{r/q'}\\
W(x, y) &= |u(x-y))|^{p/r'}|v(y)|^{q/r'}
\end{aligned}$$

satisfy $(UVW)(x, y) = u(x-y)v(y)w(x)$. Moreover,

$$\begin{aligned}
\|V\|_{q'} &= \left(\int_{\mathbb{R}^n} |w(x)|^r\,dx \int_{\mathbb{R}^n} |u(x-y)|^p\,dy\right)^{1/q'}\\
&= \left(\int_{\mathbb{R}^n} |w(x)|^r\,dx \int_{\mathbb{R}^n} |u(z)|^p\,dz\right)^{1/q'} = \|u\|_p^{p/q'}\,\|w\|_r^{r/q'},
\end{aligned}$$

and similarly $\|U\|_{p'} = \|v\|_q^{q/p'}\,\|w\|_r^{r/p'}$ and $\|W\|_{r'} = \|u\|_p^{p/r'}\,\|v\|_q^{q/r'}$. Combining these results, we have, by the three-function form of Hölder's inequality,

$$\begin{aligned}
\left|\int_{\mathbb{R}^n} (u * v)(x)w(x)\,dx\right| &\leq \int_{\mathbb{R}^n}\int_{\mathbb{R}^n} |u(x-y||v(y||w(x)|\,dy\,dx\\
&= \int_{\mathbb{R}^n}\int_{\mathbb{R}^n} U(x, y)V(x, y)W(x, y)\,dy\,dx\\
&\leq \|U\|_{p'}\,\|V\|_{q'}\,\|W\|_{r'} = \|u\|_p\,\|v\|_q\,\|w\|_r\,.
\end{aligned}$$

We remark that (17) holds with a constant $K = K(p, q, r, n) < 1$ included on the right side. The best (smallest) constant is

$$K(p, q, r, n) = \left(\frac{p^{1/p}q^{1/q}r^{1/r}}{(p')^{1/p'}(q')^{1/q'}(r')^{1/r'}}\right)^{n/2}.$$

See [LL] for a proof of this. ∎

2.25 COROLLARY If $(1/p) + (1/q) = 1 + (1/r)$, and if $u \in L^p(\mathbb{R}^n)$ and $v \in L^q(\mathbb{R}^n)$, then $u * v \in L^r(\mathbb{R}^n)$, and

$$\|u * v\|_r \le K(p, q, r', n)\,\|u\|_p\,\|v\|_q \le \|u\|_p\,\|v\|_q\,.$$

This is known as *Young's inequality for convolution.* It also implies Young's Theorem. When $u \in C_0(\mathbb{R}^n)$, it follows from Lemma 2.7 and the case of inequality (17) proved above, with r' in place of r.

2.26 (Proof of the General Case of Corollary 2.25 and Theorem 2.24) We remove the restriction $u \in C_0(\mathbb{R}^n)$ from the above Corollary and therefore from Young's Theorem itself. We can assume that p and q are both finite, since the only other pairs satisfying the hypotheses are $(p, q) = (1, \infty)$ and $(\infty, 1)$, and these were covered before the statement of the theorem.

Fix a simple function v in $L^q(\mathbb{R}^n)$, and regard the functions u as running through the subspace $C_0(\mathbb{R}^n)$ of $L^p(\mathbb{R}^n)$. Then convolution with v is a bounded operator, T_v say, from this dense subspace of $L^p(\mathbb{R}^n)$ to $L^r(\mathbb{R}^n)$, and the norm of T_v is at most $\|v\|_q$. By the norm density of $C_0(\mathbb{R}^n)$ in $L^p(\mathbb{R}^n)$, the operator T_v extends uniquely to one with the same norm mapping all of $L^p(\mathbb{R}^n)$ to $L^r(\mathbb{R}^n)$.

Given u in $L^p(\mathbb{R}^n)$, find a sequence $\{u_j\}$ in $C_0(\mathbb{R}^n)$ converging in L^p norm to u. Then $T_v(u_j)$ converges in L^r norm to $T_v(u)$. Pass to a subsequence, if necessary, to also get almost-everywhere convergence of $T_v(u_j)$ to $T_v(u)$. Since the simple function v also belongs to $L^{p'}$, the integrals (16) defining $u * v$ and $u_j * v$ all converge absolutely, and

$$u * v(x) = \lim_{j\to\infty} \big(u_j * v(x)\big) \quad \text{for all } x \in R^n.$$

So $T_v(u)(x)$ agrees almost everywhere with $u * v(x)$ as given in (16), and hence $\|u * v\|_r \le \|u\|_p\|v\|_q$ when u is any function in $L^p(\mathbb{R}^n)$ and v is any simple function in $L^q(\mathbb{R}^n)$.

We complete the proof with an argument passing from simple functions v to general functions in $L^q(\mathbb{R}^n)$. For any fixed u in $L^p(\mathbb{R}^n)$ convolution with u defines an operator, S_u say, with norm at most $\|u\|_p$, from the subspace of simple functions in $L^q(\mathbb{R}^n)$ to $L^r(\mathbb{R}^n)$. By the density of that subspace, the operator S_u extends uniquely to one with the same norm mapping all of $L^q(\mathbb{R}^n)$ to $L^r(\mathbb{R}^n)$.

To relate this extended operator S_u to formula (16), it suffices to deal with the case where the functions u and v are both nonnegative. Pick an increasing sequence $\{v_j\}$ of nonnegative simple functions converging in L^q norm to v. Then the sequence $\{u * v_j\}$ converges in L^r norm to $S_u(v)$. Again pass to a subsequence that converges almost everywhere to $S_u(v)$. Since the function u is nonnegative, the product sequence $\{u * v_j(x)\}$ increases for each x. So it either diverges to ∞ or converges to a finite value for $u * v(x)$. From the a.e. convergence above, the latter must happen for almost all x, and $\|u * v\|_r = \|S_u(v)\|_r \le \|u\|_p \|v\|_q$ as required. ∎

2.27 (The Space ℓ^p) It is sometimes useful to classify sequences of real or complex numbers according to their degree of summability. We denote by ℓ^p the set of doubly infinite sequences $a = \{a_i\}_{i=-\infty}^{\infty}$ for which

$$\|a\,;\ell^p\| = \begin{cases} \left(\displaystyle\sum_{i=-\infty}^{\infty} |a_i|^p \right)^{1/p} & \text{if } 0 < p < \infty \\ \displaystyle\sup_{-\infty<i<\infty} |a_i| & \text{if } p = \infty \end{cases}$$

is finite. Evidently, $\|a\,;\ell^p\| = \|f\|_p$ where f is the function defined on $\mathbb{R}$ by $f(t) = a_i$ for $i \le t < i+1$, $-\infty < i < \infty$.

If $1 \le p \le \infty$, then ℓ^p is a Banach space with norm $\|\cdot\,;\ell^p\|$. Singly infinite sequences such as $\{a_i\}_{i=0}^{\infty}$ or even finite sequences such as $\{a_i\}_{i=m}^{n}$ can be regarded as defined for $-\infty < i < \infty$ with all $a_i = 0$ for i outside the appropriate interval, and as such they determine subspaces of of ℓ^p.

Hölder's inequality, Minkowski's inequality, and Young's inequality follow for the spaces ℓ^p by the same methods used for $L^p(\mathbb{R})$. Specifically, suppose that $a = \{a_i\}_{i=-\infty}^{\infty}$ and $b = \{b_i\}_{i=-\infty}^{\infty}$.

(a) If $a \in \ell^p$ and $b \in \ell^q$, then $ab = \{a_i b_i\}_{i=-\infty}^{\infty} \in \ell^r$ where r satisfies $(1/r) = (1/p) + (1/q)$, and

$$\|ab\,;\ell^r\| \le \|a\,;\ell^p\|\,\|b\,;\ell^q\|. \qquad \text{(Hölder's Inequality)}$$

(b) If $a, b \in \ell^p$, then

$$\|a+b\,;\ell^p\| \le \|a\,;\ell^p\| + \|b\,;\ell^p\|. \qquad \text{(Minkowski's Inequality)}$$

(c) If $a \in \ell^p$ and $b \in \ell^q$ where $(1/p) + (1/q) \ge 1$, then the series $(a * b)_i$ defined by

$$(a * b)_i = \sum_{j=-\infty}^{\infty} a_{i-j} b_j, \qquad (-\infty < i < \infty),$$

converges absolutely. Moreover, the sequence $a * b$, called the *convolution* of a and b, belongs to ℓ^r, where $1 + (1/r) = (1/p) + (1/q)$, and

$$\|a * b\,;\ell^r\| \le \|a\,;\ell^p\|\,\|b\,;\ell^q\|. \qquad \text{(Young's Inequality)}$$

Note, however, that the ℓ^p spaces imbed into one another in the reverse order to the imbeddings of the spaces $L^p(\Omega)$ where Ω has finite volume. (See Theorem 2.14.) If $0 < p \le q \le \infty$, then

$$\ell^p \to \ell^q, \qquad \text{and} \qquad \left\| a\,; \ell^q \right\| \le \left\| a\,; \ell^p \right\|.$$

The latter inequality is obvious if $q = \infty$ and follows for other $q \ge p$ from summing the inequality

$$|a_i|^q = |a_i|^p |a_i|^{q-p} \le |a_i|^p \left\| a\,; \ell^\infty \right\|^{q-p} \le |a_i|^p \left\| a\,; \ell^p \right\|^{q-p}.$$

Mollifiers and Approximation by Smooth Functions

2.28 (Mollifiers) Let J be a nonnegative, real-valued function belonging to $C_0^\infty(\mathbb{R}^n)$ and having the properties

(i) $J(x) = 0$ if $|x| \ge 1$, and

(ii) $\int_{\mathbb{R}^n} J(x)\,dx = 1$.

For example, we may take

$$J(x) = \begin{cases} k \exp[-1/(1-|x|^2)] & \text{if } |x| < 1 \\ 0 & \text{if } |x| \ge 1, \end{cases}$$

where $k > 0$ is chosen so that condition (ii) is satisfied. If $\epsilon > 0$, the function $J_\epsilon(x) = \epsilon^{-n} J(x/\epsilon)$ is nonnegative, belongs to $C_0^\infty(\mathbb{R}^n)$, and satisfies

(i) $J_\epsilon(x) = 0$ if $|x| \ge \epsilon$, and

(ii) $\int_{\mathbb{R}^n} J_\epsilon(x)\,dx = 1$.

J_ϵ is called a *mollifier* and the convolution

$$J_\epsilon * u(x) = \int_{\mathbb{R}^n} J_\epsilon(x-y)u(y)\,dy, \tag{18}$$

defined for functions u for which the right side of (18) makes sense, is called a *mollification* or *regularization* of u. The following theorem summarizes some properties of mollification.

2.29 THEOREM (Properties of Mollification) Let u be a function which is defined on $\mathbb{R}^n$ and vanishes identically outside Ω.

(a) If $u \in L^1_{\text{loc}}(\mathbb{R}^n)$, then $J_\epsilon * u \in C^\infty(\mathbb{R}^n)$.

(b) If $u \in L^1_{\text{loc}}(\Omega)$ and $\operatorname{supp}(u) \Subset \Omega$, then $J_\epsilon * u \in C_0^\infty(\Omega)$ provided

$$\epsilon < \operatorname{dist}\big(\operatorname{supp}(u), \operatorname{bdry}(\Omega)\big).$$

(c) If $u \in L^p(\Omega)$ where $1 \le p < \infty$, then $J_\epsilon * u \in L^p(\Omega)$. Also

$$\|J_\epsilon * u\|_p \le \|u\|_p \quad \text{and} \quad \lim_{\epsilon \to 0+} \|J_\epsilon * u - u\|_p = 0.$$

(d) If $u \in C(\Omega)$ and if $G \Subset \Omega$, then $\lim_{\epsilon \to 0+} J_\epsilon * u(x) = u(x)$ uniformly on G.

(e) If $u \in C(\overline{\Omega})$, then $\lim_{\epsilon \to 0+} J_\epsilon * u(x) = u(x)$ uniformly on Ω.

Proof. Since $J_\epsilon(x-y)$ is an infinitely differentiable function of x and vanishes if $|y-x| \ge \epsilon$, and since for every multi-index α we have

$$D^\alpha (J_\epsilon * u)(x) = \int_{\mathbb{R}^n} D_x^\alpha J_\epsilon(x-y) u(y)\, dy,$$

conclusions (a) and (b) are valid.

If $u \in L^p(\Omega)$ where $1 < p < \infty$, then by Hölder's inequality (3),

$$\begin{aligned} |J_\epsilon * u(x)| &= \left| \int_{\mathbb{R}^n} J_\epsilon(x-y) u(y)\, dy \right| \\ &\le \left(\int_{\mathbb{R}^n} J_\epsilon(x-y)\, dy \right)^{1/p'} \left(\int_{\mathbb{R}^n} J_\epsilon(x-y) |u(y)|^p\, dy \right)^{1/p} \\ &= \left(\int_{\mathbb{R}^n} J_\epsilon(x-y) |u(y)|^p\, dy \right)^{1/p}. \end{aligned} \tag{19}$$

Hence by Fubini's Theorem 1.54

$$\begin{aligned} \int_\Omega |J_\epsilon * u(x)|^p\, dx &\le \int_{\mathbb{R}^n} \int_{\mathbb{R}^n} J_\epsilon(x-y) |u(y)|^p\, dy\, dx \\ &= \int_{\mathbb{R}^n} |u(y)|^p\, dy \int_{\mathbb{R}^n} J_\epsilon(x-y)\, dx = \|u\|_p^p. \end{aligned}$$

For $p = 1$ this inequality follows directly from (18).

Now let $\eta > 0$ be given. By Theorem 2.19 there exists $\phi \in C_0(\Omega)$ such that $\|u - \phi\|_p < \eta/3$. Thus also $\|J_\epsilon * u - J_\epsilon * \phi\|_p < \eta/3$. Now

$$\begin{aligned} |J_\epsilon * \phi(x) - \phi(x)| &= \left| \int_{\mathbb{R}^n} J_\epsilon(x-y) \big(\phi(y) - \phi(x)\big)\, du \right| \\ &\le \sup_{|y-x|<\epsilon} |\phi(y) - \phi(x)|. \end{aligned} \tag{20}$$

Since ϕ is uniformly continuous on Ω, the right side of (20) tends to zero as $\epsilon \to 0+$. Since supp (ϕ) is compact, we can ensure that $\|J_\epsilon * \phi - \phi\|_p < \eta/3$

by choosing ϵ sufficiently small. For such ϵ we have $\|J_\epsilon * u - u\|_p < \eta$ and (c) follows.

The proofs of (d) and (e) may be obtained by replacing ϕ by u in inequality (20). ∎

2.30 COROLLARY $C_0^\infty(\Omega)$ is dense in $L^p(\Omega)$ if $1 \le p < \infty$. ∎

This is an immediate consequence of conclusions (b) and (e) of the theorem and Theorem 2.19.

Precompact Sets in $L^p(\Omega)$

2.31 The following theorem plays a role in the study of L^p spaces similar to that played by the Arzela-Ascoli Theorem 1.33 in the study of spaces of continuous functions. If u is a function defined a.e. on $\Omega \subset \mathbb{R}^n$, let $\tilde{u}$ denote the zero extension of u outside Ω:

$$\tilde{u} = \begin{cases} u(x) & \text{if } x \in \Omega, \\ 0 & \text{if } x \in \mathbb{R}^n - \Omega. \end{cases}$$

2.32 THEOREM Let $1 \le p < \infty$. A bounded subset $K \subset L^p(\Omega)$ is precompact in $L^p(\Omega)$ if and only if for every number $\epsilon > 0$ there exists a number $\delta > 0$ and a subset $G \Subset \Omega$ such that for every $u \in K$ and $h \in \mathbb{R}^n$ with $|h| < \delta$ both of the following inequalities hold:

$$\int_\Omega |\tilde{u}(x+h) - \tilde{u}(x)|^p \, dx < \epsilon^p, \tag{21}$$

$$\int_{\Omega - \overline{G}} |u(x)|^p \, dx < \epsilon^p. \tag{22}$$

Proof. Let $T_h u$ denote the translate of u by h:

$$T_h u(x) = u(x+h).$$

First we assume that K is precompact in $L^p(\Omega)$. Let $\epsilon > 0$ be given. Since K has a finite $\epsilon/6$-net (Theorem 1.19), and since $C_0(\Omega)$ is dense in $L^p(\Omega)$ (Theorem 2.19), there exists a finite set S of continuous functions having compact support in Ω, such that for each $u \in K$ there exists $\phi \in S$ satisfying $\|u - \phi\|_p < \epsilon/3$. Let G be the union of the supports of the finitely many functions in S. Then $G \Subset \Omega$ and inequality (22) follows immediately. To prove inequality (21) choose a closed ball $\overline{B_r}$ of radius r centred at the origin and containing G. Note that $(T_h\phi - \phi)(x) = \phi(x+h) - \phi(x)$ is uniformly continuous and vanishes outside B_{r+1} provided $|h| < 1$. Hence

$$\lim_{|h| \to 0} \int_{\mathbb{R}^n} |T_h\phi(x) - \phi(x)|^p \, dx = 0,$$

the convergence being uniform for $\phi \in S$. For $|h|$ sufficiently small, we have $\|T_h\phi - \phi\|_p < \epsilon/3$. If $\phi \in S$ satisfies $\|u - \phi\|_p < \epsilon/3$, then also $\|T_h\tilde{u} - T_h\phi\|_p < \epsilon/3$. Hence we have for $|h|$ sufficiently small (independent of $u \in K$),

$$\|T_h\tilde{u} - \tilde{u}\|_p \le \|T_h\tilde{u} - T_h\phi\|_p + \|T_h\phi - \phi\|_p + \|\phi - u\|_p < \epsilon$$

and (21) follows. (This argument shows that translation is continuous in $L^p(\mathbb{R}^n)$.)

It is sufficient to prove the converse for the special case $\Omega = \mathbb{R}^n$, as it follows for general Ω from its application in this special case to the set $\tilde{K} = \{\tilde{u} : u \in K\}$.

Let $\epsilon > 0$ be given and choose $G \Subset \mathbb{R}^n$ such that for all $u \in K$

$$\int_{\mathbb{R}^n - \overline{G}} |u(x)|^p \, dx < \frac{\epsilon}{3}. \tag{23}$$

For any $\eta > 0$ the function $J_\eta * u$ defined as in (18) belongs to $C^\infty(\mathbb{R}^n)$ and in particular to $C(\overline{G})$. If $\phi \in C_0(\mathbb{R}^n)$, then by Hölder's inequality,

$$\begin{aligned}\left|J_\eta * \phi(x) - \phi(x)\right|^p &= \left|\int_{\mathbb{R}^n} J_\eta(y)\big(\phi(x-y) - \phi(x)\big)\, dy\right|^p \\ &\le \int_{B_\eta} J_\eta(y)\left|T_{-y}\phi(x) - \phi(x)\right|^p dy.\end{aligned}$$

Hence

$$\left\|J_\eta * \phi - \phi\right\|_p \le \sup_{h \in B_\eta} \|T_h\phi - \phi\|_p .$$

If $u \in L^p(\mathbb{R}^n)$, let $\{\phi_j\}$ be a sequence in $C_0(\mathbb{R}^n)$ converging to u in L^p norm. By 2.29(c), $\{J_\eta * \phi_j\}$ is a Cauchy sequence converging to $J_\eta * u$ in $L^p(\mathbb{R}^n)$. Since also $T_h\phi_j \to T_h u$ in $L^p(\mathbb{R}^n)$, we have

$$\left\|J_\eta * u - u\right\|_p \le \sup_{h \in B_\eta} \|T_h u - u\|_p .$$

Now (21) implies that $\lim_{|h|\to 0} \|T_h u - u\|_p = 0$ uniformly for $u \in K$. Hence $\lim_{\eta\to 0} \left\|J_\eta * u - u\right\|_p = 0$ uniformly for $u \in K$. Fix $\eta > 0$ so that

$$\int_{\overline{G}} |J_\eta * u(x) - u(x)|^p \, dx < \frac{\epsilon}{3 \cdot 2^{p-1}} \tag{24}$$

for all $u \in K$.

We show that $\{J_\eta * u : u \in K\}$ satisfies the conditions of the Arzela-Ascoli Theorem 1.33 on $\overline{G}$ and hence is precompact in $C(\overline{G})$. By (19) we have

$$|J_\eta * u(x)| \le \left(\sup_{y \in \mathbb{R}^n} J_\eta(y)\right)^{1/p} \|u\|_p$$

which is bounded uniformly for $x \in \mathbb{R}^n$ and $u \in K$ since K is bounded in $L^p(\mathbb{R}^n)$ and η is fixed. Similarly,

$$|J_\eta * u(x+h) - J_\eta * u(x)| \le \left(\sup_{y\in\mathbb{R}^n} J_\eta(y)\right)^{1/p} \|T_h u - u\|_p$$

and so $\lim_{|h|\to 0} J_\eta * u(x+h) = J_\eta * u(x)$ uniformly for $x \in \mathbb{R}^n$ and $u \in K$. Thus $\{J_\eta * u : u \in K\}$ is precompact in $C(\overline{G})$, and by Theorem 1.19 there exists a finite set $\{\psi_1, \ldots, \psi_m\}$ of functions in $C(\overline{G})$ such that if $u \in K$, then for some j, $1 \le j \le m$, and all $x \in \overline{G}$ we have

$$|\psi_j(x) - J_\eta * u(x)| < \frac{\epsilon}{3 \cdot 2^{p-1} \cdot \mathrm{vol}(\overline{G})}.$$

This, together with (23), (24), and the inequality $(|a|+|b|)^p \le 2^{p-1}(|a|^p + |b|^p)$ of Lemma 2.2, implies that

$$\begin{aligned}\int_{\mathbb{R}^n} |u(x) - \tilde{\psi}_j(x)|^p\,dx &= \int_{\mathbb{R}^n - \overline{G}} |u(x)|^p\,dx + \int_{\overline{G}} |u(x) - \psi_j(x)|^p\,dx \\ &< \frac{\epsilon}{3} + 2^{p-1}\int_{\overline{G}} \left(|u(x) - J_\eta * u(x)|^p + |J_\eta * u(x) - \psi_j(x)|^p\right) dx \\ &< \frac{\epsilon}{3} + 2^{p-1}\left(\frac{\epsilon}{3\cdot 2^{p-1}} + \frac{\epsilon}{3 \cdot 2^{p-1}\cdot \mathrm{vol}(\overline{G})}\mathrm{vol}(\overline{G})\right) = \epsilon.\end{aligned}$$

Hence K has a finite ϵ-net in $L^p(\mathbb{R}^n)$ and is precompact there by Theorem 1.19. ∎

2.33 THEOREM Let $1 \le p < \infty$ and let $K \subset L^p(\Omega)$. Suppose there exists a sequence $\{\Omega_j\}$ of subdomains of Ω having the following properties:

(i) $\Omega_j \subset \Omega_{j+1}$ for each j.

(ii) The set of restrictions to Ω_j of the functions in K is precompact in $L^p(\Omega_j)$ for each j.

(iii) For every $\epsilon > 0$ there exists j such that

$$\int_{\Omega - \Omega_j} |u(x)|^p\,dx < \epsilon \qquad \text{for every } u \in K.$$

Then K is precompact in $L^p(\Omega)$.

Proof. Let $\{u_n\}$ be a sequence in K. By (ii) there exists a subsequence $\{u_n^{(1)}\}$ whose restrictions of Ω_1 converge in $L^p(\Omega_1)$. Having selected $\{u_n^{(1)}\}, \ldots, \{u_n^{(k)}\}$, we may select a subsequence $\{u_n^{(k+1)}\}$ of $\{u_n^{(k)}\}$ whose restrictions to Ω_{k+1} converge in $L^p(\Omega_{k+1})$. The restrictions of $\{u_n^{(k+1)}\}$ to Ω_j also converge in $L^p(\Omega_j)$ for $1 \le j \le k$ by (i).

Let $v_n = u_n^{(n)}$ for $n = 1, 2, \ldots$. Clearly $\{v_n\}$ is a subsequence of $\{u_n\}$. Given $\epsilon > 0$, (iii) assures us that there exists j such that

$$\int_{\Omega - \Omega_j} |v_n(x) - v_m(x)|^p \, dx < \frac{\epsilon}{2}$$

for all $n, m = 1, 2, \ldots$. Except for the first $j - 1$ terms, $\{v_n\}$ is a subsequence of $\{u_n^{(j)}\}$, so its restrictions to Ω_j form a Cauchy sequence in $L^p(\Omega_j)$. Thus for n, m sufficiently large,

$$\int_{\Omega_j} |v_n(x) - v_m(x)|^p \, dx < \frac{\epsilon}{2},$$

and

$$\int_{\Omega} |v_n(x) - v_m(x)|^p \, dx < \epsilon.$$

Thus $\{v_n\}$ is a Cauchy sequence in $L^p(\Omega)$ and so converges there. Hence K is precompact in $L^p(\Omega)$. ∎

Uniform Convexity

2.34 As noted previously, the parallelogram law in an inner product space guarantees the uniform convexity of the corresponding norm on that space. This applies to $L^2(\Omega)$. Now we will develop certain inequalities due to Clarkson [Clk] that generalize the parallelogram law and verify the uniform convexity of $L^p(\Omega)$ for $1 < p < \infty$.

We begin by preparing three technical lemmas needed for the proof.

2.35 LEMMA If $0 < s < 1$, then $f(t) = (1 - s^t)/t$ is a decreasing function of $t > 0$.

Proof. $f'(t) = (1/t^2)\big(g(s^t) - 1\big)$ where $g(r) = r - r \ln r$. Since $0 < s^t < 1$ and since $g'(r) = -\ln r \geq 0$ for $0 < r \leq 1$, it follows that $g(s^t) < g(1) = 1$ whence $f'(t) < 0$. ∎

2.36 LEMMA If $1 < p \leq 2$ and $0 \leq t \leq 1$, then

$$\left(\frac{1+t}{2}\right)^{p'} + \left(\frac{1-t}{2}\right)^{p'} \leq \left(\frac{1}{2} + \frac{1}{2}t^p\right)^{1/(p-1)}, \tag{25}$$

where $p' = p/(p-1)$ is the exponent conjugate to p.

Proof. Since equality holds in (25) if either $p = 2$ or $t = 0$ or $t = 1$, we may assume that $1 < p < 2$ and that $0 < t < 1$. Under the transformation

$t = (1-s)/(1+s)$, which maps $0 < t < 1$ onto $1 > s > 0$, (25) reduces to the equivalent form

$$\frac{1}{2}\left((1+s)^p + (1-s)^p\right) - (1+s^{p'})^{p-1} \geq 0. \tag{26}$$

The power series expansion of the left side of (26) takes the form

$$\begin{aligned}
&\frac{1}{2}\sum_{k=0}^{\infty}\binom{p}{k}s^k + \frac{1}{2}\sum_{k=0}^{\infty}\binom{p}{k}(-s)^k - \sum_{k=0}^{\infty}\binom{p-1}{k}s^{p'k} \\
&= \sum_{k=0}^{\infty}\binom{p}{2k}s^{2k} - \sum_{k=0}^{\infty}\binom{p-1}{k}s^{p'k} \\
&= \sum_{k=1}^{\infty}\left[\binom{p}{2k}s^{2k} - \binom{p-1}{2k-1}s^{p'(2k-1)} - \binom{p-1}{2k}s^{2p'k}\right],
\end{aligned}$$

where

$$\binom{p}{0} = 1 \quad \text{and} \quad \binom{p}{k} = \frac{p(p-1)(p-2)\cdots(p-k+1)}{k!}, \quad k \geq 1.$$

The latter series certainly converges for $0 \leq s < 1$. We prove (26) by showing that each term of the series is positive for $0 < s < 1$. The kth term (in square brackets above) can be written in the form

$$\begin{aligned}
&\frac{p(p-1)(2-p)(3-p)\cdots(2k-1-p)}{(2k)!}s^{2k} \\
&\quad - \frac{(p-1)(2-p)\cdots(2k-1-p)}{(2k-1)!}s^{p'(2k-1)} + \frac{(p-1)(2-p)\cdots(2k-p)}{(2k)!}s^{2kp'} \\
&= \frac{(2-p)\cdots(2k-p)}{(2k-1)!}s^{2k}\left[\frac{p(p-1)}{2k(2k-p)} - \frac{p-1}{2k-p}s^{p'(2k-1)-2k} + \frac{p-1}{2k}s^{2kp'-2k}\right] \\
&= \frac{(2-p)\cdots(2k-p)}{(2k-1)!}s^{2k}\left[\frac{1-s^{(2k-p)/(p-1)}}{(2k-p)/(p-1)} - \frac{1-s^{2k/(p-1)}}{2k/(p-1)}\right].
\end{aligned}$$

The first factor is positive since $p < 2$; the factor in the square brackets is positive by Lemma 2.35 since $0 < (2k-p)/(p-1) < 2k/(p-1)$. Thus (26) and hence (25) is established. ∎

2.37 LEMMA Let $z, w \in \mathbb{C}$. If $1 < p \leq 2$ and $p' = p/(p-1)$, then

$$\left|\frac{z+w}{2}\right|^{p'} + \left|\frac{z-w}{2}\right|^{p'} \leq \left(\frac{1}{2}|z|^p + \frac{1}{2}|w|^p\right)^{1/(p-1)}. \tag{27}$$

If $2 \le p < \infty$, then

$$\left|\frac{z+w}{2}\right|^p + \left|\frac{z-w}{2}\right|^p \le \frac{1}{2}|z|^p + \frac{1}{2}|w|^p. \tag{28}$$

Proof. Since (27) obviously holds if $z = 0$ or $w = 0$ and is symmetric in z and w, we can assume that $|z| \ge |w| > 0$. If $w/z = re^{i\theta}$ where $0 \le r \le 1$ and $0 \le \theta < 2\pi$, then (27) can be rewritten in the form

$$\left|\frac{1+re^{i\theta}}{2}\right|^{p'} + \left|\frac{1-re^{i\theta}}{2}\right|^{p'} \le \left(\frac{1}{2} + \frac{1}{2}r^p\right)^{1/(p-1)}. \tag{29}$$

If $\theta = 0$, then (29) is just the result of Lemma 2.36. We complete the proof of (29) by showing that for fixed r, $0 < r \le 1$, the function

$$f(\theta) = |1 + re^{i\theta}|^{p'} + |1 - re^{i\theta}|^{p'}$$

has a maximum value for $0 \le \theta < 2\pi$ at $\theta = 0$. Since

$$f(\theta) = (1 + r^2 + 2r\cos\theta)^{p'/2} + (1 + r^2 - 2r\cos\theta)^{p'/2},$$

satisfies $f(2\pi - \theta) = f(\pi - \theta) = f(\theta)$, we need consider f only on the interval $0 \le \theta \le \pi/2$. Since $p' \ge 2$, on that interval

$$f'(\theta) = -p'r\sin\theta\big[(1 + r^2 + 2r\cos\theta)^{(p'/2)-1} - (1 + r^2 - 2r\cos\theta)^{(p'/2)-1}\big] \le 0.$$

Thus the maximum value of f does indeed occur at $\theta = 0$ and (29), and therefore also (27), is proved.

If $2 \le p < \infty$, then $1 < p' \le 2$, and we have by interchanging p and p' in (27) and using Lemma 2.2,

$$\begin{aligned}
\left|\frac{z+w}{2}\right|^p + \left|\frac{z-w}{2}\right|^p &\le \left(\frac{1}{2}|z|^{p'} + \frac{1}{2}|w|^{p'}\right)^{1/(p'-1)} \\
&= \left(\frac{1}{2}|z|^{p'} + \frac{1}{2}|w|^{p'}\right)^{p/p'} \\
&\le 2^{(p/p')-1}\left[\left(\frac{1}{2}\right)^{p/p'}|z|^p + \left(\frac{1}{2}\right)^{p/p'}|w|^p\right] \\
&= \frac{1}{2}|z|^p + \frac{1}{2}|w|^p,
\end{aligned}$$

so that (28) is also proved. ∎

2.38 THEOREM (Clarkson's Inequalities) Let $u, v \in L^p(\Omega)$. For $1 < p < \infty$ let $p' = p/(p-1)$. If $2 \le p < \infty$, then

$$\left\| \frac{u+v}{2} \right\|_p^p + \left\| \frac{u-v}{2} \right\|_p^p \le \frac{1}{2} \|u\|_p^p + \frac{1}{2} \|v\|_p^p, \tag{30}$$

$$\left\| \frac{u+v}{2} \right\|_p^{p'} + \left\| \frac{u-v}{2} \right\|_p^{p'} \ge \left(\frac{1}{2} \|u\|_p^p + \frac{1}{2} \|v\|_p^p \right)^{p'-1}. \tag{31}$$

If $1 < p \le 2$, then

$$\left\| \frac{u+v}{2} \right\|_p^{p'} + \left\| \frac{u-v}{2} \right\|_p^{p'} \le \left(\frac{1}{2} \|u\|_p^p + \frac{1}{2} \|v\|_p^p \right)^{p'-1}, \tag{32}$$

$$\left\| \frac{u+v}{2} \right\|_p^p + \left\| \frac{u-v}{2} \right\|_p^p \ge \frac{1}{2} \|u\|_p^p + \frac{1}{2} \|v\|_p^p. \tag{33}$$

Proof. For $2 \le p < \infty$, (30) is obtained by using $z = u(x)$ and $w = v(x)$ in (28) and integrating over Ω. To prove (32) for $1 < p \le 2$ we first note that $\left\| |u|^{p'} \right\|_{p-1} = \|u\|_p^{p'}$ for any $u \in L^p(\Omega)$. Using the reverse Minkowski inequality (7) corresponding to the exponent $p - 1 < 1$, and (27) with $z = u(x)$ and $w = v(x)$, we obtain

$$\begin{aligned}
\left\| \frac{u+v}{2} \right\|_p^{p'} + \left\| \frac{u-v}{2} \right\|_p^{p'} &= \left\| \left| \frac{u+v}{2} \right|^{p'} \right\|_{p-1} + \left\| \left| \frac{u-v}{2} \right|^{p'} \right\|_{p-1} \\
&\le \left[\int_\Omega \left(\left| \frac{u(x)+v(x)}{2} \right|^{p'} + \left| \frac{u(x)-v(x)}{2} \right|^{p'} \right)^{p-1} dx \right]^{1/(p-1)} \\
&\le \left[\int_\Omega \left(\frac{1}{2} |u(x)|^p + \frac{1}{2} |v(x)|^p \right) dx \right]^{p'-1} \\
&= \left(\frac{1}{2} \|u\|_p^p + \frac{1}{2} \|v\|_p^p \right)^{p'-1}
\end{aligned}$$

which is (32).

Inequality (31) is proved for $2 \le p < \infty$ by the same method used to prove (32) except that the direct Minkowski inequality (5), corresponding to $p - 1 \ge 1$, is used in place of the reverse inequality, and in place of (27) is used the inequality

$$\left(\left| \frac{\xi + \eta}{2} \right|^{p'} + \left| \frac{\xi - \eta}{2} \right|^{p'} \right)^{p-1} \ge \frac{1}{2} |\xi|^p + \frac{1}{2} |\eta|^p,$$

which is obtained from (27) by replacing p by p', z by $\xi + \eta$, and w by $\xi - \eta$. Finally, (33) can be obtained from a similar revision of (28).

We remark that if $p = 2$, all four Clarkson inequalities reduce to the parallelogram law

$$\|u + v\|_2^2 + \|u - v\|_2^2 = 2\,\|u\|_2^2 + 2\,\|v\|_2^2 . \blacksquare$$

2.39 THEOREM If $1 < p < \infty$, then $L^p(\Omega)$ is uniformly convex.

Proof. Let $u, v \in L^p(\Omega)$ satisfy $\|u\|_p = \|v\|_p = 1$ and $\|u - v\|_p \geq \epsilon$ where $0 < \epsilon < 2$. If $2 \leq p < \infty$, then (30) implies that

$$\left\| \frac{u+v}{2} \right\|_p^p \leq 1 - \frac{\epsilon^p}{2^p}.$$

If $1 < p \leq 2$, then (32) implies that

$$\left\| \frac{u+v}{2} \right\|_p^{p'} \leq 1 - \frac{\epsilon^{p'}}{2^{p'}}.$$

In either case there exists $\delta = \delta(\epsilon) > 0$ such that $\|(u+v)/2\|_p \leq 1 - \delta$. $\blacksquare$

See [BKC] for sharper information on L^p geometry.

2.40 COROLLARY $L^p(\Omega)$ is reflexive if $1 < p < \infty$. $\blacksquare$

This is a consequence of uniform convexity via Theorem 1.21. We will give a direct proof after calculating the normed dual of $L^p(\Omega)$.

The Normed Dual of $L^p(\Omega)$

2.41 (Linear Functionals) Let $1 \leq p \leq \infty$ and let p' be the exponent conjugate to p. Each element $v \in L^{p'}(\Omega)$ defines a linear functional L_v on $L^p(\Omega)$ via

$$L_v(u) = \int_\Omega u(x)v(x)\,dx, \qquad u \in L^p(\Omega).$$

By Hölder's inequality $|L_v(u)| \leq \|u\|_p \|v\|_{p'}$, so that $L_v \in [L^p(\Omega)]'$ and

$$\|L_v; [L^p(\Omega)]'\| \leq \|v\|_{p'} .$$

Equality must hold above. If $1 < p \leq \infty$, let $u(x) = |v(x)|^{p'-2}\overline{v(x)}$ if $v(x) \neq 0$ and $u(x) = 0$ otherwise. Then $u \in L^p(\Omega)$ and $L_v(u) = \|u\|_p \|v\|_{p'}$.

Now suppose $p = 1$ so $p' = \infty$. If $\|v\|_{p'} = 0$, let $u(x) = 0$. Otherwise let $0 < \epsilon < \|v\|_\infty$ and let A be a measurable subset of Ω such that $0 < \mu(A) < \infty$

and $|v(x)| > \|v\|_\infty - \epsilon$ on A. Let $u(x) = \overline{v(x)}/|v(x)|$ on A and $u(x) = 0$ elsewhere. Then $u \in L^1(\Omega)$ and $L_v(u) \geq \|u\|_1 (\|v\|_\infty - \epsilon)$. Thus we have shown that

$$\|L_v; [L^p(\Omega)]'\| = \|v\|_{p'}, \tag{34}$$

so that the operator $\mathcal{L}$ mapping v to L_v is an isometric isomorphism of $L^{p'}(\Omega)$ onto a subspace of $[L^p(\Omega)]'$.

2.42 It is natural to ask if the range if the isomorphism $\mathcal{L}$ is all of $[L^p(\Omega)]'$. That is, is every continuous linear functional on $L^p(\Omega)$ of the form L_v for some $v \in L^{p'}(\Omega)$? We will show that such is the case if $1 \leq p < \infty$. For $p = 2$, this is an immediate consequence of the Riesz Representation Theorem 1.12 for Hilbert spaces. For general p a direct proof can be based on the Radon-Nikodym Theorem 1.52 (see [Ru2] or Theorem 8.19). We will give a more elementary proof based on a variational argument and uniform convexity. We will use a limiting argument to obtain the case $p = 1$ from the case $p > 1$.

2.43 LEMMA Let $1 < p < \infty$. If $L \in [L^p(\Omega)]'$, and $\|L; [L^p(\Omega)]'\| = 1$, then there exists a unique $w \in L^p(\Omega)$ such that $\|w\|_p = L(w) = 1$. Dually, if $w \in L^p(\Omega)$ is given and $\|w\|_p = 1$, then there exists a unique $L \in [L^p(\Omega)]'$ such that $\|L; [L^p(\Omega)]'\| = L(w) = 1$.

Proof. First assume that $L \in [L^p(\Omega)]'$ is given and $\|L; [L^p(\Omega)]'\| = 1$. Then there exists a sequence $\{w_n\} \in L^p(\Omega)$ satisfying $\|w_n\|_p = 1$ and such that $\lim_{n\to\infty} |L(w_n)| = 1$. We may assume that $|L(w_n)| > 1/2$ for each n, and, replacing w_n by a suitable multiple of w_n by a complex number of unit modulus, that $L(w_n)$ is real and positive. Let $\epsilon > 0$. By the definition of uniform convexity, there exists a positive number $\delta > 0$ such that if u and v belong to the unit ball of $L^p(\Omega)$ and if $\|(u+v)/2\|_p > 1-\delta$, then $\|u - v\|_p < \epsilon$. On the other hand, there exist an integer N such that $L(w_n) > 1 - \delta$ for all $n > N$. When $m > N$ also, we have that $L\big((w_m + w_n)/2\big) > 1 - \delta$, and then $\|w_m - w_n\|_p < \epsilon$. Therefore $\{w_n\}$ is a Cauchy sequence in $L^p(\Omega)$ and so converges to a limit w in that space. Clearly, $\|w\|_p = 1$ and $L(w) = \lim_{n\to\infty} L(w_n) = 1$. For uniqueness, if there were two candidates v and w, then the sequence $\{v, w, v, w, \ldots\}$ would have to converge, forcing $v = w$.

Now suppose $w \in L^p(\Omega)$ is given and $\|w\|_p = 1$. As noted in Paragraph 2.41 the functional L_v defined by

$$L_v(u) = \int_\Omega u(x)v(x)\,dx, \qquad u \in L^p(\Omega), \tag{35}$$

where

$$v(x) = \begin{cases} |w(x)|^{p-2}\overline{w(x)} & \text{if } w(x) \neq 0 \\ 0 & \text{otherwise} \end{cases} \tag{36}$$

satisfies $L_v(w) = \|w\|_p^p = 1$ and $\|L_v; [L^p(\Omega)]'\| = \|v\|_{p'} = \|w\|_p^{p/p'} = 1$. It remains to be shown, therefore, that if $L_1, L_2 \in [L^p(\Omega)]'$ satisfy $\|L_1\| = \|L_2\| = 1$ and $L_1(w) = L_2(w) = 1$, then $L_1 = L_2$. Suppose not. Then there exists $u \in L^p(\Omega)$ such that $L_1(u) \neq L_2(u)$. Replacing u by a suitable multiple of u, we may assume that $L_1(u) - L_2(u) = 2$. Then replacing u by its sum with a suitable multiple of w, we can arrange that $L_1(u) = 1$ and $L_2(u) = -1$. If $t > 0$, then $L(w + tu) = 1 + t$. Since $\|L_1\| = 1$, therefore $\|w + tu\|_p \geq 1 + t$. Similarly, $L_2(w - tu) = 1 + t$ and so $\|w - tu\|_p \geq 1 + t$. If $1 < p \leq 2$, Clarkson's inequality (33) gives

$$1 + t^p \|u\|_p^p = \left\| \frac{(w + tu) + (w - tu)}{2} \right\|_p^p + \left\| \frac{(w + tu) - (w - tu)}{2} \right\|_p^p$$

$$\geq \frac{1}{2} \|w + tu\|_p^p + \frac{1}{2} \|w - tu\|_p^p \geq (1 + t)^p,$$

which is not possible for all $t > 0$. Similarly, if $2 \leq p < \infty$, Clarkson's inequality (31) gives

$$1 + t^{p'} \|u\|_p^{p'} = \left\| \frac{(w + tu) + (w - tu)}{2} \right\|_p^{p'} + \left\| \frac{(w + tu) - (w - tu)}{2} \right\|_p^{p'}$$

$$\geq \left(\frac{1}{2} \|w + tu\|_p^p + \frac{1}{2} \|w - tu\|_p^p \right)^{p'-1} \geq (1 + t)^{p'},$$

which is also not possible for all $t > 0$. Thus no such u can exist, and $L_1 = L_2$. ∎

2.44 THEOREM (The Riesz Representation Theorem for $L^p(\Omega)$) Let $1 < p < \infty$ and let $L \in [L^p(\Omega)]'$. Then there exists $v \in L^{p'}(\Omega)$ such that for all $u \in L^p(\Omega)$

$$L(u) = L_v(u) = \int_\Omega u(x)v(x)\, dx.$$

Moreover, $\|v\|_{p'} = \|L; [L^p(\Omega)]'\|$. Thus $[L^p(\Omega)]' \cong L^{p'}(\Omega)$; $[L^p(\Omega)]'$ is isometrically isomorphic to $L^{p'}(\Omega)$.

Proof. If $L = 0$ we may take $v = 0$. Thus we can assume $L \neq 0$, and, without loss of generality, that $\|L; [L^p(\Omega)]'\| = 1$. By Lemma 2.43 there exists $w \in L^p(\Omega)$ with $\|w\|_p = 1$ such that $L(w) = 1$. Let v be given by (36). Then L_v, defined by (35), satisfies $\|L_v; [L^p(\Omega)]'\| = 1$ and $L_v(w) = 1$. By Lemma 2.43 again, we have $L = L_v$. Since $\|v\|_{p'} = 1$, the proof is complete. ∎

2.45 THEOREM (The Riesz Representation Theorem for $L^1(\Omega)$) Let $L \in [L^1(\Omega)]'$. Then there exists $v \in L^\infty(\Omega)$ such that for all $u \in L^1(\Omega)$

$$L(u) = \int_\Omega u(x)v(x)\, dx$$

and $\|v\|_\infty = \|L; [L^1(\Omega)]'\|$. Thus $[L^1(\Omega)]' \cong L^\infty(\Omega)$.

Proof. Once again we assume that $L \neq 0$ and $\|L; [L^1(\Omega)]'\| = 1$. Let us suppose, for the moment, that Ω has finite volume. If $1 < p < \infty$, then by Theorem 2.14 we have $L^p(\Omega) \subset L^1(\Omega)$ and

$$|L(u)| \leq \|u\|_1 \leq \left(\mathrm{vol}(\Omega)\right)^{1-(1/p)} \|u\|_p$$

for any $u \in L^p(\Omega)$. Hence $L \in [L^p(\Omega)]'$ and by Theorem 2.44 there exists $v_p \in L^{p'}(\Omega)$ such that

$$L(u) = \int_\Omega u(x)v_p(x)\,dx, \qquad u \in L^p(\Omega) \tag{37}$$

and

$$\left\|v_p\right\|_{p'} \leq \left(\mathrm{vol}(\Omega)\right)^{1-(1/p)}. \tag{38}$$

Since $C_0^\infty(\Omega)$ is dense in $L^p(\Omega)$ for $1 < p < \infty$, and since for any p, q satisfying $1 < p, q < \infty$ and any $\phi \in C_0^\infty(\Omega)$ we have

$$\int_\Omega \phi(x)v_p(x)\,dx = L(\phi) = \int_\Omega \phi(x)v_q(x)\,dx,$$

it follows that $v_p = v_q$ a.e. on Ω. Hence we may replace v_p in (37) with a function v belonging to $L^p(\Omega)$ for each p, $1 < p < \infty$, and satisfying, following (38)

$$\|v\|_{p'} \leq \left(\mathrm{vol}(\Omega)\right)^{1-(1/p)} = \left(\mathrm{vol}(\Omega)\right)^{1/p'}.$$

It follows by Theorem 2.14 again that $v \in L^\infty(\Omega)$ and

$$\|v\|_\infty \leq \lim_{p'\to\infty} \left(\mathrm{vol}(\Omega)\right)^{1/p'} = 1. \tag{39}$$

The argument of Paragraph 2.41 shows that there must be equality in (39).

Even if Ω does not have finite volume, we can still write $\Omega = \bigcup_{j=1}^\infty G_j$, where $G_j = \{x \in \Omega : j-1 \leq |x| < j\}$ has finite volume. The sets G_j are mutually disjoint. Let χ_j be the characteristic function of G_j. If $u_j \in L^1(G_j)$, let $\tilde{u}_j$ denote the zero extension of u_j outside G_j. Let $L_j(u_j) = L(\tilde{u}_j)$. Then $L_j \in [L^1(G_j)]'$ and $\|L_j; [L^1(G_j)]'\| \leq 1$. By the finite volume case considered above, there exists $v_j \in L^\infty(G_j)$ such that $\left\|v_j\right\|_{\infty,G_j} \leq 1$ and

$$L_j(u_j) = \int_{G_j} u_j(x)v_j(x)\,dx = \int_\Omega \tilde{u}_j(x)v(x)\,dx,$$

where $v(x) = v_j(x)$ for $x \in G_j$, $j = 1, 2, \ldots$, so that $\|v\|_\infty \le 1$. If $u \in L^1(\Omega)$, we put $u = \sum_{j=1}^{\infty} \chi_j u$; the series is norm convergent in $L^1(\Omega)$ by dominated convergence. Since

$$L\left(\sum_{j=1}^{k} \chi_j u\right) = \sum_{j=1}^{k} L_j(\chi_j u) = \int_\Omega \sum_{j=1}^{k} \chi_j(x) u(x) v(x)\, dx,$$

we obtain, passing to the limit by dominated convergence,

$$L(u) = \int_\Omega u(x) v(x)\, dx.$$

It then follows, as in the finite volume case, that $\|v\|_\infty = 1$. ∎

2.46 THEOREM (Reflexivity of $L^p(\Omega)$) $L^p(\Omega)$ is reflexive if and only if $1 < p < \infty$.

Proof. Let $X = L^p(\Omega)$, where $1 < p < \infty$. Since $X' \cong L^{p'}(\Omega)$, we have

$$X'' \cong [L^{p'}(\Omega)]' \cong L^p(\Omega).$$

That is, for every element $w \in X''$ there exists $u \in L^p(\Omega) = X$ such that $w(v) = v(u) = Ju(v)$ for all $v \in X'$, where J is the natural isometric isomorphism of x into X''. (See Paragraph 1.14.) Since the range of J is therefore all of X'', X is reflexive. ∎

Since $L^1(\Omega)$ is separable while its dual, which is isometrically isomorphic to $L^\infty(\Omega)$ is not separable, neither $L^1(\Omega)$ nor $L^\infty(\Omega)$ can be reflexive.

2.47 The Riesz Representation Theorem cannot hold for the space $L^\infty(\Omega)$ in a form analogous to Theorem 2.44, for if so, then the argument of Theorem 2.46 would show that $L^1(\Omega)$ was reflexive. The dual of $L^\infty(\Omega)$ is larger than $L^1(\Omega)$. It may be identified with a space of absolutely continuous, finitely additive set functions of bounded total variation on Ω. See, for example, [Y, p 118] for details.

Mixed-Norm L^p Spaces

2.48 It is sometimes useful to consider L^p type norms of functions on $\mathbb{R}^n$ involving different exponents in different coordinate directions. Given a measurable function u on $\mathbb{R}^n$ and an index vector $\mathbf{p} = (p_1, \ldots, p_n)$ where $0 < p_i \le \infty$ for $1 \le i \le n$, we can calculate the number $\|u\|_{\mathbf{p}}$ by calculating first the L^{p_1}-norm of $u(x_1, x_2, \ldots, x_n)$ with respect to the variable x_1, and then the L^{p_2}-norm of the result with respect to the variable x_2, and so on, finishing with the L^{p_n}-norm with respect to x_n:

$$\|u\|_{\mathbf{p}} = \left\| \cdots \left\| \|u\|_{L^{p_1}(dx_1)} \right\|_{L^{p_2}(dx_2)} \cdots \right\|_{L^{p_n}(dx_n)}$$

where

$$\|f\|_{L^q(dt)} = \begin{cases} \left[\int_{-\infty}^{\infty} |f(\dots,t,\dots)|^q \, dt\right]^{1/q} & \text{if } 0 < q < \infty \\ \operatorname*{ess\,sup}_t |f(\dots,t,\dots)| & \text{if } q = \infty. \end{cases}$$

Of course, $\|\cdot\|_{L^q(dt)}$ is not a norm unless $q \geq 1$. For instance, if all the numbers p_i are finite, then

$$\|u\|_{\mathbf{p}} = \left[\int_{-\infty}^{\infty} \cdots \left[\int_{-\infty}^{\infty} \left[\int_{-\infty}^{\infty} |u(x_1,\dots,x_n)|^{p_1} \, dx_1\right]^{p_2/p_1} dx_2\right]^{p_3/p_2} dx_3 \cdots dx_n\right]^{1/p}$$

We will denote by $L^{\mathbf{p}} = L^{\mathbf{p}}(\mathbb{R}^n)$ the set of (equivalence classes of almost everywhere equal) functions u for which $\|u\|_{\mathbf{p}} < \infty$; this is a Banach space with norm $\|\cdot\|_{\mathbf{p}}$ if all $p_i \geq 1$. The standard reference for information on these *mixed-norm* spaces is [BP]. All that we require about mixed norms in this book are two elementary results, a version of Hölder's inequality, and an inequality concerning the effect on mixed norms of permuting the order in which the L^{p_i}-norms are calculated.

2.49 (Hölder's Inequality for Mixed Norms) Let $0 < p_i \leq \infty$ and let $0 < q_i \leq \infty$ for $1 \leq i \leq n$. If $u \in L^{\mathbf{p}}$ and $v \in L^{\mathbf{q}}$, then $uv \in L^{\mathbf{r}}$ where

$$\frac{1}{r_i} = \frac{1}{p_i} + \frac{1}{q_i}, \qquad 1 \leq i \leq n, \tag{40}$$

and we have Hölder's inequality:

$$\|uv\|_{\mathbf{r}} \leq \|u\|_{\mathbf{p}} \|v\|_{\mathbf{q}}$$

This inequality can be proved by simply applying the (scalar) version of Hölder's inequality given in Corollary 2.5 one variable at a time. As in Corollary 2.5, p_i and q_i are allowed to be less than 1 in this form of Hölder's inequality. The n equations (40) are usually summarized with the convenient abuse of notation

$$\frac{1}{\mathbf{r}} = \frac{1}{\mathbf{p}} + \frac{1}{\mathbf{q}}.$$

The above form of Hölder's inequality can be iterated to provide a version for a product of k functions:

$$\left\|\prod_{j=1}^{k} u_j\right\|_{\mathbf{r}} \leq \prod_{j=1}^{k} \|u_j\|_{\mathbf{p}_j} \qquad \text{where} \quad \frac{1}{\mathbf{r}} = \sum_{j=1}^{k} \frac{1}{\mathbf{p}_j}.$$

2.50 (Permuted Mixed Norms) The definition of $\|u\|_{\mathbf{p}}$ requires the successive L^{p_i}-norms to be calculated in the order of appearance of the variables in the argument of u. This order can be changed by permuting the arguments and associated indices. If σ is a permutation of the set $\{1, 2, \dots, n\}$, denote $\sigma x = (x_{\sigma(1)}, x_{\sigma(2)}, \dots, x_{\sigma(n)})$, and let $\sigma\mathbf{p}$ be defined similarly. Let σu be defined by $\sigma u(\sigma x) = u(x)$, that is, $\sigma u(x) = u(\sigma^{-1}x)$. Then $\|\sigma u\|_{\sigma\mathbf{p}}$ is called a permuted mixed norm of u. For example, if $n = 2$ and $\sigma\{1, 2\} = \{2, 1\}$, then

$$\|u\|_{\mathbf{p}} = \left[\int_{-\infty}^{\infty}\left[\int_{-\infty}^{\infty}|u(x_1, x_2)|^{p_1}\,dx_1\right]^{p_2/p_1} dx_2\right]^{1/p_2}$$

$$\|\sigma u\|_{\sigma\mathbf{p}} = \left[\int_{-\infty}^{\infty}\left[\int_{-\infty}^{\infty}|u(x_1, x_2)|^{p_2}\,dx_2\right]^{p_1/p_2} dx_1\right]^{1/p_1}.$$

Note that $\|u\|_{\mathbf{p}}$ and $\|\sigma u\|_{\sigma\mathbf{p}}$ involve the same L^{p_i}-norms with respect to the same variables; only the order of evaluation of those norms has been changed. The question of comparing the sizes of these mixed norms naturally arises.

2.51 THEOREM (The Permutation Inequality for Mixed Norms) Given an index vector $\mathbf{p}$, let σ_* and σ^* be permutations of $\{1, 2, \dots, n\}$ having components in nondecreasing order and nonincreasing order respectively:

$$p_{\sigma_*(1)} \le p_{\sigma_*(2)} \le \cdots \le p_{\sigma_*(n)},$$

$$p_{\sigma^*(1)} \ge p_{\sigma^*(2)} \ge \cdots \ge p_{\sigma^*(n)}.$$

Then for any permutation σ of $\{1, 2, \dots, n\}$ and any function u we have

$$\|\sigma_* u\|_{\sigma_*\mathbf{p}} \le \|\sigma u\|_{\sigma\mathbf{p}} \le \|\sigma^* u\|_{\sigma^*\mathbf{p}}.$$

Proof. Since any permutation can be decomposed into a product of special permutations each of which transposes two adjacent elements and leaves the rest unmoved, proving the inequality reduces to demonstrating the special case: if $p_1 \le p_2 < \infty$, then

$$\left[\int_{-\infty}^{\infty}\left[\int_{-\infty}^{\infty}|u|^{p_1}\,dx_1\right]^{p_2/p_1} dx_2\right]^{1/p_2} \le \left[\int_{-\infty}^{\infty}\left[\int_{-\infty}^{\infty}|u|^{p_2}\,dx_2\right]^{p_1/p_2} dx_1\right]^{1/p_1}.$$

But this is just a version of Minkowski's inequality for integrals (Theorem 2.9), namely

$$\left\|\int_{-\infty}^{\infty}|v(x_1, x_2)|\,dx_1\right\|_{L^r(dx_2)} \le \int_{-\infty}^{\infty}\|v(x_1, \cdot)\|_{L^r(dx_2)}\,dx_1$$

applied to $v = |u|^{p_1}$ with $r = p_2/p_1$. The case where $p_2 = \infty$ is easier. ∎

2.52 REMARK Similar permutation inequalities hold for mixed norm ℓ^p spaces and for hybrid mixtures of ℓ^p and L^q norms. We will use such inequalities in Chapter 7.

The Marcinkiewicz Interpolation Theorem

2.53 (Distribution Functions) Let Ω be a domain in $\mathbb{R}^n$ and u be a measurable function defined on Ω. For $t \geq 0$, let

$$\Omega_{u,t} = \{x \in \Omega \,:\, |u(x)| > t\}.$$

We define the *distribution function* of u to be

$$\delta_u(t) = \mu\left(\Omega_{u,t}\right),$$

where μ is the Lebesgue measure on $\mathbb{R}^n$. Evidently δ_u is nonincreasing for $t \geq 0$ and if $|u(x)| \leq |v(x)|$ a.e. on Ω, then $\delta_u(t) \leq \delta_v(t)$ for $t \geq 0$.

Since $|u(x)| > t$ implies $|u(x)| > t + (1/k)$ for some integer $k > 0$, we have $\Omega_{u,t} = \bigcup_{k=1}^{\infty} \Omega_{u,t+(1/k)}$ and it follows that δ_u is right continuous on the interval $[0, \infty)$. Similarly, if $|u(x)|$ is an increasing limit of $\{|u_j(x)|\}$ at each x, then $|u(x)| > t$ implies $|u_j(x)| > t$ for some j and so $\Omega_{u,t} = \bigcup_{j=1}^{\infty} \Omega_{u_j,t}$. Hence $\lim_{j\to\infty} \delta_{u_j}(t) = \delta_u(t)$.

If $|u(x) + v(x)| > t$, then either $|u(x)| > t/2$ or $|v(x)| > t/2$ (or both), so that $\Omega_{u+v,t} \subset \Omega_{u,t/2} \cup \Omega_{v,t/2}$ and hence

$$\delta_{u+v}(t) \leq \delta_u(t/2) + \delta_v(t/2). \tag{41}$$

Now suppose $u \in L^p(\Omega)$ for some p satisfying $0 < p < \infty$. For $t > 0$ we have

$$\|u\|_p^p = \int_\Omega |u(x)|^p\,dx \geq \int_{\Omega_{u,t}} |u(x)|^p\,dx \geq t^p \mu\left(\Omega_{u,t}\right),$$

from which we obtain *Chebyshev's inequality*

$$\delta_u(t) = \mu\left(\Omega_{u,t}\right) \leq t^{-p}\,\|u\|_p^p. \tag{42}$$

2.54 LEMMA If $0 < p < \infty$, then

$$\|u\|_p^p = \int_\Omega |u(x)|^p\,dx = p\int_0^\infty t^{p-1}\delta_u(t)\,dt. \tag{43}$$

Proof. First suppose $|u|$ is a simple function, say

$$|u(x)| = a_j \quad \text{on} \quad A_j \subset \Omega, \qquad 1 \le j \le k,$$

where $0 < a_1 < a_2 < \cdots < a_k$ and $A_i \cap A_j$ is empty for $i \neq j$. Then

$$\delta_u(t) = \begin{cases} \sum_{i=1}^{k} \mu(A_i) & \text{if } t < a_1 \\ \sum_{i=j}^{k} \mu(A_i) & \text{if } a_{j-1} \le t < a_j, \quad (2 \le j \le k) \\ 0 & \text{if } t \ge a_k. \end{cases}$$

Therefore,

$$\begin{aligned} p\int_0^\infty t^{p-1}\delta_u(t)\,dt &= p\left(\int_0^{a_1} + \sum_{j=2}^{k}\int_{a_{j-1}}^{a_j} + \int_{a_k}^{\infty}\right) t^{p-1}\delta_u(t)\,dt \\ &= a_1^p \sum_{j=1}^{k}\mu(A_j) + \sum_{j=2}^{k}\left(a_j^p - a_{j-1}^p\right)\sum_{i=j}^{k}\mu(A_i) \\ &= \sum_{j=1}^{k} a_j^p \mu(A_j) = \|u\|_p^p, \end{aligned}$$

so (43) holds for simple functions. By Theorem 1.44, if u is measurable, then $|u|$ is a limit of a monotonically increasing sequence of measurable simple functions. Equation (43) now follows by monotone convergence. ∎

2.55 (Weak L^p Spaces) If u is a measurable function on Ω, let

$$[u]_p = [u]_{p,\Omega} = \left(\sup_{t>0} t^p \delta_u(t)\right)^{1/p}.$$

We define the space weak-$L^p(\Omega)$ as follows:

$$\text{weak-}L^p(\Omega) = \{u : [u]_p < \infty\}.$$

It is easily checked that $[cu]_p = |c|[u]_p$ for complex c, but $[\cdot]_p$ is not a norm on weak-$L^p(\Omega)$ because it does not satisfy the triangle inequality. However, by (41)

$$\begin{aligned} [u+v]_p &= \left(\sup_{t>0} t^p \delta_{u+v}(t)\right)^{1/p} \\ &\le \left(2^p \sup_{t>0}\left(\frac{t}{2}\right)^p \delta_u(t/2) + 2^p \sup_{t>0}\left(\frac{t}{2}\right)^p \delta_v(t/2)\right)^{1/p} \\ &= 2\left([u]_p + [v]_p\right), \end{aligned}$$

so weak-$L^p(\Omega)$ is a vector space and the "open balls" $B_r(u) = \{v \in \text{weak-}L^p(\Omega) : [v - u]_p < r\}$ do generate a topology on weak-$L^p(\Omega)$ with respect to which weak-$L^p(\Omega)$ is a topological vector space. A functional $[\cdot]$ with the properties of a norm except that the triangle inequality is replaced with a weaker version of the form $[u + v] \le K([u] + [v])$ for some constant $K > 1$ is called a *quasi-norm.*

Chebyshev's inequality (42) shows that $[u]_p \le \|u\|_p$ so that $L^p(\Omega) \subset \text{weak-}L^p(\Omega)$. The inclusion is strict since, if $x_0 \in \Omega$ it is easily shown that $u(x) = |x - x_0|^{-n/p}$ belongs to weak-$L^p(\Omega)$ but not to $L^p(\Omega)$.

2.56 (Strong and Weak Type Operators) A operator F mapping a vector space X of measurable functions into another such space Y is called *sublinear* if, for all $u, v \in X$ and scalars c,

$$|F(u + v)| \le |F(u)| + |F(v)|, \qquad \text{and}$$
$$|T(cu)| = |c||T(u)|.$$

A linear operator from X into Y is certainly sublinear. We will be especially concerned with operators from L^p spaces on a domain Ω in $\mathbb{R}^n$ into $L^q(\Omega')$ or weak-$L^q(\Omega')$ where Ω' is a domain in $\mathbb{R}^k$ with k not necessarily equal to n.

We distinguish two important classes of sublinear operators:

(a) F is of *strong type* (p, q), where $1 \le p \le \infty$ and $1 \le q \le \infty$, if F maps $L^p(\Omega)$ into $L^q(\Omega')$ and there exists a constant K such that for all $u \in L^p(\Omega)$,

$$\|F(u)\|_{q,\Omega'} \le K \|u\|_{p,\Omega} .$$

(b) F is of *weak type* (p, q), where $1 \le p \le \infty$ and $1 \le q < \infty$, if F maps $L^p(\Omega)$ into weak-$L^q(\Omega')$ and there exists a constant K such that for all $u \in L^p(\Omega)$,

$$[F(u)]_{q,\Omega'} \le K \|u\|_{p,\Omega} .$$

We also say that F is of weak type (p, ∞) if F is of strong type (p, ∞).

Strong type (p, q) implies weak type (p, q) but not conversely unless $q = \infty$.

2.57 The following theorem has its origins in the work of Marcinkiewicz [Mk] and was further developed by Zygmund [Z]. It is valid in more general contexts than stated here, but we only need it for operators between L^p spaces on domains in $\mathbb{R}^n$ and only state it in this context. It will form one of the cornerstones on which our proof of the Sobolev imbedding theorem will rest. In that context it will only be used for linear operators.

Because the Marcinkiewicz theorem involves an operator on a vector space containing two different L^p spaces, say X and Y, (over the same domain) it is convenient to consider its domain to be the sum of those spaces, that is the vector space consisting of sums $u + v$ where $u \in X$ and $v \in Y$.

There are numerous proofs of the Marcinkiewicz theorem in the literature. See, for example, [St] and [SW]. Our proof is based on Folland [Fo].

2.58 THEOREM (The Marcinkiewicz Interpolation Theorem) Let $1 \le p_1 \le q_1 < \infty$ and $1 \le p_2 \le q_2 \le \infty$, with $q_1 < q_2$. Suppose the numbers p and q satisfy

$$\frac{1}{p} = \frac{1-\theta}{p_1} + \frac{\theta}{p_2}, \qquad \frac{1}{q} = \frac{1-\theta}{q_1} + \frac{\theta}{q_2},$$

where $0 < \theta < 1$. Let Ω and Ω' be domains in $\mathbb{R}^n$ and $\mathbb{R}^k$, respectively; k may or may not be equal to n. Let F be a sublinear operator from $L^{p_1}(\Omega) + L^{p_2}(\Omega)$ into the space of measurable functions on Ω'. If F is of weak type (p_1, q_1) and also of weak type (p_2, q_2), then F is of strong type (p, q). That is, if

$$[F(u)]_{q_j,\Omega'} \le K_j \, \|u\|_{p_j,\Omega}, \qquad j = 1, 2,$$

then

$$\|F(u)\|_{q,\Omega'} \le K \, \|u\|_{p,\Omega},$$

where the constant K depends only on $p, p_1, q_1, p_2, q_2, K_1$, and K_2.

Proof. First consider the case where $q_1 < q < q_2 < \infty$ so that p_1 and p_2 are necessarily both finite. The conditions satisfied by p and q imply that $(1/p, 1/q)$ is an interior point of the line segment joining (p_1^{-1}, q_1^{-1}) and (p_2^{-1}, q_2^{-1}) in the (p, q)-plane. Let c be the extended real number equal to q/p times the slope of that line segment;

$$c = \frac{p_1(q_1 - q)}{q_1(p_1 - p)} = \frac{p_2(q_2 - q)}{q_2(p_2 - p)}. \tag{44}$$

Given any $T > 0$, a measurable function u on Ω can be written as a sum of a "small" part $u_{S,T}$ and a "big" part $u_{B,T}$ defined as follows:

$$u_{S,T}(x) = \begin{cases} u(x) & \text{if } |u(x)| \le T \\ T\dfrac{u(x)}{|u(x)|} & \text{if } |u(x)| > T, \end{cases}$$

$$u_{B,T}(x) = u(x) - u_{S,T}(x) = \begin{cases} 0 & \text{if } |u(x)| \le T \\ u(x)\left(1 - \dfrac{T}{|u(x)|}\right) & \text{if } |u(x)| > T. \end{cases}$$

Since $|u_{S,T}(x)| \le T$ and $|u_{B,T}(x)| = \max\{0, |u(x)| - T\}$ for all $x \in \Omega$, the distribution functions of $u_{S,T}$ and $u_{B,T}$ are given by

$$\delta_{u_{S,T}}(t) = \begin{cases} \delta_u(t) & \text{if } t < T \\ 0 & \text{if } t \ge T, \end{cases}$$

$$\delta_{u_{B,T}}(t) = \delta_u(t + T).$$

It follows, using (43), that

$$\int_\Omega |u_{S,T}(x)|^{p_2}\,dx = p_2\int_0^\infty t^{p_2-1}\delta_{u_{S,T}}(t)\,dt = p_2\int_0^T t^{p_2-1}\delta_u(t)\,dt$$

$$\begin{aligned}\int_\Omega |u_{B,T}(x)|^{p_1}\,dx &= p_1\int_0^\infty t^{p_1-1}\delta_{u_{B,T}}(t)\,dt = p_1\int_0^\infty t^{p_1-1}\delta_u(t+T)\,dt\\ &= p_1\int_T^\infty (t-T)^{p_1-1}\delta_u(t)\,dt \le p_1\int_T^\infty t^{p_1-1}\delta_u(t)\,dt.\end{aligned}$$

Using (43) followed by the sublinearity of F and inequality (41), we calculate

$$\begin{aligned}\int_{\Omega'} |F(u)(y)|^q\,dy &= q\int_0^\infty t^{q-1}\delta_{F(u)}(t)\,dt\\ &= 2^q q\int_0^\infty t^{q-1}\delta_{F(u)}(2t)\,dt\\ &\le 2^q q\int_0^\infty t^{q-1}\delta_{F(u_{S,T})+F(u_{B,T})}(2t)\,dt\\ &\le 2^q q\int_0^\infty t^{q-1}\delta_{F(u_{S,T})}(t)\,dt + 2^q q\int_0^\infty t^{q-1}\delta_{F(u_{B,T})}(t)\,dt. \qquad (45)\end{aligned}$$

This inequality holds for any $T > 0$; we can choose T to depend on t if we wish. In the following, let $T = t^c$ where c is given by (44). For positive s, the definition of $[\cdot]_s$ implies that $\delta_v(t) \le t^{-s}[v]_s^s$. Using this and the given estimate $[F(v)]_{q_2,\Omega'} \le K_2\,\|v\|_{p,\Omega}$ we obtain

$$\begin{aligned}\int_0^\infty t^{q-1}\delta_{F(u_{S,T})}(t)\,dt &\le \int_0^\infty t^{q-1-q_2}[F(u_{S,T})]_{q_2}^{q_2}\,dt\\ &\le \int_0^\infty t^{q-1-q_2}\left(K_2\,\|u_{S,T}\|_{p_2}\right)^{q_2}dt\\ &\le K_2^{q_2}p_2^{q_2/p_2}\int_0^\infty t^{q-1-q_2}\left[\int_0^{t^c}\tau^{p_2-1}\delta_u(\tau)\,d\tau\right]^{q_2/p_2}dt\\ &= K_2^{q_2}p_2^{q_2/p_2}I_2.\end{aligned}$$

Since $q_2 \ge p_2$ we can estimate the latter iterated integral I_2 using Minkowski's inequality for integrals, Theorem 2.9.

$$
\begin{aligned}
I_2 &= \int_0^\infty \left[\int_0^{t^c} t^{(q-1-q_2)(p_2/q_2)} \tau^{p_2-1} \delta_u(\tau)\,d\tau \right]^{q_2/p_2} dt \\
&\le \left[\int_0^\infty \left(\int_{\tau^{1/c}}^\infty t^{q-1-q_2} \left(\tau^{p_2-1}\delta_u(\tau)\right)^{q_2/p_2} dt \right)^{p_2/q_2} d\tau \right]^{q_2/p_2} \\
&= \left[\int_0^\infty \tau^{p_2-1}\delta_u(\tau) \left(\int_{\tau^{1/c}}^\infty t^{q-1-q_2}\,dt \right)^{p_2/q_2} d\tau \right]^{q_2/p_2} \\
&= \left[\frac{1}{q_2-q} \int_0^\infty \tau^{p_2-1+[(q-q_2)/c](p_2/q_2)} \delta_u(\tau)\,d\tau \right]^{q_2/p_2} \\
&= \left(\frac{1}{q_2-q} \int_0^\infty \tau^{p-1}\delta_u(\tau)\,d\tau \right)^{q_2/p_2} = \left(\frac{1}{p(q_2-q)} \|u\|_{p,\Omega}^p \right)^{q_2/p_2}.
\end{aligned}
$$

It follows that

$$
2^q q \int_0^\infty t^{q-1} \delta_{F(u_{S,T})}(t)\,dt \le 2^q q K_2^{q_2} \left(\frac{p_2}{p(q_2-q)} \|u\|_{p,\Omega}^p \right)^{q_2/p_2}. \tag{46}
$$

An entirely parallel argument using $q_1 < q$ instead of $q_2 > q$ shows that

$$
2^q q \int_0^\infty t^{q-1} \delta_{F(u_{B,T})}(t)\,dt \le 2^q q K_1^{q_1} \left(\frac{p_1}{p(q-q_1)} \|u\|_{p,\Omega}^p \right)^{q_1/p_1}. \tag{47}
$$

If $\|u\|_{p,\Omega} = 1$, we therefore have

$$
\|F(u)\|_{q,\Omega'} \le K = 2q^{1/q} \left[\left(\frac{p_2 K_2^{p_2}}{p(q_2-q)} \right)^{q_2/p_2} + \left(\frac{p_1 K_1^{p_1}}{p(q-q_1)} \right)^{q_1/p_1} \right]^{1/q}.
$$

By the homogeneity of F, if $u \ne 0$ in $L^p(\Omega)$, then

$$
\begin{aligned}
\|F(u)\|_{q,\Omega'} &= \left\| F\left(\|u\|_{p,\Omega} \frac{u}{\|u\|_{p,\Omega}} \right) \right\|_{q,\Omega'} \\
&= \|u\|_{p,\Omega} \left\| F\left(\frac{u}{\|u\|_{p,\Omega}} \right) \right\|_{q,\Omega'} \le K \|u\|_{p,\Omega}.
\end{aligned}
$$

Now we examine the case where $q_2 = \infty$. It is possible to choose T (depending on t) in the above argument to ensure that $\delta_{F(u_{S,T})}(t) = 0$ for all $t > 0$. If $p_2 = \infty$, the appropriate choice is $T = t/K_2$ for then

$$
\left\|F(u_{S,T})\right\|_{\infty,\Omega'} \le K_2 \left\|u_{S,T}\right\|_{\infty,\Omega} \le K_2 T = t,
$$

and $\delta_{F(u_{S,T})}(t) = 0$. If $p_2 < \infty$, the appropriate choice is

$$T = \left(\frac{t}{K_2 \left(p_2 \|u\|_{p,\Omega}^p / p \right)^{1/p_2}} \right)^c ,$$

where $c = p_2/(p_2 - p)$, the limit as $q_2 \to \infty$ of the value of c used in the earlier part of this proof. For this choice of T,

$$\begin{aligned} \|F(u_{S,T})\|_{\infty,\Omega'}^{p_2} &\le K_2^{p_2} \|u_{S,T}\|_{p_2}^{p_2} = K_2^{p_2} p_2 \int_0^T t^{p_2-1} \delta_{u_{S,T}}(t)\, dt \\ &\le K_2^{p_2} p_2 T^{p_2-p} \int_0^T t^{p-1} \delta_u(t)\, dt \\ &\le K_2^{p_2} p_2 T^{p_2-p} \int_0^\infty t^{p-1} \delta_u(t)\, dt \\ &= K_2^{p_2} p_2 T^{p_2-p} (1/p) \|u\|_{p,\Omega}^p = t^{p_2}, \end{aligned}$$

and again $\delta_{F(u_{S,T})}(t) = 0$. In either of these cases the first term in (45) is zero and an estimate similar to (47) holds for the second term provided $p_1 < p_2$.

If $q_2 = \infty$ and $p_2 < p_1 < \infty$ we can instead assure that the second term in (45) is zero by choosing T to force $\delta_{F(u_{B,t})}(t) = 0$ and obtain an estimate similar to (46) for the first term.

There remains one case to be considered: $q_1 < q < q_2 = \infty$, $p_1 = p = p_2 < \infty$. In this case it follows directly from the definition of $[\cdot]_s$ that

$$t^{q_1} \delta_{F(u)}(t) \le [F(u)]_{q_1}^{q_1} \le K_1^{q_1} \|u\|_{p,\Omega}^{q_1},$$

and hence $\delta_{F(u)} \le \left(K_1 \|u\|_{p,\Omega} / t \right)^{q_1}$. On the other hand, $\delta_{F(u)}(t) = 0$ if we have $t \ge T = K_2 \|u\|_{p,\Omega} \ge \|F(u)\|_{\infty,\Omega'}$. Thus

$$\begin{aligned} \|F(u)\|_{q,\Omega'}^q &= q \int_0^\infty t^{q-1} \delta_{F(u)}(t)\, dt = q \int_0^T t^{q-1} \delta_{F(u)}(t)\, dt \\ &\le q \left(K_1 \|u\|_{p,\Omega} \right)^{q_1} \int_0^T t^{q-1-q_1}\, dt = K^q \|u\|_{p,\Omega}^{q_1}, \end{aligned}$$

where K is a finite constant because $q_1 < q$. This completes the proof. ∎

3

THE SOBOLEV SPACES $W^{m,p}(\Omega)$

In this chapter we introduce Sobolev spaces of integer order and establish some of their most important properties. These spaces are defined over an arbitrary domain $\Omega \subset \mathbb{R}^n$ and are vector subspaces of various Lebesgue spaces $L^p(\Omega)$.

Definitions and Basic Properties

3.1 (The Sobolev Norms) We define a functional $\|\cdot\|_{m,p}$, where m is a positive integer and $1 \le p \le \infty$, as follows:

$$\|u\|_{m,p} = \left(\sum_{0 \le |\alpha| \le m} \|D^\alpha u\|_p^p \right)^{1/p} \quad \text{if} \quad 1 \le p < \infty, \tag{1}$$

$$\|u\|_{m,\infty} = \max_{0 \le |\alpha| \le m} \|D^\alpha u\|_\infty \tag{2}$$

for any function u for which the right side makes sense, $\|\cdot\|_p$ being, of course, the norm in $L^p(\Omega)$. In some situations where confusion of domains may occur we will use $\|u\|_{m,p,\Omega}$ in place of $\|u\|_{m,p}$. Evidently (1) or (2) defines a norm on any vector space of functions on which the right side takes finite values provided functions are identified in the space if they are equal almost everywhere in Ω.

3.2 (Sobolev Spaces) For any positive integer m and $1 \le p \le \infty$ we consider three vector spaces on which $\|\cdot\|_{m,p}$ is a norm:

(a) $H^{m,p}(\Omega) \equiv$ the completion of $\{u \in C^m(\Omega) : \|u\|_{m,p} < \infty\}$ with respect to the norm $\|\cdot\|_{m,p}$,

(b) $W^{m,p}(\Omega) \equiv \{u \in L^p(\Omega) : D^\alpha u \in L^p(\Omega)$ for $0 \le |\alpha| \le m\}$, where $D^\alpha u$ is the weak (or distributional) partial derivative of Paragraph 1.62, and

(c) $W_0^{m,p}(\Omega) \equiv$ the closure of $C_0^\infty(\Omega)$ in the space $W^{m,p}(\Omega)$.

Equipped with the appropriate norm (1) or (2) these are called *Sobolev spaces* over Ω. Clearly $W^{0,p}(\Omega) = L^p(\Omega)$, and if $1 \le p < \infty$, $W_0^{0,p}(\Omega) = L^p(\Omega)$ because $C_0^\infty(\Omega)$ is dense in $L^p(\Omega)$. (See Paragraph 2.30.) For any m, we have the obvious chain of imbeddings

$$W_0^{m,p}(\Omega) \to W^{m,p}(\Omega) \to L^p(\Omega).$$

We will show in Theorem 3.17 that $H^{m,p}(\Omega) = W^{m,p}(\Omega)$ for every domain Ω. This result, published in 1964 by Meyers and Serrin [MS] ended much confusion about the relationship of these spaces that existed in the literature before that time. It is surprising that this elementary result remained undiscovered for so long.

The spaces $W^{m,p}(\Omega)$ were introduced by Sobolev [So1,So2]. Many related spaces were being studied by other writers, in particular Morrey [Mo] and Deny and Lions [DL]. Many different symbols ($W^{m,p}$, $H^{m,p}$, $P^{m,p}$, L_p^m, etc.) have been used to denote these spaces and their variants, and before they became generally associated with the name of Sobolev they were sometimes referred to under other names, for example, as Beppo Levi spaces.

Numerous generalizations and specializions of the basic spaces $W^{m,p}(\Omega)$ have been constructed. Much of this literature originated in the Soviet Union. In particular, there are extensions that allow arbitrary real values of m (see Chapter 7) which are interpreted as corresponding to fractional orders of differentiation. There are weighted spaces that introduce weight functions into the L^p norms; see Kufner [Ku]. There are spaces of vector fields that are annihilated by differential operators like curl and divergence; see [DaL]. Other generalizations involve different orders of differentiation and different L^p norms in different coordinate directions (anisotropic spaces — see [BIN1, BIN2]), and Orlicz-Sobolev spaces (see Chapter 8) modeled on the generalizations of L^p spaces known as Orlicz spaces. Finally, there has been much work on the interaction between Sobolev spaces and differential geometry [Hb] and a flurry of recent activity on Sobolev spaces on metric spaces [Hn, HK].

We will not be able to investigate the most of these generalizations here.

3.3 THEOREM $W^{m,p}(\Omega)$ is a Banach space.

Proof. Let $\{u_n\}$ be a Cauchy sequence in $W^{m,p}(\Omega)$. Then $\{D^\alpha u\}$ is a Cauchy sequence in $L^p(\Omega)$ for $0 \le |\alpha| \le m$. Since $L^p(\Omega)$ is complete there exist functions u and u_α, $0 \le |\alpha| \le m$, such that $u_n \to u$ and $D^\alpha u_n \to u_\alpha$ in $L^p(\Omega)$ as

$n \to \infty$. Now $L^p(\Omega) \subset L^1_{\mathrm{loc}}(\Omega)$ and so u_n determines a distribution $T_{u_n} \in \mathscr{D}'(\Omega)$ as in Paragraph 1.58. For any $\phi \in \mathscr{D}(\Omega)$ we have

$$|T_{u_n}(\phi) - T_u(\phi)| \le \int_\Omega |u_n(x) - u(x)||\phi(x)|\,dx \le \|\phi\|_{p'}\,\|u_n - u\|_p$$

by Hölder's inequality, where p' is the exponent conjugate to p. Therefore $T_{u_n}(\phi) \to T_u(\phi)$ for every $\phi \in \mathscr{D}(\Omega)$ as $n \to \infty$. Similarly, $T_{D^\alpha u_n}(\phi) \to T_{u_\alpha}(\phi)$ for every $\phi \in \mathscr{D}(\Omega)$. It follows that

$$T_{u_\alpha}(\phi) = \lim_{n\to\infty} T_{D^\alpha u_n}(\phi) = \lim_{n\to\infty} (-1)^{|\alpha|} T_{u_n}(D^\alpha \phi) = (-1)^{|\alpha|} T_u(D^\alpha \phi)$$

for every $\phi \in \mathscr{D}(\Omega)$. Thus $u_\alpha = D^\alpha u$ in the distributional sense on Ω for $0 \le |\alpha| \le m$, whence $u \in W^{m,p}(\Omega)$. Since $\lim_{n\to\infty} \|u_n - u\|_{m,p} = 0$, the space $W^{m,p}(\Omega)$ is complete. ∎

3.4 COROLLARY $H^{m,p}(\Omega) \subset W^{m,p}(\Omega)$.

Proof. Distributional and classical partial derivatives coincide whenever the latter exist and are continuous on Ω. Therefore the set

$$S = \{\phi \in C^m(\Omega) : \|\phi\|_{m,p} < \infty\}$$

is contained in $W^{m,p}(\Omega)$. Since $W^{m,p}(\Omega)$ is complete, the identity operator on S extends to an isometric isomorphism between $H^{m,p}(\Omega)$, the completion of S, and the closure of S in $W^{m,p}(\Omega)$. We can identify $H^{m,p}(\Omega)$ with this closure. ∎

3.5 Several important properties of the spaces $W^{m,p}(\Omega)$ can be easily obtained by regarding $W^{m,p}(\Omega)$ as a closed subspace of an L^p space on a union of disjoint copies of Ω.

Given integers $n \ge 1$ and $m \ge 0$, let $N \equiv N(n, m)$ be the number of multi-indices $\alpha = (\alpha_1, \ldots, \alpha_n)$ such that $|\alpha| \le m$. For each such multi-index α let Ω_α be a copy of Ω in a different copy of $\mathbb{R}^n$, so that the N domains Ω_α are *de facto* mutually disjoint. Let $\Omega^{(m)}$ be the union of these N domains; $\Omega^{(m)} = \bigcup_{|\alpha|\le m} \Omega_\alpha$. Given a function u in $W^{m,p}(\Omega)$, let U be the function on $\Omega^{(m)}$ that coincides with $D^\alpha u$ on Ω_α. It is easy to check that the map P taking u to U is an isometry from $W^{m,p}(\Omega)$ into $L^p(\Omega^{(m)})$. Since $W^{m,p}(\Omega)$ is complete, the range W of the isometry P is a closed subspace of $L^p(\Omega^{(m)})$. It follows that W is separable if $1 \le p < \infty$, and is uniformly convex and reflexive if $1 < p < \infty$. The same conclusions must therefore hold for $W^{m,p}(\Omega) = P^{-1}(W)$.

3.6 THEOREM $W^{m,p}(\Omega)$ is separable if $1 \le p < \infty$, and is uniformly convex and reflexive if $1 < p < \infty$. In particular, $W^{m,2}(\Omega)$ is a separable Hilbert space with inner product

$$(u, v)_m = \sum_{0\le|\alpha|\le m} (D^\alpha u, D^\alpha v),$$

where $(u, v) = \int_\Omega u(x)\overline{v(x)}\,dx$ is the inner product on $L^2(\Omega)$. ∎

Duality and the Spaces $W^{-m,p'}(\Omega)$

3.7 In this section we shall take, for fixed Ω, m, and p, the number N, the spaces $L^p(\Omega^{(m)})$ and W, and the operator P to be specified as in Paragraph 3.5. We also define

$$\langle u, v\rangle = \int_\Omega u(x)v(x)\,dx$$

for any functions u, v for which the right side makes sense. For given p let us agree that p' always denotes the conjugate exponent:

$$p' = \begin{cases} \infty & \text{if } p = 1 \\ p/(p-1) & \text{if } 1 < p < \infty \\ 1 & \text{if } p = \infty. \end{cases}$$

First we extend the Riesz Representation Theorem to the space $W^{m,p}(\Omega)$. Then, we identify the dual of $W_0^{m,p}(\Omega)$ with a subspace of $\mathscr{D}'(\Omega)$. Finally, we show that if $1 < p < \infty$, the dual of $W_0^{m,p}(\Omega)$ can also be identified with the completion of $L^{p'}(\Omega)$ with respect to a norm weaker than the usual $L^{p'}$ norm.

3.8 (The Dual of $L^p(\Omega^{(m)})$) To every $L \in (L^p(\Omega^{(m)}))'$, where $1 \le p < \infty$, there corresponds a unique $v \in L^{p'}(\Omega^{(m)})$ such that for every $u \in L^p(\Omega^{(m)})$,

$$L(u) = \int_{\Omega^{(m)}} u(x)v(x)\,dx = \sum_{|\alpha|\le m}\int_{\Omega_\alpha} u_\alpha(x)v_\alpha(x)\,dx = \sum_{|\alpha|\le m}\langle u_\alpha, v_\alpha\rangle,$$

where u_α and v_α are the restrictions of u and v, respectively, to Ω_α. Moreover, $\left\|L\,;\,\left(L^p(\Omega^{(m)})\right)'\right\| = \left\|v\,;\,L^{p'}(\Omega^{(m)})\right\|$. Thus $(L^p(\Omega^{(m)}))' \equiv L^{p'}(\Omega^{(m)})$.

This is valid because $L^p(\Omega^{(m)})$ is, after all, an L^p space, albeit one defined on an unusual domain.

3.9 THEOREM (The Dual of $W^{m,p}(\Omega)$) Let $1 \le p < \infty$. For every $L \in \left(W^{m,p}(\Omega)\right)'$ there exist elements $v \in L^{p'}(\Omega^{(m)})$ such that if the restriction of v to Ω_α is v_α, we have for all $u \in W^{m,p}(\Omega)$

$$L(u) = \sum_{0\le|\alpha|\le m}\langle D^\alpha u, v_\alpha\rangle. \tag{3}$$

Moreover

$$\|L;\left(W^{m,p}(\Omega)\right)'\| = \inf\|v; L^{p'}(\Omega^{(m)})\| = \min\|v; L^{p'}(\Omega^{(m)})\|, \tag{4}$$

the infimum being taken over, and *attained on* the set of all $v \in L^{p'}(\Omega^{(m)})$ for which (3) holds for every $u \in W^{m,p}(\Omega)$.

If $1 < p < \infty$, the element $v \in L^{p'}(\Omega^{(m)})$ satisfying (3) and (4) is unique.

Proof. A linear functional L^* is defined as follows on the range W of the operator P defined in Paragraph 3.5:

$$L^*(Pu) = L(u), \qquad u \in W^{m,p}(\Omega).$$

Since P is an isometric isomorphism, $L^* \in W'$ and

$$\|L^*; W'\| = \left\| L; \left(W^{m,p}(\Omega)\right)' \right\|.$$

By the Hahn-Banach Theorem 1.13 there exists a norm preserving extension $\hat{L}$ of L^* to all of $L^p(\Omega^{(m)})$, and, as observed in Paragraph 3.8 there exists $v \in L^{p'}(\Omega^{(m)})$ such that if $u \in L^p(\Omega^{(m)})$, then

$$\hat{L}(u) = \sum_{0\le|\alpha|\le m} \langle u_\alpha, v_\alpha \rangle.$$

Thus, for $u \in W^{m,p}(\Omega)$ we obtain

$$L(u) = L^*(Pu) = \hat{L}(Pu) = \sum_{0\le|\alpha|\le m} \langle D^\alpha u, v_\alpha \rangle.$$

Moreover,

$$\|L; \left(W^{m,p}(\Omega)\right)'\| = \|L^*; W'\| = \|\hat{L}; (L^p(\Omega^{(m)}))'\| = \|v; L^{p'}(\Omega^{(m)})\|.$$

Now (4) must hold because any element $v \in L^{p'}(\Omega^{(m)})$ for which (3) holds for every $u \in W^{m,p}(\Omega)$ corresponds to an extension L of L^* and so will have norm $\|v; L^{p'}(\Omega^{(m)})\|$ not less than $\|L; \left(W^{m,p}(\Omega)\right)'\|$.

The uniqueness of v if $1 < p < \infty$ follows from the uniform convexity of $L^p(\Omega^{(m)})$ and $L^{p'}(\Omega^{(m)})$ by an argument similar to that in Lemma 2.43. ∎

3.10 If $1 \le p < \infty$ every element L of $\left(W^{m,p}(\Omega)\right)'$ is an extension to $W^{m,p}(\Omega)$ of a distribution $T \in \mathscr{D}'(\Omega)$. To see what form this distribution takes, suppose L is given by (3) for some $v \in L^{p'}(\Omega^{(m)})$ and define T and T_{v_α} on $\mathscr{D}(\Omega)$ by

$$T = \sum_{0\le|\alpha|\le m} (-1)^{|\alpha|} D^\alpha T_{v_\alpha}, \quad T_{v_\alpha}(\phi) = \langle \phi, v_\alpha \rangle, \quad 0 \le |\alpha| \le m, \tag{5}$$

For every $\phi \in \mathscr{D}(\Omega) \subset W^{m,p}(\Omega)$ we have $T(\phi) = \sum_{0\le|\alpha|\le m} T_{v_\alpha}(D^\alpha \phi) = L(\phi)$ so that L is clearly an extension of T. Moreover, by (4)

$$\|L; \left(W^{m,p}(\Omega)\right)'\| = \min\left\{\|v; L^{p'}(\Omega^{(m)})\| \;:\; L \text{ extends } T \text{ given by (5)}\right\}.$$

These remarks also hold for $L \in \left(W_0^{m,p}(\Omega)\right)'$ since any such functional possesses a norm-preserving extension to $W^{m,p}(\Omega)$.

3.11 Now suppose T is any element of $\mathscr{D}'(\Omega)$ having the form (5) for some $v \in L^{p'}(\Omega^{(m)})$, where $1 \le p' \le \infty$. Then T possesses (possibly non-unique) continuous extensions to $W^{m,p}(\Omega)$. However, T possesses a unique continuous extension to $W_0^{m,p}(\Omega)$. To see this, for $u \in W_0^{m,p}(\Omega)$ let $\{\phi_n\}$ be a sequence in $C_0^\infty(\Omega) = \mathscr{D}(\Omega)$ converging to u in norm in $W_0^{m,p}(\Omega)$. Then

$$\begin{aligned}|T(\phi_k) - T(\phi_n)| &\le \sum_{0\le|\alpha|\le m} |T_{v_\alpha}(D^\alpha \phi_k - D^\alpha \phi_n)| \\ &\le \sum_{0\le|\alpha|\le m} \|D^\alpha(\phi_k - \phi_n)\|_p \, \|v_\alpha\|_{p'} \\ &\le \|\phi_k - \phi_n\|_{m,p} \, \|v; L^{p'}(\Omega^{(m)})\| \to 0 \quad \text{as } k, n \to \infty.\end{aligned}$$

Thus $\{T(\phi_n)\}$ is a Cauchy sequence in $\mathbb{C}$ and so converges to a limit that we can denote by $L(u)$ since it is clear that if also $\{\psi_n\} \subset \mathscr{D}(\Omega)$ and $\|\psi_n - u\|_{m,p} \to 0$, then $T(\phi_n) - T(\psi_n) \to 0$ as $n \to \infty$. The functional L thus defined is linear and belongs to $\left(W_0^{m,p}(\Omega)\right)'$, for if $u = \lim_{n\to\infty} \phi_n$ as above, then

$$|L(u)| = \lim_{n\to\infty} |T(\phi_n)| \le \lim_{n\to\infty} \|\phi_n\|_{m,p} \, \|v; L^{p'}(\Omega^{(m)})\| = \|u\|_{m,p} \, \|v; L^{p'}(\Omega^{(m)})\|.$$

We have therefore proved the following theorem.

3.12 THEOREM (The Normed Dual of $W_0^{m,p}(\Omega)$) If $1 \le p < \infty$, p' is the exponent conjugate to p, and $m \ge 1$, the dual space $\left(W_0^{m,p}(\Omega)\right)'$ is isometrically isomorphic to the Banach space $W^{-m,p'}(\Omega)$ consisting of those distributions $T \in \mathscr{D}'(\Omega)$ that satisfy (5) and having norm

$$\|T\| = \min\left\{\|v; L^{p'}(\Omega^{(m)})\| \ : \ v \text{ satisfies (5)}\right\}. \quad \blacksquare$$

The completeness of this space is a consequence of the isometric isomorphism. Evidently $W^{-m,p'}(\Omega)$ is separable and reflexive if $1 < p < \infty$.

When $W_0^{m,p}(\Omega)$ is a proper subset of $W^{m,p}(\Omega)$, continuous linear functionals on $W^{m,p}(\Omega)$ are not fully determined by their restrictions to $C_0(\Omega)$, and so are not determined by distributions T given by (5).

3.13 (The $(-m, p')$ norm on $L^{p'}(\Omega)$) There is another way of characterizing the dual of $W_0^{m,p}(\Omega)$ if $1 < p < \infty$. Each element $v \in L^{p'}(\Omega)$ determines an element L_v of $\left(W_0^{m,p}(\Omega)\right)'$ by means of $L_v(u) = \langle u, v\rangle$, because

$$|L_v(u)| = |\langle u, v\rangle| \le \|v\|_{p'} \, \|u\|_p \le \|v\|_{p'} \, \|u\|_{m,p}.$$

We define the $(-m, p')$-norm of $v \in L^{p'}(\Omega)$ to be the norm of L_v, that is

$$\|v\|_{-m,p'} = \|L_v; \left(W_0^{m,p}(\Omega)\right)'\| = \sup_{u\in W_0^{m,p}(\Omega),\ \|u\|_{m,p}\le 1} |\langle u, v\rangle|.$$

Clearly $\|v\|_{-m,p'} \le \|v\|_{p'}$ and for any $u \in W_0^{m,p}(\Omega)$ and $v \in L^{p'}(\Omega)$ we have

$$|\langle u, v\rangle| = \|u\|_{m,p}\left|\left\langle \frac{u}{\|u\|_{m,p}}, v\right\rangle\right| \le \|u\|_{m,p}\,\|v\|_{-m,p'}, \tag{6}$$

which is a generalization of Hölder's inequality.

Let $V = \{L_v : v \in L^{p'}(\Omega)\}$, which is a vector subspace of $\left(W_0^{m,p}(\Omega)\right)'$. We show that V is dense in $\left(W_0^{m,p}(\Omega)\right)'$. To this end it is sufficient to show that if $F \in \left(W_0^{m,p}(\Omega)\right)''$ satisfies $F(L_v) = 0$ for every $L_v \in V$, then $F = 0$ in $\left(W_0^{m,p}(\Omega)\right)''$. But since $W_0^{m,p}(\Omega)$ is reflexive, there exists $f \in W_0^{m,p}(\Omega)$ corresponding to $F \in \left(W_0^{m,p}(\Omega)\right)''$ such that $\langle f, v\rangle = L_v(f) = F(L_v) = 0$ for every $v \in L^{p'}(\Omega)$. But then $f(x)$ must be zero a.e. in Ω. Hence $f = 0$ in $W_0^{m,p}(\Omega)$ and $F = 0$ in $\left(W_0^{m,p}(\Omega)\right)''$.

Let $H^{-m,p'}(\Omega)$ denote the completion of $L^{p'}(\Omega)$ with respect to the norm $\|\cdot\|_{-m,p'}$. Then we have

$$H^{-m,p'}(\Omega) \equiv \left(W_0^{m,p}(\Omega)\right)' \equiv W^{-m,p'}(\Omega).$$

In particular, corresponding to each $v \in H^{-m,p'}(\Omega)$, there exists a distribution $T_v \in W^{-m,p'}(\Omega)$ such that $T_v(\phi) = \lim_{n\to\infty}\langle \phi, v_n\rangle$ for every $\phi \in \mathscr{D}(\Omega)$ and every sequence $\{v_n\} \subset L^{p'}(\Omega)$ for which $\lim_{n\to\infty}\|v_n - v\|_{-m,p'} = 0$. Conversely, any $T \in W^{-m,p'}(\Omega)$ satisfies $T = T_v$ for some such v. Moreover, by (6), $|T_v(\phi)| \le \|\phi\|_{m,p}\,\|v\|_{-m,p'}$.

3.14 A similar argument to that above shows that the dual space $\left(W^{m,p}(\Omega)\right)'$ can be characterized for $1 < p < \infty$ as the completion of $L^{p'}(\Omega)$ with respect to the norm

$$\|v\|^*_{-m,p'} = \sup_{u\in W^{m,p}(\Omega),\ \|u\|_{m,p}\le 1} |\langle u, v\rangle|.$$

Approximation by Smooth Functions on Ω

We wish to prove that $\{\phi \in C^\infty(\Omega) : \|\phi\|_{m,p} < \infty\}$ is dense in $W^{m,p}(\Omega)$. For this we require the following existence theorem for infinitely differentiable *partitions of unity*.

3.15 THEOREM (Partitions of Unity) Let A be an arbitrary subset of $\mathbb{R}^n$ and let $\mathscr{O}$ be a collection of open sets in $\mathbb{R}^n$ which cover A, that is, $A \subset \bigcup_{U\in\mathscr{O}} U$. Then there exists a collection Ψ of functions $\psi \in C_0^\infty(\mathbb{R}^n)$ having the following properties:

(i) For every $\psi \in \Psi$ and every $x \in \mathbb{R}^n$, $0 \le \psi(x) \le 1$.

(ii) If $K \Subset A$, all but finitely many $\psi \in \Psi$ vanish identically on K.

(iii) For every $\psi \in \Psi$ there exists $U \in \mathcal{O}$ such that $\text{supp}(\psi) \subset U$.

(iv) For every $x \in A$, we have $\sum_{\psi \in \Psi} \psi(x) = 1$.

Such a collection Ψ is called a *C^∞-partition of unity for A subordinate to $\mathcal{O}$*.

Proof. Since the proof can be found in many texts, we give only an outline of it. First suppose that A is compact. Then there is a finite collection of sets in $\mathcal{O}$ that cover A, say $A \subset \bigcup_{j=1}^N U_j$. Compact sets $K_1 \subset U_1, \ldots, K_N \subset U_N$ can then be constructed so that $A \subset \bigcup_{j=1}^N K_j$. For each j a nonnegative-valued function $\phi_j \in C_0^\infty(U_j)$ can be found such that $\phi_j(x) > 0$ for $x \in K_j$. A function ϕ in $C^\infty(\mathbb{R}^n)$ can then be constructed so that $\phi(x) > 0$ on $\mathbb{R}^n$ and $\phi(x) = \sum_{j=1}^N \phi_j(x)$ for $x \in A$. Now $\Psi = \{\psi_n : \psi_j(x) = \phi_j(x)/\phi(x), 1 \le j \le N\}$ has the required properties. If A is an arbitrary open set. Then $A = \bigcup_{j=1}^\infty A_j$, where

$$A_j = \{x \in A : |x| \le j \text{ and } \text{dist}(x, \text{bdry}\, A) \ge 1/j\}$$

is compact. Taking $A_0 = A_{-1} = \emptyset$, for each $j \ge 1$ the collection

$$\mathcal{O}_j = \{U \cap (\text{interior of } A_{j+1} \cap A_{j-2}^c) : U \in \mathcal{O}\}$$

covers A_j and so there exists a finite C^∞-partition of unity Ψ_j for A_j subordinate to $\mathcal{O}_j$. The sum $\sigma(x) = \sum_{j=1}^\infty \sum_{\phi \in \Psi_j} \phi(x)$ involves only finitely many nonzero terms at each $x \in A$. The collection $\Psi = \{\psi : \psi(x) = \phi(x)/\sigma(x)$ for some ϕ in some Ψ_j if $x \in A$, $\psi(x) = 0$ if $x \notin A\}$ has the prescribed properties.
Finally, if A is arbitrary, then $A \subset B$ where B is the union of all $U \in \mathcal{O}$ and is an open set. Any partition of unity for B will do for A as well. ∎

3.16 LEMMA (Mollification in $W^{m,p}(\Omega)$) Let J_ϵ be defined as in Paragraph 2.28 and let $1 \le p < \infty$ and $u \in W^{m,p}(\Omega)$. If Ω' is a subdomain with compact closure in Ω, then $\lim_{\epsilon \to 0+} J_\epsilon * u = u$ in $W^{m,p}(\Omega')$.

Proof. Let $\epsilon < \text{dist}(\Omega', \text{bdry}\,\Omega)$ and $\tilde{u}$ be the zero extension of u outside Ω. If $\phi \in \mathcal{D}(\Omega')$,

$$\begin{aligned}\int_{\Omega'} J_\epsilon * u(x) D^\alpha \phi(x)\,dx &= \int_{\mathbb{R}^n}\int_{\mathbb{R}^n} \tilde{u}(x-y) J_\epsilon(y) D^\alpha \phi(x)\,dx\,dy \\ &= (-1)^{|\alpha|} \int_{\mathbb{R}^n}\int_{\Omega'} D_x^\alpha u(x-y) J_\epsilon(y) \phi(x)\,dx\,dy \\ &= (-1)^{|\alpha|} \int_{\Omega'} J_\epsilon * D^\alpha u(x) \phi(x)\,dx.\end{aligned}$$

Thus $D^\alpha J_\epsilon * u = J_\epsilon * D^\alpha u$ in the distributional sense in Ω'. Since $D^\alpha u \in L^p(\Omega)$ for $0 \le |\alpha| \le m$ we have by Theorem 2.29(c)

$$\lim_{\epsilon \to 0+} \|D^\alpha J_\epsilon * u - D^\alpha u\|_{p,\Omega'} = \lim_{\epsilon \to 0+} \|J_\epsilon * D^\alpha u - D^\alpha u\|_{p,\Omega'} = 0.$$

Thus $\lim_{\epsilon \to 0+} \|J_\epsilon u - u\|_{m,p,\Omega'} = 0$. ∎

3.17 THEOREM (H = W) (See [MS].) If $1 \le p < \infty$, then

$$H^{m,p}(\Omega) = W^{m,p}(\Omega).$$

Proof. By Corollary 3.4 it is sufficient to show that $W^{m,p}(\Omega) \subset H^{m,p}(\Omega)$, that is, that $\{\phi \in C^m(\Omega) : \|\phi\|_{m,p} < \infty\}$ is dense in $W^{m,p}(\Omega)$. If $u \in W^{m,p}(\Omega)$ and $\epsilon > 0$, we in fact show that there exists $\phi \in C^\infty(\Omega)$ such that $\|\phi - u\|_{m,p} < \epsilon$, so that $C^\infty(\Omega)$ is dense in $W^{m,p}(\Omega)$. For $k = 1, 2, \ldots$ let

$$\Omega_k = \{x \in \Omega : |x| < k \text{ and } \operatorname{dist}(x, \operatorname{bdry}\Omega) > 1/k,$$

and let $\Omega_0 = \Omega_{-1} = \emptyset$, the empty set. Then

$$\mathcal{O} = \{U_k : U_k = \Omega_{k+1} \cap (\overline{\Omega_{k-1}})^c,\ k = 1, 2, \ldots\}$$

is a collection of open subsets of Ω that covers Ω. Let Ψ be a C^∞-partition or unity for Ω subordinate to $\mathcal{O}$. Let ψ_k denote the sum of the finitely many functions $\psi \in \Psi$ whose supports are contained in U_k. Then $\psi_k \in C_0^\infty(U_k)$ and $\sum_{k=1}^\infty \psi_k(x) = 1$ on Ω.

If $0 < \epsilon < 1/(k+1)(k+2)$, then $J_\epsilon * (\psi_k u)$ has support in the intersection $V_k = \Omega_{k+2} \cap (\overline{\Omega_{k-2}})^c \Subset \Omega$. Since $\psi_k u \in W^{m,p}(\Omega)$ we may choose ϵ_k, satisfying $0 < \epsilon_k < 1/(k+1)(k+2)$, such that

$$\left\| J_{\epsilon_k} * (\psi_k u) - \psi_k u \right\|_{m,p,\Omega} = \left\| J_{\epsilon_k} * (\psi_k u) - \psi_k u \right\|_{m,p,V_k} < \epsilon/(2^k).$$

Let $\phi = \sum_{k=1}^\infty J_{\epsilon_k} * (\psi_k u)$. On any $\Omega' \Subset \Omega$ only finitely many terms in the sum can be nonzero. Thus $\phi \in C^\infty(\Omega)$. For $x \in \Omega_k$, we have

$$u(x) = \sum_{j=1}^{k+2} \psi_j(x) u(x), \qquad \text{and} \qquad \phi(x) = \sum_{j=1}^{k+2} J_{\epsilon_j} * (\psi_j u)(x).$$

Thus

$$\|u - \phi\|_{m,p,\Omega_k} \le \sum_{j=1}^{k+2} \left\| J_{\epsilon_j} * (\psi_j u) - \psi_j u \right\|_{m,p,\Omega} < \epsilon.$$

By the monotone convergence theorem 1.48, $\|u - \phi\|_{m,p,\Omega} < \epsilon$. ∎

3.18 EXAMPLE Theorem 3.17 can not be extended to the case $p = \infty$. For instance, if $\Omega = \{x \in \mathbb{R} : -1 < x < 1$, and $u(x) = |x|$, then $u'(x) = x/|x|$ for $x \ne 0$ and so $u \in W^{1,\infty}(\Omega)$. But $u \notin H^{1,\infty}(\Omega)$. In fact, if $0 < \epsilon < 1/2$, there exists no function $\phi \in C^1(\Omega)$ such that $\left\|\phi' - u'\right\|_\infty < \epsilon$.

Approximation by Smooth Functions on $\mathbb{R}^n$

3.19 Having shown that an element of $W^{m,p}(\Omega)$ can always be approximated by functions smooth on Ω we now ask whether the approximation can in fact be done with bounded functions having bounded derivatives of all orders, or at least of all orders up to and including at least m. That is, we are asking whether, for any values of $k \geq m$, the space $C^k(\overline{\Omega})$ is dense in $W^{m,p}(\Omega)$. The following example shows that the answer may be negative.

3.20 EXAMPLE Let $\Omega = \{(x, y) \in \mathbb{R}^2 : 0 < |x| < 1, 0 < y < 1\}$. Then the function defined on Ω by

$$u(x, y) = \begin{cases} 1 & \text{if } x > 0 \\ 0 & \text{if } x < 0 \end{cases}$$

evidently belongs to $W^{1,p}(\Omega)$. However, if $\epsilon > 0$ is sufficiently small, there can exist no $\phi \in C^1(\overline{\Omega})$ such that $\|u - \phi\|_{1,p,\Omega} < \epsilon$. To see this, suppose there exists such a ϕ. If $L = \{(x, y) : -1 \leq x \leq 0, 0 \leq y \leq 1\}$ and $R = \{(x, y) : 0 \leq x \leq 1, 0 \leq y \leq 1\}$, then $\overline{\Omega} = L \cup R$. We have $\|\phi\|_{1,L} \leq \|\phi\|_{p,L} < \epsilon$ and similarly $\|1 - \phi\|_{1,R} < \epsilon$ from which we obtain $\|\phi\|_{1,R} > 1 - \epsilon$. If

$$\Phi(x) = \int_0^1 \phi(x, y)\, dy,$$

then there exist a and b with $-1 \leq a < 0$ and $0 < b \leq 1$ such that $\Phi(a) < \epsilon$ and $\Phi(b) > 1 - \epsilon$. If $0 < \epsilon < 1/2$, then

$$\begin{aligned} 1 - 2\epsilon < \Phi(b) - \Phi(a) = \int_a^b \Phi'(x)\, dx &\leq \int_{\overline{\Omega}} |D_x\phi(x, y)|\, dx\, dy \\ &\leq 2^{1/p'} \|D_x\phi\|_{p,\Omega} < 2^{1/p'}\epsilon. \end{aligned}$$

Thus $1 < \epsilon(2 + 2^{1/p'})$, which is not possible for small ϵ.

The difficulty with the domain in this example is that it lies on both sides of part of its boundary, namely the line segment $x = 0$, $0 \leq y \leq 1$. We now formulate a condition on a domain Ω that prevents this from happening and guarantees that for any k and m, $C^k(\overline{\Omega})$ is dense in $W^{m,p}(\Omega)$ provided $1 \leq p < \infty$.

3.21 (The Segment Condition) We say that a domain Ω satisfies the *segment condition* if every $x \in \operatorname{bdry}\Omega$ has a neighbourhood U_x and a nonzero vector y_x such that if $z \in \overline{\Omega} \cap U_x$, then $z + ty_x \in \Omega$ for $0 < t < 1$.

If nonempty, the boundary of a domain satisfying this condition must be $(n-1)$-dimensional, and the domain cannot lie on both sides of any part of its boundary.

3.22 THEOREM If Ω satisfies the segment condition, then the set of restrictions to Ω of functions in $C_0^\infty(\mathbb{R}^n)$ is dense in $W^{m,p}(\Omega)$ for $1 \leq p < \infty$.

Proof. Let f be a fixed function in $C_0^\infty(\mathbb{R}^n)$ satisfying

(i) $f(x) = 1$ if $|x| \le 1$,

(ii) $f(x) = 0$ if $|x| \ge 2$,

(iii) $|D^\alpha f(x)| \le M$ (constant) for all x and $0 \le |\alpha| \le m$.

For $\epsilon > 0$ let $f_\epsilon(x) = f(\epsilon x)$. Then $f_\epsilon(x) = 1$ for $|x| \le 1/\epsilon$ and also $|D^\alpha f_\epsilon(x)| \le M\epsilon^{|\alpha|} \le M$ if $\epsilon \le 1$. If $u \in W^{m,p}(\Omega)$, then $u_\epsilon = f_\epsilon u$ belongs to $W^{m,p}(\Omega)$ and has bounded support. For $0 < \epsilon \le 1$ and $|\alpha| \le m$

$$|D^\alpha u_\epsilon(x)| = \left| \sum_{\beta \le \alpha} \binom{\alpha}{\beta} D^\beta u(x) D^{\alpha-\beta} f_\epsilon(x) \right| \le M \sum_{\beta \le \alpha} \binom{\alpha}{\beta} |D^\beta u(x)|.$$

Therefore, setting $\Omega_\epsilon = \{x \in \Omega : |x| > 1/\epsilon\}$, we have

$$\begin{aligned} \|u - u_\epsilon\|_{m,p,\Omega} &= \|u - u_\epsilon\|_{m,p,\Omega_\epsilon} \\ &\le \|u\|_{m,p,\Omega_\epsilon} + \|u_\epsilon\|_{m,p,\Omega_\epsilon} \le \text{const}\, \|u\|_{m,p,\Omega_\epsilon}. \end{aligned}$$

The right side approaches zero as $\epsilon \to 0+$. Thus any $u \in W^{m,p}(\Omega)$ can be approximated in that space by functions with bounded supports.

We now, therefore, assume that $K = \{x \in \Omega : u(x) \neq 0\}$ is bounded. The set $F = \overline{K} - \left(\bigcup_{x \in \text{bdry}\,\Omega} U_x \right)$ is thus compact and contained in Ω, $\{U_x\}$ being the collection of open sets referred to in the definition of the segment condition. There exists an open set U_0 such that $F \Subset U_0 \Subset \Omega$. Since $\overline{K}$ is compact, there exists finitely many of the sets U_x, let us rename them $U_1, \dots, U_k$, such that $\overline{K} \subset U_0 \cup U_1 \cup \cdots \cup U_k$. Moreover, there are other open sets $V_0, V_1, \dots, V_k$ such that $V_j \Subset U_j$ for $0 \le j \le k$ but still $\overline{K} \subset V_0 \cup V_1 \cup \cdots \cup V_k$.

Let Ψ be a C^∞-partition of unity subordinate to $\{V_j : 0 \le j \le k\}$, and let ψ_j be the sum of the finitely many functions $\psi \in \Psi$ whose supports lie in V_j. Let $u_j = \psi_j u$. Suppose that for each j we can find $\phi_j \in C_0^\infty(\mathbb{R}^n)$ such that

$$\left\| u_j - \phi_j \right\|_{m,p,\Omega} < \epsilon/(k+1). \tag{7}$$

Then, putting $\phi = \sum_{j=0}^k \phi_j$, we would obtain

$$\|u - \phi\|_{m,p,\Omega} \le \sum_{j=0}^k \left\| u_j - \phi_j \right\|_{m,p,\Omega} < \epsilon.$$

A function $\phi_0 \in C_0^\infty(\mathbb{R}^n)$ satisfying (7) for $j = 0$ can be found via Lemma 3.16 since $\operatorname{supp}(u_0) \subset V_0 \Subset \Omega$. It remains, therefore, to find ϕ_j satisfying (7) for $1 \le j \le k$. For fixed such j we extend u_j to be identically zero outside Ω. Thus $u_j \in W^{m,p}(\mathbb{R}^n - \Gamma)$, where $\Gamma = \overline{V_j} \cap \text{bdry}\,\Omega$. Let y be the nonzero vector associated with the set U_j in the definition of the segment condition. (See Fig. 1.) Let $\Gamma_t = \{x - ty : x \in \Gamma\}$, where t is so chosen that

$$0 < t < \min\{1, \operatorname{dist}(V_j, \mathbb{R}^n - U_j)/|y|\}.$$

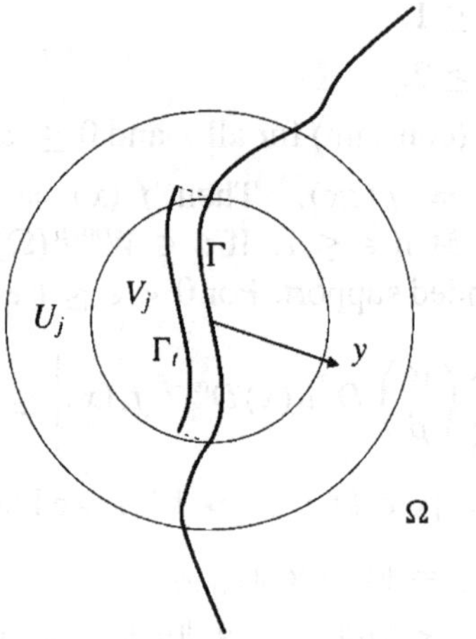

Fig. 1

Then $\Gamma_t \subset U_j$ and $\Gamma_t \cap \overline{\Omega}$ is empty by the segment condition. Let us define $u_{j,t}(x) = u_j(x + ty)$. Then $u_{j,t} \in W^{m,p}(\mathbb{R}^n - \Gamma_t)$. Translation is continuous in $L^p(\Omega)$ (see the proof of Theorem 2.32) so $D^\alpha u_{j,t} \to D^\alpha u_j$ in $L^p(\Omega)$ as $t \to 0+$ for $|\alpha| \le m$. Thus $u_{j,t} \to u_j$ in $W^{m,p}(\Omega)$ as $t \to 0+$, and so it is sufficient to find $\phi_j \in C_0^\infty(\mathbb{R}^n)$ such that $\|u_{j,t} - \phi_j\|_{m,p}$ is sufficiently small. However, $\Omega \cap U_j \Subset \mathbb{R}^n - \Gamma_t$, and so by Lemma 3.16 we can take $\phi_j = J_\delta * u_{j,t}$ for suitably small $\delta > 0$. This completes the proof. ∎

3.23 COROLLARY $W_0^{m,p}(\mathbb{R}^n) = W^{m,p}(\mathbb{R}^n)$.

Approximation by Functions in $C_0^\infty(\Omega)$

3.24 Corollary 3.23 suggests the question: For what domains Ω is it true that $W^{m,p}(\Omega) = W_0^{m,p}(\Omega)$, that is, when is $C_0^\infty(\Omega)$ dense in $W^{m,p}(\Omega)$? A partial answer to this question can be formulated in terms of the nature of the distributions belonging to $W^{-m,p'}(\mathbb{R}^n)$. The approach below is due to Lions [Lj]. Throughout this discussion we assume $1 < p < \infty$ and p' is the conjugate exponent $p' = p/(p-1)$.

3.25 **((m, p')-Polar sets)** Let F be a closed subset of $\mathbb{R}^n$. A distribution $T \in \mathscr{D}'(\mathbb{R}^n)$ is said to have support in F ($\operatorname{supp}(T) \subset F$) provided that $T(\phi) = 0$ for every $\phi \in \mathscr{D}(\mathbb{R}^n - F)$. We say that the closed set F is (m, p')-polar if the only distribution $T \in W^{-m,p'}(\mathbb{R}^n)$ having support in F is the zero distribution $T = 0$.

If F has positive measure, it cannot be (m, p')-polar because the characteristic function of any compact subset of F having positive measure belongs to $L^{p'}(\mathbb{R}^n)$ and hence to $W^{-m,p'}(\mathbb{R}^n)$.

We shall show later that if $mp > n$, then $W^{m,p}(\mathbb{R}^n) \to C(\mathbb{R}^n)$ in the sense that if $u \in W^{m,p}(\mathbb{R}^n)$, then there exists $v \in C(\mathbb{R}^n)$ such that $u(x) = v(x)$ a.e. in $\mathbb{R}^n$ and

$$|v(x)| \le \text{const } \|u\|_{m,p},$$

the constant being independent of x and u. It follows that the Dirac distribution δ_x given by $\delta_x(\phi) = \phi(x)$ belongs to $\left(W^{m,p}(\mathbb{R}^n)\right)' = \left(W_0^{m,p}(\mathbb{R}^n)\right)' = W^{-m,p'}(\mathbb{R}^n)$. Hence, if $mp > n$ a set F cannot be (m, p')-polar unless it is empty.

Since $W^{m+1,p}(\Omega) \to W^{m,p}(\Omega)$ any bounded linear functional on the latter space is also bounded on the former. Thus $W^{-m,p'}(\Omega) \subset W^{-m-1,p'}(\Omega)$ and, in particular, any $(m+1, p')$-polar set is also (m, p')-polar. The converse is, of course, generally not true.

3.26 (Zero Extensions) If function u is defined on Ω let $\tilde{u}$ denote the zero extension of u to the complement Ω^c of Ω in $\mathbb{R}^n$:

$$\tilde{u}(x) = \begin{cases} u(x) & \text{if } x \in \Omega, \\ 0 & \text{if } x \in \Omega^c. \end{cases}$$

The following lemma shows that the mapping $u \mapsto \tilde{u}$ maps $W_0^{m,p}(\Omega)$ (isometrically) into $W^{m,p}(\mathbb{R}^n)$.

3.27 LEMMA Let $u \in W_0^{m,p}(\Omega)$. If $|\alpha| \le m$, then $D^\alpha \tilde{u} = \widetilde{D^\alpha u}$ in the distributional sense in $\mathbb{R}^n$. Hence $\tilde{u} \in W^{m,p}(\mathbb{R}^n)$.

Proof. Let $\{\phi_j\}$ be a sequence in $C_0^\infty(\Omega)$ converging to u in $W_0^{m,p}(\Omega)$. If $\psi \in \mathscr{D}(\mathbb{R}^n)$, then for $|\alpha| \le m$

$$\begin{aligned}(-1)^{|\alpha|} \int_{\mathbb{R}^n} \tilde{u}(x) D^\alpha \psi(x)\, dx &= (-1)^{|\alpha|} \int_\Omega u(x) D^\alpha \psi(x)\, dx \\ &= \lim_{j\to\infty} (-1)^{|\alpha|} \int_\Omega \phi_j(x) D^\alpha \psi(x)\, dx \\ &= \lim_{j\to\infty} \int_\Omega D^\alpha \phi_j(x) \psi(x)\, dx \\ &= \int_{\mathbb{R}^n} \widetilde{D^\alpha u}(x) \psi(x)\, dx.\end{aligned}$$

Thus $D^\alpha \tilde{u} = \widetilde{D^\alpha u}$ in the distributional sense in $\mathbb{R}^n$ and these locally integrable functions are equal a.e. in $\mathbb{R}^n$. It follows that $\|\tilde{u}\|_{m,p,\mathbb{R}^n} = \|u\|_{m,p,\Omega}$. ∎

We can now give a necessary and sufficient condition that the mapping $u \mapsto \tilde{u}$ carries $W_0^{m,p}(\Omega)$ *onto* $W^{m,p}(\mathbb{R}^n)$.

3.28 THEOREM $C_0^\infty(\Omega)$ is dense in $W^{m,p}(\mathbb{R}^n)$ if and only if the complement Ω^c of Ω is (m, p')-polar.

Proof. First suppose $C_0^\infty(\Omega)$ is dense in $W^{m,p}(\mathbb{R}^n)$. Let $T \in W^{-m,p'}(\mathbb{R}^n)$ have support in Ω^c. If $u \in W^{m,p}(\mathbb{R}^n)$, then there exists a sequence $\{\phi_j\} \subset C_0^\infty(\Omega)$ converging to u in $W^{m,p}(\mathbb{R}^n)$. Hence $T(u) = \lim_{j\to\infty} T(\phi_j) = 0$ and so $T = 0$. Thus Ω^c is (m, p')-polar.

Conversely, suppose that $C_0^\infty(\Omega)$ is not dense in $W^{m,p}(\mathbb{R}^n)$. Then there exists $u \in W^{m,p}(\mathbb{R}^n)$ and a constant $k > 0$ such that for all $\phi \in C_0^\infty(\Omega)$ we have $\|u - \phi\|_{m,p,\mathbb{R}^n} \geq k$. The Hahn-Banach theorem 1.13 can be used to show that there exists $T \in W^{-m,p'}(\mathbb{R}^n)$ such that $T(\phi) = 0$ for all $u \in C_0^\infty(\Omega)$ but $T(u) \neq 0$. Since $\operatorname{supp}(T) \subset \Omega^c$ but $T \neq 0$, Ω^c cannot be (m, p')-polar. ∎

As a final preparation for our investigation of the possible identity of $W_0^{m,p}(\Omega)$ and $W^{m,p}(\Omega)$ we establish a distributional analog of the fact, obvious for differentiable functions, that the vanishing of first derivatives over a rectangle implies constancy on that rectangle. We extend this first to distributions (in Corollary 3.30) and then to locally integrable functions.

3.29 LEMMA Let $B = (a_1, b_1) \times \cdots \times (a_n, b_n)$ be an open rectangular box in $\mathbb{R}^n$ and let $\phi \in \mathscr{D}(B)$. If $\int_B \phi(x)\,dx = 0$, then $\phi(x) = \sum_{j=1}^n \phi_j(x)$, where $\phi_j \in \mathscr{D}(B)$ and

$$\int_{a_j}^{b_j} \phi_j(x_1, \ldots, x_j, \ldots, x_n)\,dx_j = 0 \tag{8}$$

for every fixed $(x_1, \ldots, x_{j-1}, x_{j+1}, \ldots, x_n) \in \mathbb{R}^{n-1}$.

Proof. For $1 \leq j \leq n$ select functions $u_j \in C_0^\infty(a_j, b_j)$ such that $\int_{a_j}^{b_j} u_j(t)\,dt = 1$. For $2 \leq j \leq n$, let

$$B_j = (a_j, b_j) \times (a_{j+1}, b_{j+1}) \times \cdots \times (a_n, b_n),$$

$$\psi_j(x_j, \ldots, x_n) = \int_{a_1}^{b_1} dt_1 \int_{a_2}^{b_2} dt_2 \cdots \int_{a_{j-1}}^{b_{j-1}} \phi(t_1, \ldots, t_{j-1}, x_j, \ldots, x_n)\,dt_{j-1},$$

$$\omega_j(x) = u_1(x_1) \cdots u_{j-1}(x_{j-1}) \psi_j(x_j, \ldots, x_n).$$

Then $\psi_j \in \mathscr{D}(B_j)$ and $\omega_j \in \mathscr{D}(B)$. Moreover

$$\int_{B_j} \psi_j(x_j, \ldots, x_n)\,dx_j \cdots dx_n = \int_B \phi(x)\,dx = 0.$$

Let $\phi_1 = \phi - \omega_2$, $\phi_j = \omega_j - \omega_{j+1}$ if $2 \leq j \leq n-1$, and $\phi_n = \omega_n$. Clearly $\phi_j \in \mathscr{D}(B)$ for $1 \leq j \leq n$, and $\phi = \sum_{j=1}^n \phi_j$. Finally,

$$\int_{a_1}^{b_1} \phi_1(x_1, \dots, x_n)\, dx_1$$
$$= \int_{a_1}^{b_1} \phi(x_1, \dots, x_n)\, dx_1 - \psi_2(x_2, \dots, x_n) \int_{a_1}^{b_1} u_1(x_1)\, dx_1 = 0$$

$$\int_{a_j}^{b_j} \phi_j(x_1, \dots, x_n)\, dx_j$$
$$= u_1(x_1) \cdots u_{j-1}(x_{j-1})$$
$$\times \left(\int_{a_j}^{b_j} \psi_j(x_1, \dots, x_n)\, dx_j - \psi_{j+1}(x_{j+1}, \dots, x_n) \int_{a_j}^{b_j} u_j(x_j)\, dx_j \right)$$
$$= 0, \qquad 2 \le j \le n-1,$$

$$\int_{a_n}^{b_n} \phi_n(x_1, \dots, x_n)\, dx_n = u_1(x_1) \cdots u_{n-1}(x_{n-1}) \int_{a_n}^{b_n} \psi_n(x_n)\, dx_n$$
$$= u_1(x_1) \cdots u_{n-1}(x_{n-1}) \int_B \phi(x)\, dx = 0. \quad ∎$$

3.30 COROLLARY If $T \in \mathscr{D}'(B)$ and $D_j T = 0$ for $1 \le j \le n$, then there exists a constant k such that for all $\phi \in \mathscr{D}(B)$,

$$T(\phi) = k \int_B \phi(x)\, dx.$$

Proof. First note that if $\int_B \phi(x)\, dx = 0$, then $T(\phi) = 0$, for, by the above lemma we may write $\phi = \sum_{j=1}^n \phi_j$, where $\phi_j \in \mathscr{D}(B)$ satisfies (8), and hence $\phi_j = D_j \theta_j$, where θ_j defined by

$$\theta_j(x) = \int_{a_j}^{x_j} \phi_j(x_1, \dots, x_{j-1}, t, x_{j+1}, \dots, x_n)\, dt$$

belongs to $\mathscr{D}(B)$. Thus $T(\phi) = \sum_{j=1}^n T(D_j \theta_j) = -\sum_{j=1}^n (D_j T)(\theta_j) = 0$. Now suppose $T \ne 0$. Then there exists $\phi_0 \in \mathscr{D}(B)$ such that $T(\phi_0) = k_1 \ne 0$. Thus $\int_B \phi_0(x)\, dx = k_2 \ne 0$ and $T(\phi_0) = k \int_B \phi_0(x)\, dx$, where $k = k_1/k_2$. If $\phi \in \mathscr{D}(B)$ is arbitrary, let $K(\phi) = \int_B \phi(x)\, dx$. Then

$$\int_B \left(\phi(x) - \frac{K(\phi)}{k_2} \phi_0(x) \right) dx = 0$$

and so $T\big(\phi - [K(\phi)/k_2]\phi_0\big) = 0$. It follows that

$$T(\phi) = \frac{T(\phi_0)}{k_2} K(\phi) = kK(\phi) = k \int_B \phi(x)\, dx. \qquad \blacksquare$$

Note that this corollary can be extended to any connected set $\Omega \in \mathbb{R}^n$ via a partition of unity for Ω subordinate to some open cover of Ω by open rectangular boxes that are contained in Ω. We do not, however, require this extension.

The following lemma shows that different locally integrable functions on an open set Ω determine different distributions on Ω.

3.31 LEMMA Let $u \in L^1_{\text{loc}}(\Omega)$ satisfy $\int_\Omega u(x)\phi(x)\, dx = 0$ for every ϕ in $\mathscr{D}(\Omega)$. Then $u(x) = 0$ a.e. in Ω.

Proof. If $\psi \in C_0(\Omega)$, then for sufficiently small positive ϵ, the mollifier $J_\epsilon * \psi$ belongs to $\mathscr{D}(\Omega)$. By Lemma 2.29, $J_\epsilon * \psi \to \psi$ uniformly on Ω as $\epsilon \to 0+$. Hence $\int_\Omega u(x)\psi(x)\, dx = 0$ for every $\psi \in C_0(\Omega)$.

Let $K \Subset \Omega$ and let $\epsilon > 0$. Let χ_K be the characteristic function of K. Then $\int_K |u(x)|\, dx < \infty$. There exists $\delta > 0$ such that for any measurable set $A \subset K$ with $\mu(A) < \delta$ we have $\int_A |u(x)|\, dx < \epsilon/2$ (see, for example, [Ru2, p. 124]). By Lusin's theorem 1.42(f) there exists $\psi \in C_0(\mathbb{R}^n)$ with $|\psi(x)| \le 1$ for all x, such that

$$\mu\big(\{x \in \mathbb{R}^n : \psi(x) \ne \chi_K(x)\operatorname{sgn}\overline{u(x)}\}\big) < \delta.$$

Here

$$\operatorname{sgn} v(x) = \begin{cases} v(x)/|v(x)| & \text{if } v(x) \ne 0 \\ 0 & \text{if } v(x) = 0. \end{cases}$$

Hence

$$\begin{aligned} \int_K |u(x)|\, dx &= \int_\Omega u(x)\chi_K(x)\operatorname{sgn}\overline{u(x)}\, dx \\ &= \int_\Omega u(x)\psi(x)\, dx + \int_\Omega u(x)\big(\chi_K(x)\operatorname{sgn}\overline{u(x)} - \psi(x)\big)\, dx \\ &\le 0 + 2\int_{\{x\in\Omega:\psi(x)\ne\chi_K(x)\operatorname{sgn}\overline{u(x)}\}} |u(x)|\, dx < \epsilon. \end{aligned}$$

Since ϵ is arbitrary, $u(x) = 0$ a.e. in K for each such K, and hence a.e. in Ω. ∎

3.32 COROLLARY If B is a rectangular box as in Lemma 3.29 and u in $L^1_{\text{loc}}(B)$ possesses weak derivatives $D_j u = 0$ for $1 \le j \le n$, then for some constant k, $u(x) = k$ a.e. in B.

Proof. By Corollary 3.30, since $D_j T_u = 0$ for $1 \le j \le n$, we have

$$\int_B u(x)\phi(x)\, dx = T_u(\phi) = k \int_B \phi(x)\, dx.$$

Hence $u(x) - k = 0$ a.e. in B. ∎

3.33 THEOREM Let $m \geq 1$.

(a) If $W^{m,p}(\Omega) = W_0^{m,p}(\Omega)$, then Ω^c is (m, p')-polar.

(b) If Ω^c is both $(1, p)$-polar and (m, p')-polar, then $W^{m,p}(\Omega) = W_0^{m,p}(\Omega)$.

Proof. (a) Assume $W^{m,p}(\Omega) = W_0^{m,p}(\Omega)$. We deduce first that Ω^c must have measure zero. If not, there would exist some finite open rectangle $B \subset \mathbb{R}^n$ which intersects both Ω and Ω^c in sets of positive measure. Let u be the restriction to Ω of a function in $C_0^\infty(\mathbb{R}^n)$ which is identically one on $B \cap \Omega$. Then $u \in W^{m,p}(\Omega)$ and so $u \in W_0^{m,p}(\Omega)$. By Lemma 3.27, the zero extension $\tilde{u}$ of u to $\mathbb{R}^n$ belongs to $W^{m,p}(\mathbb{R}^n)$ and $D_j\tilde{u} = \widetilde{D_j u}$ in the distributional sense in $\mathbb{R}^n$ for $1 \leq j \leq n$. Now $\widetilde{D_j u}$ is identically zero on B and so $D_j\tilde{u} = 0$ as a distribution on B. By Corollary 3.32, $\tilde{u}$ must have a constant value a.e. in B. Since $\tilde{u} = 1$ on $B \cap \Omega$ and $\tilde{u} = 0$ on $B \cap \Omega^c$, we have a contradiction. Thus Ω^c has measure zero.

Now if $v \in W^{m,p}(\mathbb{R}^n)$ and u is the restriction of v to Ω, then u belongs to $W^{m,p}(\Omega)$ and hence, by assumption, also to $W_0^{m,p}(\Omega)$. By Lemma 3.27, $\tilde{u} \in W^{m,p}(\mathbb{R}^n)$ and can be approximated by elements of $C_0^\infty(\Omega)$. But $v(x) = \tilde{u}(x)$ on Ω, that is, a.e. in $\mathbb{R}^n$. Hence v and $\tilde{u}$ have the same distributional derivatives, and coincide in $W^{m,p}(\mathbb{R}^n)$. Therefore $C_0^\infty(\Omega)$ is dense in $W^{m,p}(\mathbb{R}^n)$ and Ω^c is (m, p')-polar by Theorem 3.28.

(b) Now assume Ω^c is $(1, p)$-polar and (m, p')-polar. Let $u \in W^{m,p}(\Omega)$. We show that $u \in W_0^{m,p}(\Omega)$. Since $\tilde{u} \in L^p(\mathbb{R}^n)$, the distribution $T_{D_j\tilde{u}}$, corresponding to $D_j\tilde{u}$, belongs to $W^{-1,p}(\mathbb{R}^n)$. Since $\widetilde{D_j u} \in L^p(\mathbb{R}^n) \subset H^{-1,p}(\mathbb{R}^n)$ (see Paragraph 3.13), therefore $T_{\widetilde{D_j u}} \in W^{-1,p}(\mathbb{R}^n)$. Hence $T_{D_j\tilde{u}-\widetilde{D_j u}} \in W^{-1,p}(\mathbb{R}^n)$. But $D_j\tilde{u} - \widetilde{D_j u} = 0$ on Ω so $\operatorname{supp}\left(T_{D_j\tilde{u}-\widetilde{D_j u}}\right) \subset \Omega^c$. Since Ω^c is $(1,p)$-polar, $D_j\tilde{u} = \widetilde{D_j u}$ in the distributional sense on $\mathbb{R}^n$, whence $D_j\tilde{u} \in L^p(\mathbb{R}^n)$ and $\tilde{u} \in W^{m,p}(\mathbb{R}^n)$. Since Ω^c is (m, p')-polar, $C_0^\infty(\Omega)$ is dense in $W^{m,p}(\mathbb{R}^n)$, and thus $u \in W_0^{m,p}(\Omega)$. ∎

3.34 If (m, p')-polarity implies $(1, p)$-polarity, then Theorem 3.33 amounts to the assertion that (m, p')-polarity of Ω^c is necessary and sufficient for the equality of $W^{m,p}(\Omega)$ and $W_0^{m,p}(\Omega)$. This is certainly the case if $p = 2$.

The following two lemmas develop properties of polarity. The first of these shows that it is a local property.

3.35 LEMMA $F \subset \mathbb{R}^n$ is (m, p')-polar if and only if $F \cap K$ is (m, p')-polar for every compact set $K \subset \mathbb{R}^n$.

Proof. Clearly the (m, p')-polarity of F implies that of $F \cap K$ for every compact K. We need only prove the converse.

Let $T \in W^{-m,p'}(\mathbb{R}^n)$ be given by $T = \sum_{0\leq|\alpha|\leq m}(-1)^{|\alpha|}D^\alpha T_{v_\alpha}$, where sequence $\{v_\alpha\} \subset L^{p'}(\mathbb{R}^n)$. Suppose T has support in F. We must show that $T = 0$. Let $f \in C_0^\infty(\mathbb{R}^n)$ satisfy $f(x) = 1$ if $|x| \leq 1$ and $f(x) = 0$ if $|x| \geq 2$. For $\epsilon > 0$, let

$f_\epsilon(x) = f(\epsilon x)$ so that $D^\alpha f_\epsilon(x) = \epsilon^{|\alpha|} D^\alpha f(\epsilon x) \to 0$ uniformly in x as $\epsilon \to 0+$. Then $f_\epsilon T \in W^{-m,p'}(\mathbb{R}^n)$ by induction on m, and for any $\phi \in \mathscr{D}(\mathbb{R}^n)$ we have

$$
\begin{aligned}
|T(\phi) - f_\epsilon T(\phi)| &= |T(\phi) - T(f_\epsilon \phi)| \\
&= \left| \sum_{0\le|\alpha|\le m} \int_{\mathbb{R}^n} v_\alpha(x) D^\alpha \big[\phi(x)\big(1 - f_\epsilon(x)\big)\big]\, dx \right| \\
&= \left| \sum_{0\le|\alpha|\le m} \sum_{\beta\le\alpha} \binom{\alpha}{\beta} \int_{\mathbb{R}^n} v_\alpha(x) D^\beta \phi(x) D^{\alpha-\beta}\big(1 - f_\epsilon(x)\big)\, dx \right| \\
&\le \sum_{\beta\le\alpha} \int_{\mathbb{R}^n} |w_\beta(x) D^\beta \phi(x)|\, dx \le \|\phi\|_{m,p} \left\| w\, ;\, L^{p'}(\Omega^{(m)}) \right\|,
\end{aligned}
$$

where

$$
\begin{aligned}
w_\beta(x) &= \sum_{|\alpha|\le m,\ \beta\le\alpha} \binom{\alpha}{\beta} v_\alpha(x) D^{\alpha-\beta}\big(1 - f_\epsilon(x)\big) \\
&= v_\beta(x)\big(1 - f_\epsilon(x)\big) - \sum_{|\alpha|\le m,\ \beta\le\alpha,\beta\ne\alpha} \binom{\alpha}{\beta} v_\alpha(x) D^{\alpha-\beta} f_\epsilon(x).
\end{aligned}
$$

Since $f_\epsilon(x) = 1$ for $|x| \le 1/\epsilon$, we have $\lim_{\epsilon\to0+} \left\| w_\beta \right\|_{p'} = 0$. Thus $f_\epsilon T \to T$ in $W^{-m,p'}(\mathbb{R}^n)$ as $\epsilon \to 0+$. But $f_\epsilon T = 0$ by assumption since it has compact support in K. Thus $T = 0$. ∎

3.36 LEMMA If $p' < q'$ (that is, $p > q$) and $f \subset \mathbb{R}^n$ is (m, p')-polar, then F is also (m, q')-polar.

Proof. Let $K \subset \mathbb{R}^n$ be compact. By the previous lemma it is sufficient to show that $F \cap K$ is (m, q')-polar. Let G be an open, bounded set in $\mathbb{R}^n$ containing K. By Theorem 2.14, $W_0^{m,p}(G) \to W_0^{m,q}(G)$, so that $W^{-m,q'}(G) \subset W^{-m,p'}(G)$. Any distribution $T \in W^{-m,q'}(\mathbb{R}^n)$ having support in $K \cap F$ also belongs to $W^{-m,q'}(G)$ and so to $W^{-m,p'}(G)$. Since $K \cap F$ is (m, p')-polar, $T = 0$. Thus $K \cap F$ is also (m, q')-polar. ∎

3.37 THEOREM Let $m \ge 1$ and $p \ge 2$. Then $W^{m,p}(\Omega) = W_0^{m,p}(\Omega)$ if and only of Ω^c is (m, p')-polar.

Proof. Since $p' \le 2$, Ω^c is (m, p)-polar and therefore also $(1, p)$-polar. The result now follows by Theorem 3.33.

3.38 The Sobolev Imbedding Theorem 4.12 can be used to extend the previous theorem to cover certain values of $p < 2$. If $(m-1)p < n$, the imbedding theorem gives

$$
W^{m,p}(\mathbb{R}^n) \to W^{1,q}(\mathbb{R}^n), \qquad q = \frac{np}{n-(m-1)p},
$$

which in turn implies that $W^{-1,q'}(\mathbb{R}^n) \subset W^{-m,p'}(\mathbb{R}^n)$. If also $p \geq 2n/(n+m-1)$, then $q' \leq p$ and so by Lemma 3.36, Ω^c is $(1, p)$-polar if it is (m, p')-polar. Note that $2n/(n+m-1) < 2$ if $m > 1$. If, on the other hand, $(m-1)p \geq n$, then $mp > n$, and, as pointed out in Paragraph 3.25, Ω^c cannot be (m, p')-polar unless it is empty, in which case it is trivially $(1, p)$-polar.

The only values of p for which we do not know that the (m, p')-polarity of Ω^c implies $(1, p)$-polarity and hence is equivalent to the identity of $W^{m,p}(\Omega)$ and $W_0^{m,p}(\Omega)$, are given by $1 \leq p \leq \min\{n/(m-1), 2n/(n+m-1)\}$.

3.39 Whenever $W_0^{m,p}(\Omega) \neq W^{m,p}(\Omega)$, the former space is a closed subspace of the latter. In the Hilbert space case, $p = 2$, we may consider the space $W_0^{\perp}$ consisting of all $v \in W^{m,2}(\Omega)$ such that $(v, \phi)_m = 0$ for all $\phi \in C_0^{\infty}(\Omega)$. Every $u \in W^{m,2}(\Omega)$ can be uniquely decomposed in the form $u = u_0 + v$, where $u_0 \in W_0^{m,2}(\Omega)$ and $v \in W_0^{\perp}$. Integration by parts shows that any $v \in W_0^{\perp}$ must satisfy

$$\sum_{0 \leq |\alpha| \leq m} (-1)^{|\alpha|} D^{2\alpha} v(x) = 0$$

in the weak sense, and hence a.e. in Ω.

Coordinate Transformations

3.40 Let Φ be a one-to-one transformation of a domain $\Omega \subset \mathbb{R}^n$ onto a domain $G \in \mathbb{R}^n$, having inverse $\Psi = \Phi^{-1}$. We say that Φ is *m-smooth* if, when we write $y = \Phi(x)$ and $x = \Psi(y)$ in the form

$$\begin{aligned} y_1 &= \phi_1(x_1, \ldots, x_n), & x_1 &= \psi_1(y_1, \ldots, y_n), \\ y_2 &= \phi_2(x_1, \ldots, x_n), & x_2 &= \psi_2(y_1, \ldots, y_n), \\ &\vdots & &\vdots \\ y_n &= \phi_n(x_1, \ldots, x_n), & x_n &= \psi_n(y_1, \ldots, y_n), \end{aligned}$$

then $\phi_1, \ldots, \phi_n$ belong to $C^m(\overline{\Omega})$ and $\psi_1, \ldots, \psi_n$ belong to $C^m(\overline{G})$.

If u is a measurable function on Ω, we define a measurable function Au on G by

$$Au(y) = u\big(\Psi(y)\big). \tag{9}$$

Suppose that Φ is 1-smooth so that there exist constants $0 < c < C$ such that for all $x \in \Omega$

$$c \leq |\det \Phi'(x)| \leq C, \tag{10}$$

where Φ' denotes the Jacobian matrix $\partial(y_1, \ldots, y_n)/\partial(y_1, \ldots, y_n)$. Since smooth functions are dense in L^p spaces, the operator A defined by (9) transforms $L^p(\Omega)$ boundedly onto $L^p(G)$ and has a bounded inverse; in fact, for $1 \leq p < \infty$,

$$c^{1/p} \|u\|_{p,\Omega} \leq \|Au\|_{p,G} \leq C^{1/p} \|u\|_{p,\Omega}.$$

We establish a similar result for Sobolev spaces.

3.41 THEOREM Let Φ be m-smooth, where $m \geq 1$. The operator A defined by (9) transforms $W^{m,p}(\Omega)$ boundedly onto $W^{m,p}(G)$ and has a bounded inverse.

Proof. We show that the inequality $\|Au\|_{m,p,G} \leq \text{const } \|u\|_{m,p,\Omega}$ holds for every $u \in W^{m,p}(\Omega)$, the constant depending only on the transformation Φ. The reverse inequality $\|Au\|_{m,p,G} \geq \text{const } \|u\|_{m,p,\Omega}$ (with a different constant) can be established similarly, using the inverse operator A^{-1}. By Theorem 3.17 for given $u \in W^{m,p}(\Omega)$, there exists a sequence $\{u_j\} \subset C^\infty(\Omega)$ converging to u in $W^{m,p}(\Omega)$-norm. For such smooth u_j it is readily checked by induction on $|\alpha|$ that

$$D^\alpha(Au_j)(y) = \sum_{\beta \leq \alpha} M_{\alpha\beta}(y) A(D^\beta u_j)(y), \tag{11}$$

where $M_{\alpha\beta}$ is a polynomial of degree not exceeding $|\beta|$ in derivatives of orders not exceeding $|\alpha|$ of the various components of Ψ. If $\theta \in \mathscr{D}(G)$ integration by parts gives

$$(-1)^{|\alpha|} \int_G (Au_j)(y) D^\alpha \theta(y)\, dy = \sum_{\beta \leq \alpha} \int_G A(D^\beta u_j)(y) M_{\alpha\beta}(y)\theta(y)\, dy, \tag{12}$$

or, replacing y by $\Phi(x)$ and expressing the integrals over Ω,

$$\begin{aligned} &(-1)^{|\alpha|} \int_\Omega u_j(x)(D^\alpha\theta)\big(\Phi(x)\big)|\det \Phi'(x)|\, dx \\ &\quad = \sum_{\beta \leq \alpha} \int_\Omega D^\beta u_j(x) M_{\alpha\beta}\big(\Phi(x)\big)\theta\big(\Phi(x)\big)|\det \Phi'(x)|\, dx. \end{aligned} \tag{13}$$

Since $D^\beta u_j \to u$ in $L^p(\Omega)$ for $|\beta| \leq m$, we can take the limit through (13) as $n \to \infty$ and hence obtain (12) with u replacing u_j. Thus (11) holds in the weak sense for any $u \in W^{m,p}(\Omega)$. Therefore

$$\begin{aligned} \int_G |D^\alpha(Au)(y)|^p\, dy &\leq \left(\sum_{\beta \leq \alpha} 1\right)^p \max_{|\beta| \leq |\alpha|} \left(\sup_{y \in G} |M_{\alpha\beta}| \int_G |(D^\beta u)|\big(\Psi(y)\big)|^p\, dy\right) \\ &\leq \text{const } \max_{|\beta| \leq |\alpha|} \int_\Omega |D^\beta u(x)|^p\, dx, \end{aligned}$$

from which it follows that $\|Au\|_{m,p,G} \leq \text{const } \|u\|_{m,p,\Omega}$. ∎

Of special importance in later chapters is the case of the above theorem corresponding to nonsingular linear transformations Φ or, more generally, affine transformations (compositions of nonsingular linear transformations and translations). For such transformations $\det \Phi'(x)$ is a nonzero constant.

4

THE SOBOLEV IMBEDDING THEOREM

4.1 The imbedding characteristics of Sobolev spaces are essential in their uses in analysis, especially in the study of differential and integral operators. The most important imbedding results for Sobolev spaces are often gathered together into a single "theorem" called *the Sobolev Imbedding Theorem* although they are of several different types and can require different methods of proof. The core results are due to Sobolev [So2] but our statement (Theorem 4.12) also includes refinements due to others, in particular Morrey [Mo] and Gagliardo [Ga1].

Most of the imbeddings hold for domains $\Omega \subset \mathbb{R}^n$ satisfying some form of "cone condition" that enables us to derive pointwise estimates for the value of a function at the vertex of a truncated cone from suitable averages of the values of the function and its derivatives over the cone. Some of the imbeddings require stronger geometric hypotheses which, roughly speaking, force Ω to have an $(n-1)$-dimensional boundary that is locally the graph of a Lipschitz continuous function and which, like the segment condition described in Paragraph 3.21, requires Ω to lie on only one side of its boundary. We will discuss these geometric properties of domains prior to the statement of the imbedding theorem itself.

4.2 (Targets of the Imbeddings) The Sobolev imbedding theorem asserts the existence of imbeddings of $W^{m,p}(\Omega)$ (or $W_0^{m,p}(\Omega)$) into Banach spaces of the following types:

(i) $W^{j,q}(\Omega)$, where $j \le m$, and in particular $L^q(\Omega)$,

(ii) $W^{j,q}(\Omega_k)$, where, for $1 \le k < n$, Ω_k is the intersection of Ω with a

k-dimensional plane in $\mathbb{R}^n$.

(iii) $C_B^j(\Omega)$, the space of functions having bounded, continuous derivatives up to order j on Ω (see Paragraph 1.27) normed by

$$\left\| u\,;\, C_B^j(\Omega) \right\| = \max_{0\le|\alpha|\le j} \sup_{x\in\Omega} |D^\alpha u(x)|.$$

(iv) $C^j(\overline{\Omega})$, the closed subspace of $C_B^j(\Omega)$ consisting of functions having bounded, uniformly continuous derivatives up to order j on Ω (see Paragraph 1.28) with the same norm as $C_B^j(\Omega)$:

$$\left\| \phi\,;\, C^j(\overline{\Omega}) \right\| = \max_{0\le\alpha\le j} \sup_{x\in\Omega} |D^\alpha \phi(x)|.$$

This space is smaller than $C_B^j(\Omega)$ in that its elements must be uniformly continuous on Ω. For example, the function u of Example 3.20 belongs to $C_B^1(\Omega)$ but certainly not to $C^1(\overline{\Omega})$ for the domain Ω of that example.

(v) $C^{j,\lambda}(\overline{\Omega})$, the closed subspace of $C^j(\overline{\Omega})$ consisting of functions whose derivatives up to order j satisfy Hölder conditions of exponent λ in Ω (see Paragraph 1.29). The norm on $C^{j,\lambda}(\overline{\Omega})$ is

$$\left\| \phi\,;\, C^{j,\lambda}(\overline{\Omega}) \right\| = \left\| \phi\,;\, C^j(\overline{\Omega}) \right\| + \max_{0\le|\alpha|\le j} \sup_{\substack{x,y\in\Omega \\ x\ne y}} \frac{|D^\alpha\phi(x) - D^\alpha\phi(y)|}{|x-y|^\lambda}.$$

Since elements of $W^{m,p}(\Omega)$ are, strictly speaking, not functions defined everywhere on Ω, but rather equivalence classes of such functions defined and equal up to sets of measure zero, we must clarify what is meant by imbeddings of types (ii)–(v). What is intended for imbeddings into the continuous function spaces (types (iii)-(v)) is that the "equivalence class" $u \in W^{m,p}(\Omega)$ should contain an element that belongs to the continuous function space that is the target of the imbedding and is bounded in that space by a constant times $\|u\|_{m,p,\Omega}$. Hence, for example, existence of the imbedding

$$W^{m,p}(\Omega) \to C^j(\overline{\Omega})$$

means that each $u \in W^{m,p}(\Omega)$ when considered as a function, can be redefined on a subset of Ω having measure zero to produce a new function $u^* \in C^j(\overline{\Omega})$ such that $u^* = u$ in $W^{m,p}(\Omega)$ (i.e. u^* and u belong to the same "equivalence class" in $W^{m,p}(\Omega)$) and

$$\left\| u^*\,;\, C^j(\overline{\Omega}) \right\| \le K\, \|u\|_{m,p,\Omega}$$

with K independent of u.

Even more care is necessary in interpreting imbeddings into spaces of type (ii):

$$W^{m,p}(\Omega) \to W^{j,q}(\Omega_k)$$

where Ω_k is the intersection of Ω with a plane of dimension $k < n$. Each element of $W^{m,p}(\Omega)$ is, by Theorem 3.17, a limit in that space of a sequence $\{u_i\}$ of functions in $C^\infty(\Omega)$. The functions u_i have *traces on* Ω_k (that is, restrictions to Ω_k) that belong to $C^\infty(\Omega_k)$. The above imbedding signifies that these traces converge in $W^{j,q}(\Omega_k)$ to a function u^* that is independent of the choice of $\{u_i\}$ and satisfies

$$\left\|u^*\right\|_{j,q,\Omega_k} \le K\,\|u\|_{m,p,\Omega}$$

with K independent of u.

4.3 Let us note as a point of interest, though of no particular use to us later, that the imbedding $W^{m,p}(\Omega) \to W^{j,q}(\Omega)$ is equivalent to the simple containment $W^{m,p}(\Omega) \subset W^{j,q}(\Omega)$. Certainly the former implies the latter. To verify the converse, suppose $W^{m,p}(\Omega) \subset W^{j,q}(\Omega)$, and let I be the linear operator taking $W^{m,p}(\Omega)$ into $W^{j,q}(\Omega)$ defined by $Iu = u$ for $u \in W^{m,p}(\Omega)$. If $u_k \to u$ in $W^{m,p}(\Omega)$ (and hence in $L^p(\Omega)$) and $Iu_k \to v$ in $W^{j,q}(\Omega)$ (and hence in $L^q(\Omega)$), then, passing to a subsequence if necessary, we have by Corollary 2.17 that $u_k(x) \to u(x)$ a.e. on Ω, $u_k(x) = Iu_k(x) \to v(x)$ a.e. on Ω. Thus $u(x) = v(x)$ a.e. on Ω, that is, $Iu = v$, and I is continuous by the closed graph theorem of functional analysis.

Geometric Properties of Domains

4.4 (Some Definitions) Many properties of Sobolev spaces defined on a domain Ω, and in particular the imbedding properties of these spaces, depend on regularity properties of Ω. Such regularity is normally expressed in terms of geometric or analytic conditions that may or may not be satisfied by a given domain. We specify below several such conditions and consider their relationships. First we make some definitions.

Let v be a nonzero vector in $\mathbb{R}^n$, and for each $x \ne 0$ let $\angle(x, v)$ be the angle between the position vector x and v. For given such v, $\rho > 0$, and κ satisfying $0 < \kappa \le \pi$, the set

$$C = \{x \in \mathbb{R}^n : x = 0 \text{ or } 0 < |x| \le \rho,\ \angle(x, v) \le \kappa/2\}$$

is called a *finite cone* of height ρ, axis direction v and aperture angle κ with vertex at the origin. Note that $x + C = \{x + y : y \in C\}$ is a finite cone with vertex at x but the same dimensions and axis direction as C and is obtained by parallel translation of C.

Given n linearly independent vectors $y_1, \ldots, y_n \in \mathbb{R}^n$, the set

$$P = \left\{ \sum_{j=1}^{n} \lambda_j y_j \ : 0 \le \lambda_j \le 1,\ 1 \le j \le n \right\}$$

is a *parallelepiped* with one vertex at the origin. Similarly, $x + P$ is a parallel translate of P having one vertex at x. The *centre* of $x + P$ is the point given by $c(x + P) = x + (1/2)(y_1 + \cdots + y_n)$. Every parallelepiped with a vertex at x is contained in a finite cone with vertex at x and also contains such a cone.

An open cover $\mathscr{O}$ of a set $S \subset \mathbb{R}^n$ is said to be *locally finite* if any compact set in $\mathbb{R}^n$ can intersect at most finitely many members of $\mathscr{O}$. Such locally finite collections of sets must be countable, so their elements can be listed in sequence. If S is closed, then any open cover of S by sets with a uniform bound on their diameters possesses a locally finite subcover.

We now specify six regularity properties that a domain $\Omega \subset \mathbb{R}^n$ may possess. We denote by Ω_δ the set of points in Ω within distance δ of the boundary of Ω:

$$\Omega_\delta = \{x \in \Omega : \operatorname{dist}(x, \operatorname{bdry}\Omega) < \delta\}.$$

4.5 (The Segment Condition) As defined in Paragraph 3.21, a domain Ω satisfies the *segment condition* if every $x \in \operatorname{bdry}\Omega$ has a neighbourhood U_x and a nonzero vector y_x such that if $z \in \overline{\Omega} \cap U_x$, then $z + ty_x \in \Omega$ for $0 < t < 1$. Since the boundary of Ω is necessarily closed, we can replace its open cover by the neighbourhoods U_x with a locally finite subcover $\{U_1, U_2, \ldots\}$ with corresponding vectors $y_1, y_2, \ldots$ such that if $x \in \overline{\Omega} \cap U_j$ for some j, then $x + ty_j \in \Omega$ for $0 < t < 1$.

4.6 (The Cone Condition) Ω satisfies the *cone condition* if there exists a finite cone C such that each $x \in \Omega$ is the vertex of a finite cone C_x contained in Ω and congruent to C. Note that C_x need not be obtained from C by parallel translation, but simply by rigid motion.

4.7 (The Weak Cone Condition) Given $x \in \Omega$, let $R(x)$ consist of all points $y \in \Omega$ such that the line segment from x to y lies in Ω; thus $R(x)$ is a union of rays and line segments emanating from x. Let

$$\Gamma(x) = \{y \in R(x) : |y - x| < 1\}.$$

We say that Ω satisfies the *weak cone condition* if there exists $\delta > 0$ such that

$$\mu_n(\Gamma(x)) \ge \delta \qquad \text{for all } x \in \Omega,$$

where μ_n is the Lebesgue measure in $\mathbb{R}^n$. Clearly the cone condition implies the weak cone condition, but there are many domains satisfying the weak cone condition that do not satisfy the cone condition.

4.8 (The Uniform Cone Condition) Ω satisfies the *uniform cone condition* if there exists a locally finite open cover $\{U_j\}$ of the boundary of Ω and a corresponding sequence $\{C_j\}$ of finite cones, each congruent to some fixed finite cone C, such that

(i) There exists $M < \infty$ such that every U_j has diameter less then M.

(ii) $\Omega_\delta \subset \bigcup_{j=1}^{\infty} U_j$ for some $\delta > 0$.

(iii) $Q_j \equiv \bigcup_{x\in\Omega\cap U_j}(x + C_j) \subset \Omega$ for every j.

(iv) For some finite R, every collection of $R + 1$ of the sets Q_j has empty intersection.

4.9 (The Strong Local Lipschitz Condition) Ω satisfies the *strong local Lipschitz condition* if there exist positive numbers δ and M, a locally finite open cover $\{U_j\}$ of bdry Ω, and, for each j a real-valued function f_j of $n - 1$ variables, such that the following conditions hold:

(i) For some finite R, every collection of $R + 1$ of the sets U_j has empty intersection.

(ii) For every pair of points $x, y \in \Omega_\delta$ such that $|x - y| < \delta$, there exists j such that

$$x, y \in V_j \equiv \left\{x \in U_j : \text{dist}(x, \text{bdry}\, U_j) > \delta\right\}.$$

(iii) Each function f_j satisfies a Lipschitz condition with constant M: that is, if $\xi = (\xi_1, \ldots, \xi_{n-1})$ and $\rho = (\rho_1, \ldots, \rho_{n-1})$ are in $\mathbb{R}^{n-1}$, then

$$|f(\xi) - f(\rho)| \le M|\xi - \rho|.$$

(iv) For some Cartesian coordinate system $(\zeta_{j,1}, \ldots, \zeta_{j,n})$ in U_j, $\Omega \cap U_j$ is represented by the inequality

$$\zeta_{j,n} < f_j(\zeta_{j,1}, \ldots, \zeta_{j,n-1}).$$

If Ω is bounded, the rather complicated set of conditions above reduce to the simple condition that Ω should have a locally Lipschitz boundary, that is, that each point x on the boundary of Ω should have a neighbourhood U_x whose intersection with bdry Ω should be the graph of a Lipschitz continuous function.

4.10 (The Uniform C^m-Regularity Condition) Ω satisfies the *uniform C^m-regularity condition* is there exists a locally finite open cover $\{U_j\}$ of bdry Ω, and a corresponding sequence $\{\Phi_j\}$ of m-smooth transformations (see Paragraph 3.40) with Φ_j taking U_j onto the ball $B = \{y \in \mathbb{R}^n : |y| < 1$ and having inverse $\Psi_j = \Phi_j^{-1}$, such that:

(i) For some finite R. every collection of $R + 1$ of the sets U_j has empty intersection.

(ii) For some $\delta > 0$, $\Omega_\delta \subset \bigcup_{j=1}^{\infty} \Psi_j\left(\{y \in \mathbb{R}^n : |y| < \frac{1}{2}\}\right)$.

(iii) For each j, $\Phi_j(U_j \cap \Omega) = \{y \in B : y_n > 0\}$.

(iv) If $(\phi_{j,1}, \ldots, \phi_{j,n})$ and $(\psi_{j,1}, \ldots, \psi_{j,n})$ are the components of Φ_j and Ψ_j, then there is a finite constant M such that for every α with $0 < |\alpha| \le m$, every i, $1 \le i \le n$, and every j we have

$$
\begin{aligned}
|D^\alpha \phi_{j,i}(x)| &\le M, \qquad \text{for } x \in U_j, \\
|D^\alpha \psi_{j,i}(y)| &\le M, \qquad \text{for } y \in B.
\end{aligned}
$$

4.11 Except for the cone condition and the weak cone condition, the other conditions defined above all require that the boundary of Ω be $(n-1)$-dimensional and that Ω lie on only one side of its boundary. The domain Ω of Example 3.20 satisfies the cone condition (and therefore the weak cone condition), but none of the other four conditions. Among those four we have:

$$
\begin{aligned}
&\text{the uniform } C^m\text{-regularity condition } (m \ge 2) \\
&\implies \text{the strong local Lipschitz condition} \\
&\implies \text{the uniform cone condition} \\
&\implies \text{the segment condition.}
\end{aligned}
$$

Also,

$$
\begin{aligned}
&\text{the uniform cone condition} \\
&\implies \text{the cone condition} \\
&\implies \text{the weak cone condition}
\end{aligned}
$$

Typically, most of the imbeddings of $W^{m,p}(\Omega)$ have been proven for domains satisfying the cone condition. Exceptions are the imbeddings into spaces $C^j(\overline{\Omega})$ and $C^{j,\lambda}(\overline{\Omega})$ of uniformly continuous functions which, as suggested by Example 3.20, require that Ω lie on one side of its boundary. These imbeddings are usually proved for domains satisfying the strong local Lipschitz condition. It should be noted, however, that Ω need not satisfy any of these conditions for appropriate imbeddings of $W_0^{m,p}(\Omega)$ to be valid.

4.12 THEOREM (The Sobolev Imbedding Theorem) Let Ω be a domain in $\mathbb{R}^n$ and, for $1 \le k \le n$, let Ω_k be the intersection of Ω with a plane of dimension k in $\mathbb{R}^n$. (If $k = n$, then $\Omega_k = \Omega$.) Let $j \ge 0$ and $m \ge 1$ be integers and let $1 \le p < \infty$.

PART I Suppose Ω satisfies the cone condition.

Case A If either $mp > n$ or $m = n$ and $p = 1$, then

$$W^{j+m,p}(\Omega) \to C_B^j(\Omega). \tag{1}$$

Moreover, if $1 \le k \le n$, then

$$W^{j+m,p}(\Omega) \to W^{j,q}(\Omega_k) \qquad \text{for } p \le q \le \infty, \tag{2}$$

and, in particular,

$$W^{m,p}(\Omega) \to L^q(\Omega) \qquad \text{for } p \le q \le \infty.$$

Case B If $1 \le k \le n$ and $mp = n$, then

$$W^{j+m,p}(\Omega) \to W^{j,q}(\Omega_k), \qquad \text{for } p \le q < \infty, \tag{3}$$

and, in particular,

$$W^{m,p}(\Omega) \to L^q(\Omega), \qquad \text{for } p \le q < \infty.$$

Case C If $mp < n$ and either $n - mp < k \le n$ or $p = 1$ and $n - m \le k \le n$, then

$$W^{j+m,p}(\Omega) \to W^{j,q}(\Omega_k), \qquad \text{for } p \le q \le p* = kp/(n - mp). \tag{4}$$

In particular,

$$W^{m,p}(\Omega) \to L^q(\Omega), \qquad \text{for } p \le q \le p* = np/(n - mp). \tag{5}$$

The imbedding constants for the imbeddings above depend only on n, m, p, q, j, k, and the dimensions of the cone C in the cone condition.

PART II Suppose Ω satisfies the strong local Lipschitz condition. (See Paragraph 4.9.) Then the target space $C_B^j(\Omega)$ of the imbedding (1) can be replaced with the smaller space $C^j(\overline{\Omega})$, and the imbedding can be further refined as follows: If $mp > n > (m - 1)p$, then

$$W^{j+m,p}(\Omega) \to C^{j,\lambda}(\overline{\Omega}) \qquad \text{for } 0 < \lambda \le m - (n/p), \tag{6}$$

and if $n = (m-1)p$, then

$$W^{j+m,p}(\Omega) \to C^{j,\lambda}(\overline{\Omega}) \qquad \text{for } 0 < \lambda < 1. \tag{7}$$

Also, if $n = m-1$ and $p = 1$, then (7) holds for $\lambda = 1$ as well.

PART III All of the imbeddings in Parts A and B are valid for *arbitrary* domains Ω if the W-space undergoing the imbedding is replaced with the corresponding W_0-space.

4.13 REMARKS

1. Imbeddings (1)–(4) are essentially due to Sobolev [So1, So2], although his original proof did not cover the all cases. Imbeddings (6)–(7) originate in the work of Morrey [Mo].
2. Imbeddings (2)–(4) involving traces of functions on planes of lower dimension can be extended in a reasonable manner to apply to traces on more general smooth manifolds. For example, see Theorem 5.36.
3. If Ω_k (or Ω) has finite volume, then imbeddings (2)–(4) also hold for $1 \le q < p$ in addition to the values of q asserted in the statement of the theorem. This follows from Theorem 2.14. It will be shown in Theorem 6.43 that no imbedding of the form $W^{m,p}(\Omega) \to L^q(\Omega)$ where $q < p$ is possible unless Ω has finite volume.
4. Part III of the theorem is an immediate consequence of Parts I and II applied to $\mathbb{R}^n$ because, by Lemma 3.27, the operator of zero extension of functions outside Ω maps $W_0^{m,p}(\Omega)$ isometrically into $W^{m,p}(\mathbb{R}^n)$.
5. More generally, suppose there exists an operator E mapping $W^{m,p}(\Omega)$ into $W^{m,p}(\mathbb{R}^n)$ such that $Eu(x) = u(x)$ a.e. in Ω and such that $\|Eu\|_{m,p,\mathbb{R}^n} \le K_1 \|u\|_{m,p,\Omega}$. (Such an operator is called an (m,p)-extension operator for Ω. If the imbedding theorem has already been proved for $\mathbb{R}^n$, then it must hold for the domain Ω as well. For example, if $W^{m,p}(\mathbb{R}^n) \to L^q(\mathbb{R}^n)$, and $u \in W^{m,p}(\Omega)$, then

   $$\|u\|_{q,\Omega} \le \|Eu\|_{q,\mathbb{R}^n} \le K_2 \|Eu\|_{m,p,\mathbb{R}^n} \le K_2 K_1 \|u\|_{m,p,\Omega}.$$

   In Chapter 5 we will establish the existence of such extension operators, but only for domains satisfying conditions stronger than the cone condition, so we will not use such a technique to prove Theorem 4.12.
6. It is sufficient to prove imbeddings (1)–(4), (6)–(7) for the special case $j = 0$, as the general case follows by applying this special case to derivative $D^\alpha u$ of u for $|\alpha| \le j$. For example, if the imbedding $W^{m,p}(\Omega) \to L^q(\Omega)$ has been proven, then for any $u \in W^{j+m,p}(\Omega)$ we have $D^\alpha u \in W^{m,p}(\Omega)$

for $|\alpha| \le j$, whence $D^\alpha u \in L^q(\Omega)$. Thus $u \in W^{j,q}(\Omega)$ and

$$\|u\|_{j,q} = \left(\sum_{|\alpha| \le j} \|D^\alpha u\|_{0,q}^q \right)^{1/q}$$
$$\le K_1 \left(\sum_{|\alpha| \le j} \|D^\alpha u\|_{m,p}^p \right)^{1/p} \le K_2 \|u\|_{j+m,p}.$$

7. The authors have shown that all of Part I can be proved for domains satisfying only the weak cone condition instead of the cone condition. See [AF1].

4.14 (Strategy for Proving the Imbedding Theorem) We use two overlapping methods to prove the imbeddings in Part I of Theorem 4.12. The first, potential theoretic in nature, was used by Sobolev. It works when $p > 1$, and gives the right order of growth of imbedding constants as $q \to \infty$ when $mp = n$; this will be useful in Chapter 7. Here we use the potential method to prove Case A and the imbeddings in Cases B and C for $p > 1$. The other approach is based on a combinatorial-averaging argument due to Gagliardo [Ga1]. We will use it to establish Cases B and C for $p = 1$, though it could be adapted (with a bit more difficulty) to prove all of Part I. (See, in particular, Theorem 5.10 and the Remark following that theorem.)

Part II of the theorem follows by sharpening certain estimates used in obtaining Case A of Part I.

The entire proof of Theorem 4.12 is fairly lengthy and is broken down into several lemmas. Throughout we use K, and occasionally K_1, K_2, ..., to represent various constants that can depend on parameters of the spaces being imbedded. The values of these constants can change from line to line. While stated for the cone condition, the potential method works verbatim under the weak cone condition as well.

Imbeddings by Potential Arguments

4.15 LEMMA (A Local Estimate) Let domain $\Omega \subset \mathbb{R}^n$ satisfy the cone condition. There exists a constant K depending on m, n, and the dimensions ρ and κ of the cone C specified in the cone condition for Ω such that for every $u \in C^\infty(\Omega)$, every $x \in \Omega$, and every r satisfying $0 < r \le \rho$, we have

$$|u(x)| \le K \left(\sum_{|\alpha| \le m-1} r^{|\alpha|-n} \int_{C_{x,r}} |D^\alpha u(y)|\, dy + \sum_{|\alpha| = m} \int_{C_{x,r}} |D^\alpha u(y)| |x-y|^{m-n}\, dy \right), \tag{8}$$

where $C_{x,r} = \{y \in C_x : |x - y| \le r\}$. Here $C_x \subset \Omega$ is a cone congruent to C having vertex at x.

Proof. We apply Taylor's formula with integral remainder,

$$f(1) = \sum_{j=0}^{m-1} \frac{1}{j!} f^{(j)}(0) + \frac{1}{(m-1)!} \int_0^1 (1-t)^{m-1} f^{(m)}(t)\, dt$$

to the function $f(t) = u\big(tx + (1-t)y\big)$, where $x \in \Omega$ and $y \in C_{x,r}$. Noting that

$$f^{(j)}(t) = \sum_{|\alpha|=j} \frac{j!}{\alpha!} D^\alpha u\big(tx + (1-t)y\big)(x-y)^\alpha,$$

where $\alpha! = \alpha_1! \cdots \alpha_n!$ and $(x-y)^\alpha = (x_1 - y_1)^{\alpha_1} \cdots (x_n - y_n)^{\alpha_n}$, we obtain

$$\begin{aligned} |u(x)| \le &\sum_{|\alpha| \le m-1} \frac{1}{\alpha!} |D^\alpha u(y)| |x-y|^{|\alpha|} \\ &+ \sum_{|\alpha|=m} \frac{m}{\alpha!} |x-y|^m \int_0^1 (1-t)^{m-1} \big|D^\alpha u\big(tx + (1-t)y\big)\big|\, dt. \end{aligned}$$

If C has volume $c\rho^n$, then $C_{x,r}$ has volume cr^n. Integration of y over $C_{x,r}$ leads to

$$\begin{aligned} &cr^n |u(x)| \\ &\le \sum_{|\alpha| \le m-1} \frac{r^{|\alpha|}}{\alpha!} \int_{C_{x,r}} |D^\alpha u(y)|\, dy \\ &\quad + \sum_{|\alpha|=m} \frac{m}{\alpha!} \int_{C_{x,r}} |x-y|^m\, dy \int_0^1 (1-t)^{m-1} \big|D^\alpha u\big(tx + (1-t)y\big)\big|\, dt. \end{aligned}$$

In the final (double) integral we first change the order of integration, then substitute $z = tx + (1-t)y$, so that $z - x = (1-t)(y-x)$ and $dz = (1-t)^n\, dy$, to obtain, for that integral,

$$\int_0^1 (1-t)^{-n-1}\, dt \int_{C_{x,(1-t)r}} |z-x|^m |D^\alpha u(z)|\, dz.$$

A second change of order of integration now gives for the above integral

$$\begin{aligned} &\int_{C_{x,r}} |x-z|^m |D^\alpha u(z)|\, dz \int_0^{1-(|z-x|/r)} (1-t)^{-n-1}\, dt \\ &\le \frac{r^n}{n} \int_{C_{x,r}} |x-z|^{m-n} |D^\alpha u(z)|\, dz. \end{aligned}$$

Inequality (8) now follows immediately. ∎

4.16 (Proof of Part I, Case A of Theorem 4.12) As noted earlier, we can assume that $j = 0$. Let $u \in W^{m,p}(\Omega) \cap C^\infty(\Omega)$ and let $x \in \Omega$. We must show that

$$|u(x)| \le K \, \|u\|_{m,p} . \tag{9}$$

For $p = 1$ and $m = n$, this follows immediately from (8). For $p > 1$ and $mp > n$, we apply Hölder's inequality to (8) with $r = \rho$ to obtain

$$|u(x)| \le K \Bigg(\sum_{|\alpha| \le m-1} c^{1/p'} \rho^{|\alpha|-(n/p)} \, \|D^\alpha u\|_{p,C_{x,\rho}} + \sum_{|\alpha|=m} \|D^\alpha u\|_{p,C_{x,\rho}} \left[\int_{C_{x,\rho}} |x-y|^{(m-n)p'} \, dy \right]^{1/p'} \Bigg),$$

where c is the volume of $C_{x,1}$ and $p' = p/(p-1)$. The final integral is finite since $(m-n)p' > -n$ when $mp > n$. Thus

$$|u(x)| \le K \sum_{|\alpha| \le m} \|D^\alpha u\|_{p,C_{x,\rho}} \tag{10}$$

and (9) follows because $C_{x,\rho} \subset \Omega$.

Next observe that since any $u \in W^{m,p}(\Omega)$ is the limit of a Cauchy sequence of continuous functions by Theorem 3.17, and since (9) implies this Cauchy sequence converges to a continuous function on Ω, u must coincide with a continuous function a.e. on Ω. Thus $u \in C_B^0(\Omega)$ and imbedding (1) is proved.

Now let Ω_k denote the intersection of Ω with a k-dimensional plane H, let $\Omega_{k,\rho} = \{x \in \mathbb{R}^n : \operatorname{dist}(x, \Omega_k) < \rho\}$, and let u and all its derivatives be extended to be zero outside Ω. Since $C_{x,\rho} \subset B_\rho(x)$, the ball of radius ρ with centre at x, we have, using (10) and denoting by dx' the k-volume element in H,

$$\begin{aligned} \int_{\Omega_k} |u(x)|^p \, dx' &\le K \sum_{|\alpha| \le m} \int_{\Omega_k} dx' \int_{B_\rho(x)} |D^\alpha u(y)|^p dy \\ &= K \sum_{|\alpha| \le m} \int_{\Omega_{k,\rho}} |D^\alpha u(y)|^p dy \int_{H \cap B_\rho(y)} dx' \le K_1 \, \|u\|_{m,p,\Omega}^p , \end{aligned}$$

and $W^{m,p}(\Omega) \to L^p(\Omega_k)$. But (9) shows that $W^{m,p}(\Omega) \to L^\infty(\Omega_k)$ and so imbedding (2) follows by Theorem 2.11. ∎

Let χ_r be the characteristic function of the ball $B_r(0) = \{x \in \mathbb{R}^n : |x| < r\}$. In the following discussion we will develop estimates for convolutions of L^p functions with the kernels $\omega_m(x) = |x|^{m-n}$ and

$$\chi_r \omega_m(x) = \begin{cases} |x|^{m-n} & \text{if } |x| < r, \\ 0 & \text{if } |x| \ge r. \end{cases}$$

Observe that if $m \le n$ and $0 < r \le 1$, then

$$\chi_r(x) \le \chi_r\omega_m(x) \le \omega_m(x).$$

4.17 LEMMA Let $p \ge 1$, $1 \le k \le n$, and $n - mp < k$. There exists a constant K such that for every $r > 0$, every k-dimensional plane $H \subset \mathbb{R}^n$, and every $v \in L^p(\mathbb{R}^n)$, we have $\chi_r\omega_m * |v| \in L^p(H)$ and

$$\|\chi_r\omega_m * |v|\|_{p,H} \le Kr^{m-(n-k)/p}\,\|v\|_{p,\mathbb{R}^n}. \tag{11}$$

In particular,

$$\|\chi_1 * |v|\|_{p,H} \le \|\chi_1\omega_m * |v|\|_{p,H} \le K\,\|v\|_{p,\mathbb{R}^n}.$$

Proof. If $p > 1$, then by Hölder's inequality

$$\begin{aligned}
\chi_r\omega_m * |v|(x) &= \int_{B_r(x)} |v(y)||x-y|^{-s}|x-y|^{s+m-n}\,dy \\
&\le \left(\int_{B_r(x)} |v(y)|^p|x-y|^{-sp}\,dy\right)^{1/p}\left(\int_{B_r(x)} |x-y|^{(s+m-n)p'}\,dy\right)^{1/p'} \\
&= Kr^{s+m-(n/p)}\left(\int_{B_r(x)} |v(y)|^p|x-y|^{-sp}\,dy\right)^{1/p},
\end{aligned}$$

provided $s + m - (n/p) > 0$. If $p = 1$ the same estimate holds provided $s + m - n \ge 0$ without using Hölder's inequality.

Integrating the pth power of the above estimate over H (with volume element dx'), we obtain

$$\begin{aligned}
\|\chi_r\omega_m * |v|\|_{p,H}^p &= \int_H \left|\chi_r\omega_m * |v|(x)\right|^p dx' \\
&\le Kr^{(s+m)p-n}\int_H dx'\int_{B_r(x)} |v(y)|^p|x-y|^{-sp}\,dy \\
&\le Kr^{(s+m)p-n}\,r^{k-sp}\,\|v\|_{p,\mathbb{R}^n}^p = Kr^{mp-(n-k)}\,\|v\|_{p,\mathbb{R}^n}^p,
\end{aligned}$$

provided $k > sp$.

Since $n - mp < k$ there exists s satisfying $(n/p) - m < s < k/p$, so both estimates above are valid and (11) holds. ∎

4.18 LEMMA Let $p > 1, mp < n, n - mp < k \le n$, and $p^* = kp/(n - mp)$. There exists a constant K such that for every k-dimensional plane H in $\mathbb{R}^n$ and every $v \in L^p(\mathbb{R}^n)$, we have $\omega_m * |v| \in L^{p^*}(H)$ and

$$\|\chi_1 * |v|\|_{p^*,H} \le \|\chi_1\omega_m * |v|\|_{p^*,H} \le \|\omega_m * |v|\|_{p^*,H} \le K\,\|v\|_{p,\mathbb{R}^n}. \tag{12}$$

Proof. Only the final inequality of (12) requires proof. Since $mp < n$, for each $x \in \mathbb{R}^n$ Hölder's inequality gives

$$\int_{\mathbb{R}^n - B_r(x)} |v(y)||x-y|^{m-n}\,dy \le \|v\|_{p,\mathbb{R}^n} \left(\int_{\mathbb{R}^n - B_r(x)} |x-y|^{(m-n)p'}\,dy \right)^{1/p'}$$
$$= K_1 \|v\|_{p,\mathbb{R}^n} \left(\int_r^\infty t^{(m-n)p'+n-1}\,dt \right)^{1/p'}$$
$$= K_1 r^{m-(n/p)} \|v\|_{p,\mathbb{R}^n}.$$

If $t > 0$, choose r so that $K_1 r^{m-(n/p)} \|v\|_{p,\mathbb{R}^n} = t/2$. If

$$\omega_m * |v|(x) = \int_{\mathbb{R}^n} |v(y)||x-y|^{m-n}\,dy > t,$$

then

$$\chi_r \omega_m * |v|(x) = \int_{B_r(x)} |v(y)||x-y|^{m-n}\,dy > t/2.$$

Thus

$$\mu_k\big(\{x \in H : \omega_m * |v|(x) > t\}\big) \le \mu_k\big(\{x \in H : \chi_r \omega_m * |v|(x) > t/2\}\big)$$
$$\le \left(\frac{2}{t}\right)^p \|\chi_r \omega_m * |v|\|_{p,H}^p$$
$$\le \left(\frac{r^{(n/p)-m}}{K_1 \|v\|_{p,\mathbb{R}^n}} \right)^p K r^{mp-n+k} \|v\|_{p,\mathbb{R}^n}^p = K_2 r^k$$

by inequality (11). But $r^k = \big(2K_1 \|v\|_{p,\mathbb{R}^n}/t\big)^{p^*}$, so

$$\mu_k\big(\{x \in H : \omega_m * |v|(x) > t\}\big) \le K_2 \left(\frac{2K_1}{t} \|v\|_{p,\mathbb{R}^n} \right)^{p^*}.$$

Thus the mapping $I : v \mapsto (\omega_m * |v|)\big|_H$ is of weak type (p, p^*).

For fixed m, n, k, the values of p satisfying the conditions of this lemma constitute an open interval, so there exist p_1 and p_2 in that interval, and a number θ satisfying $0 < \theta < 1$ such that

$$\frac{1}{p} = \frac{1-\theta}{p_1} + \frac{\theta}{p_2},$$

and

$$\frac{1}{p^*} = \frac{n/k}{p} - \frac{m}{k} = \frac{1-\theta}{p_1^*} + \frac{\theta}{p_2^*}.$$

Since $p^* > p$, the Marcinkiewicz interpolation theorem 2.58 assures us that I is bounded from $L^p(\mathbb{R}^n)$ into $L^{p^*}(H)$, that is, (12) holds. ∎

4.19 (Proof of Part I, Case C of Theorem 4.12 for $p > 1$) We have $mp < n$, $n - mp < k \le n$, and $p \le q \le p* = kp/(n - mp)$. Let $u \in C^\infty(\Omega)$ and extend u and all its derivatives to be zero on $\mathbb{R}^n \sim \Omega$. Taking $r = \rho$ in Lemma 4.15 and replacing $C_{x,r}$ with the larger ball $B_1(x)$, we obtain

$$|u(x)| \le K\left(\sum_{|\alpha|\le m-1} \chi_1 * |D^\alpha u|(x) + \sum_{|\alpha|=m} \chi_1\omega_m * |D^\alpha u|(x)\right). \tag{13}$$

If $1/q = \theta/p + (1-\theta)/p^*$ where $0 \le \theta \le 1$, then by the interpolation inequality of Theorem 2.11 and Lemmas 4.17 and 4.18

$$\begin{aligned}
\|u\|_{q,\Omega_k} &\le \|u\|_{p,H}^{\theta}\, \|u\|_{p^*,H}^{1-\theta} \\
&\le K\left(\sum_{|\alpha|\le m} \|D^\alpha u\|_{p,\mathbb{R}^n}\right)^{\theta} \left(\sum_{|\alpha|\le m} \|D^\alpha u\|_{p,\mathbb{R}^n}\right)^{1-\theta} \\
&\le K\,\|u\|_{m,p,\Omega}
\end{aligned}$$

as required. ∎

4.20 (Proof of Part I, Case B of Theorem 4.12 for $p > 1$) We have $mp = n$, $1 \le k \le n$, and $p \le q < \infty$. We can select numbers p_1, p_2, and θ such that $1 < p_1 < p < p_2$, $n - mp_1 < k$, $0 < \theta < 1$, and

$$\frac{1}{p} = \frac{\theta}{p_1} + \frac{1-\theta}{p_2}, \qquad \frac{1}{q} = \frac{\theta}{p_1}.$$

As in the above proof of Case C for $p > 1$, the maps $v \mapsto (\chi_1 * |v|)\big|_H$ and $v \mapsto (\chi_1\omega_m * |v|)\big|_H$ are bounded from $L^{p_1}(\mathbb{R}^n)$ into $L^{p_1}(\mathbb{R}^k)$ and so are of weak type (p_1, p_1). As in the proof of Case A, these same maps are bounded from $L^{p_2}(\mathbb{R}^n)$ into $L^\infty(\mathbb{R}^k)$ and so are of weak type (p_2, ∞). By the Marcinkiewicz theorem again, they are bounded from $L^p(\mathbb{R}^n)$ into $L^q(\mathbb{R}^k)$ and

$$\|\chi_1 * |v|\|_{q,H} \le \|\chi_1\omega_m * |v|\|_{q,H} \le K\,\|v\|_{p,\mathbb{R}^n}$$

and the desired result follows by applying these estimates to the various terms of (13). ∎

Imbeddings by Averaging

4.21 We still need to prove the imbeddings for Cases B and C with $p = 1$. We first prove that $W^{1,1}(\Omega) \to L^{n/(n-1)}(\Omega)$ and deduce from this and the imbeddings already established for $p > 1$ that all but one of the remaining imbeddings in Cases B and C are valid. The remaining imbedding is the special case of C where $k = n - m$, $p = 1$, $p* = 1$ which will require a special proof.

We first show that any domain satisfying the cone condition is the union of finitely many subdomains each of which is a union of parallel translates of a fixed parallelepiped. Then we establish a special case of a combinatorial lemma estimating a function in terms of averages in coordinate directions. Both of these results are due to Gagliardo [Ga1] and constitute the foundation on which rests his proof of all of Cases B and C of Part I.

4.22 LEMMA (Decomposition of Ω) Let $\Omega \subset \mathbb{R}^n$ satisfy the cone condition. Then there exists a finite collection $\{\Omega_1, \dots, \Omega_N\}$ of open subsets of Ω such that $\Omega = \bigcup_{j=1}^{N} \Omega_j$, and such that to each Ω_j there corresponds a subset $A_j \subset \Omega_j$ and an open parallelepiped P_j with one vertex at 0 such that $\Omega_j = \bigcup_{x \in A_j}(x + P_j)$. If Ω is bounded and $\rho > 0$ is given, we can accomplish the above decomposition using sets A_j each satisfying $\operatorname{diam}(A_j) < \rho$.

Finally, if Ω is bounded and $\rho > 0$ is sufficiently small, then each Ω_j will satisfy the strong local Lipschitz condition.

Proof. Let C be the finite cone with vertex at 0 such that any $x \in \Omega$ is the vertex of a finite cone $C_x \subset \Omega$ congruent to C. We can select a finite number of finite cones $C_1, \dots, C_N$ each having vertex at 0 (and each having the same height as C but smaller aperture angle than C) such that any finite cone congruent to C and having vertex at 0 must contain one of the cones C_j. For each j, let P_j be an open parallelepiped with one vertex at the origin, contained in C_j, and having positive volume. Then for each $x \in \Omega$ there exists j, $1 \le j \le N$, such that

$$x + P_j \subset x + C_j \subset C_x \subset \Omega.$$

Since Ω is open and $\overline{x + P_j}$ is compact, $y + P_j \subset \Omega$ for any y sufficiently close to x. Hence every $x \in \Omega$ belongs to $y + P_j$ for some j and some $y \in \Omega$. Let $A_j = \{y \in \overline{\Omega} : y + P_j \subset \Omega\}$ and let $\Omega_j = \bigcup_{y \in A_j}(y + P_j)$. Then $\Omega = \bigcup_{j=1}^{N} \Omega_j$.

Now suppose Ω is bounded and $\rho > 0$ is given. If $\operatorname{diam}(A_j) \ge \rho$ we can decompose A_j into a finite union of sets A_{ji} each with diameter less than ρ and define the corresponding parallelepiped $P_{ji} = P_j$. We then rename the totality of such sets A_{ji} as a single finite family, which we again call $\{A_j\}$ and define Ω_j as above.

Figure 2 attempts to illustrate these notions for the domain in $\mathbb{R}^2$ considered in

Example 3.20:

$$\begin{aligned}\Omega &= \{(x, y) \in \mathbb{R}^2 : 0 < |x| < 1, 0 < y < 1\},\\ C &= \{(x, y) \in \mathbb{R}^2 : x > 0, y > 0, x^2 + y^2 < 1/4\},\\ \rho &< 0.98.\end{aligned}$$

Finally, we show that if ρ is sufficiently small, then Ω_j satisfies the strong local Lipschitz condition. For simplicity of notation, let $G = \bigcup_{x \in A}(x + P)$, where $\operatorname{diam}(A) < \rho$ and P is a fixed parallelepiped. We show that G satisfies the strong local Lipschitz condition if ρ is suitably small. For each vertex v_j of P let $Q_j = \{y = v_j + \lambda(x - v_j) : x \in P, \lambda > 0\}$ be the infinite pyramid with vertex v_j generated by P. Then $P = \bigcap Q_j$, the intersection being taken over all 2^n vertices of P. Let $G_j = \bigcup_{x \in A}(x + Q_j)$. Let δ be the distance from the centre of P to the boundary of P and let B be an arbitrary ball of radius $\sigma = \delta/2$. For any fixed $x \in G$, B cannot intersect opposite faces of $x + P$ so we may pick a vertex v_j of P with the property that $x + v_j$ is common to all faces of $x + P$ that intersect B, if any such faces exist. Then $B \cap (x + P) = B \cap (x + Q_j)$. Now let $x, y \in A$ and suppose B could intersect relatively opposite faces of $x + P$ and $y + P$, that is, there exist points a and b on opposite faces of P such that $x + a \in B$ and $y + b \in B$. Then

$$\begin{aligned}\rho \ge \operatorname{dist}(x, y) &= \operatorname{dist}(x + b, y + b)\\ &\ge \operatorname{dist}(x + b, x + a) - \operatorname{dist}(x + a, y + b)\\ &\ge 2\delta - 2\sigma = \delta.\end{aligned}$$

It follows that if $\rho < \delta$, then B cannot intersect relatively opposite faces of $x + P$ and $y + P$ for any $x, y \in A$. Thus $B \cap (x + P) = B \cap (x + Q_j)$ for some fixed j independent of $x \in A$, whence $B \cap G = B \cap G_j$.

Choose coordinates $\xi = (\xi', \xi_n) = (\xi_1, \ldots, \xi_{n-1}, \xi_n)$ in B so that the ξ_n-axis lies in the direction of the vector from the centre of P to the vertex v_j. Then $B \cap (x + Q_j)$ is specified in B by an inequality of the form $\xi_n < f_x(\xi')$ where f_x satisfies a Lipschitz condition with constant independent of x. Thus $B \cap G_j$, and hence $B \cap G$, is specified by $\xi_n < f(\xi')$, where $f(\xi') = \sup_{x \in A} f_x(\xi')$ is itself a Lipschitz continuous function. Since this can be done for a neighbourhood B of any point on the boundary of G, it follows that G satisfies the strong local Lipschitz condition. ∎

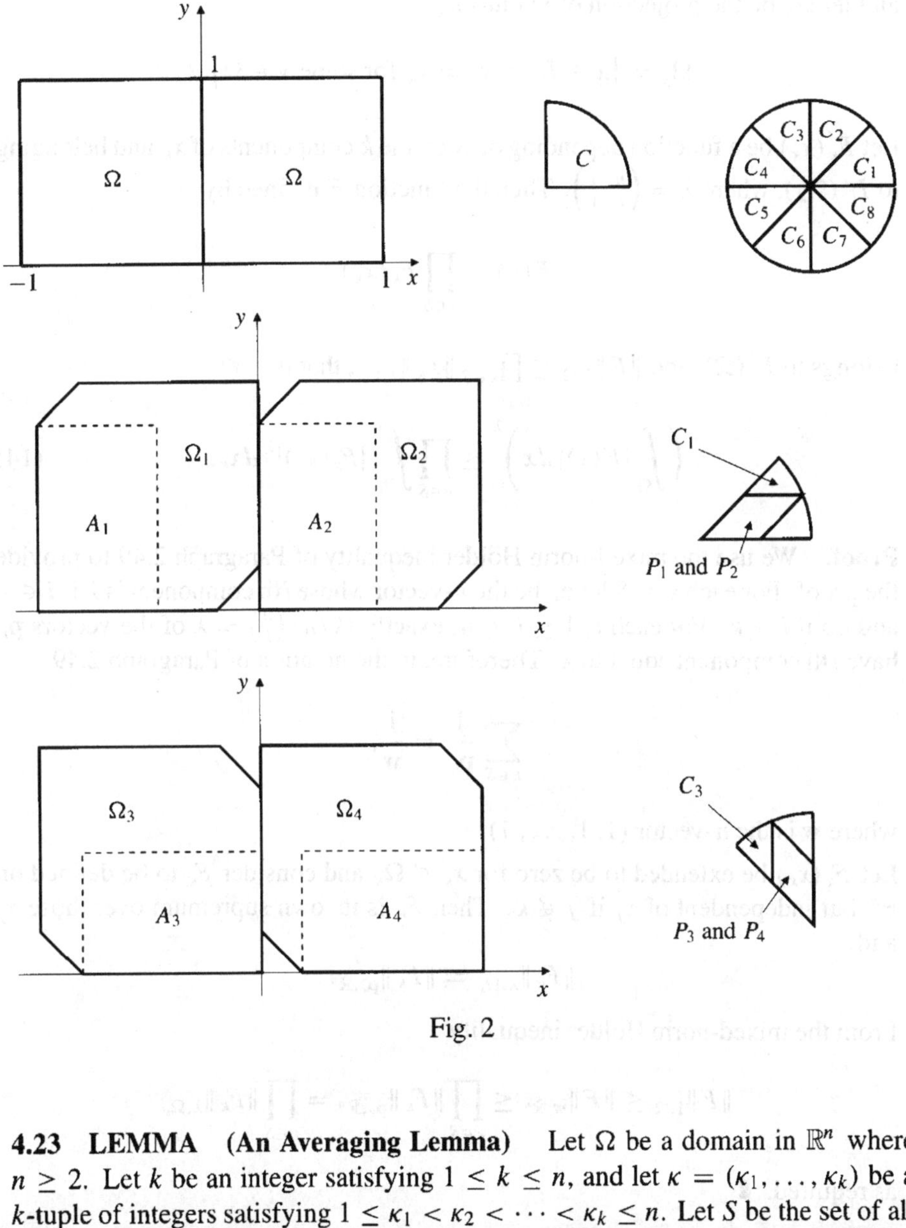

Fig. 2

4.23 LEMMA (An Averaging Lemma) Let Ω be a domain in $\mathbb{R}^n$ where $n \geq 2$. Let k be an integer satisfying $1 \leq k \leq n$, and let $\kappa = (\kappa_1, \ldots, \kappa_k)$ be a k-tuple of integers satisfying $1 \leq \kappa_1 < \kappa_2 < \cdots < \kappa_k \leq n$. Let S be the set of all $\binom{n}{k}$ such k-tuples. Given $x \in R^n$, let x_κ denote the point $(x_{\kappa_1}, \ldots, x_{\kappa_k})$ in $\mathbb{R}^k$ and let $dx_\kappa = dx_{\kappa_1} \cdots dx_{\kappa_k}$.

For $\kappa \in S$ let E_κ be the k-dimensional plane in $\mathbb{R}^n$ spanned by the coordinate axes corresponding to the components of x_κ:

$$E_\kappa = \left\{ x \in \mathbb{R}^n : x_i = 0 \text{ if } i \notin \kappa \right\},$$

and let Ω_κ be the projection of Ω onto E_κ:

$$\Omega_\kappa = \{x \in E_\kappa : x_\kappa = y_\kappa \text{ for some } y \in \Omega\}.$$

Let $F_\kappa(x_\kappa)$ be a function depending only on the k components of x_κ and belonging to $L^\lambda(\Omega_\kappa)$, where $\lambda = \binom{n-1}{k-1}$. Then the function F defined by

$$F(x) = \prod_{\kappa \in S} F_\kappa(x_\kappa)$$

belongs to $L^1(\Omega)$, and $\|F\|_{1,\Omega} \le \prod_{\kappa \in S} \|F_\kappa\|_{\lambda,\Omega_\kappa}$, that is,

$$\left(\int_\Omega |F(x)|\,dx\right)^\lambda \le \prod_{\kappa \in S} \int_{\Omega_\kappa} |F_\kappa(x_\kappa)|^\lambda\,dx_\kappa. \tag{14}$$

Proof. We use the mixed-norm Hölder inequality of Paragraph 2.49 to provide the proof. For each $\kappa \in S$ let $\mathbf{p}_\kappa$ be the n-vector whose ith component is λ if $i \in \kappa$ and ∞ if $i \notin \kappa$. For each i, $1 \le i \le n$, exactly $(k/n)\binom{n}{k} = \lambda$ of the vectors $\mathbf{p}_\kappa$ have ith component equal to λ. Therefore, in the notation of Paragraph 2.49

$$\sum_{\kappa \in S} \frac{1}{\mathbf{p}_\kappa} = \frac{1}{\mathbf{w}},$$

where $\mathbf{w}$ is the n-vector $(1, 1, \ldots, 1)$.

Let $F_\kappa(x_\kappa)$ be extended to be zero for $x_\kappa \notin \Omega_\kappa$ and consider F_κ to be defined on $\mathbb{R}^n$ but independent of x_j if $j \notin \kappa$. Then F_κ is its own supremum over those x_j and

$$\|F_\kappa\|_{\lambda,\Omega_\kappa} = \|F_\kappa\|_{\mathbf{p}_\kappa,\mathbb{R}^n}.$$

From the mixed-norm Hölder inequality

$$\|F\|_{1,\Omega} \le \|F\|_{\mathbf{w},\mathbb{R}^n} \le \prod_{\kappa \in S} \|F_\kappa\|_{\mathbf{p}_\kappa,\mathbb{R}^n} = \prod_{\kappa \in S} \|F_\kappa\|_{\lambda,\Omega_\kappa}$$

as required. ∎

4.24 LEMMA If Ω satisfies the cone condition, then $W^{1,1}(\Omega) \to L^p(\Omega)$ for $1 \le p \le n/(n-1)$.

Proof. By Lemma 4.22, Ω is a finite union of subdomains each of which is a union of parallel translates of a fixed parallelepiped. It is therefore sufficient to prove the imbedding for one such subdomain. Thus we assume $\Omega = \bigcup_{x \in A}(x + P)$ where P is a parallelepiped. There is a linear transformation of $\mathbb{R}^n$ onto itself that maps P onto a cube Q of unit edge with edges parallel to the coordinate axes. By Theorem 3.41 it is therefore sufficient to prove the lemma for $\Omega = \bigcup_{x \in A}(x + Q)$. For $x \in \Omega$ let ℓ be the intersection of Ω with the line through x parallel to the x_1-axis. Evidently ℓ contains a closed interval of length 1 containing x_1, say the interval $[\xi_1, \xi_2]$. If $f \in C^1([0, 1])$, then $|f(t_0)| \le |f(t)| + \left|\int_{t_0}^{t} f'(\tau)\, d\tau\right|$, and integrating over t over $[0, 1]$ yields

$$|f(t_0)| \le \int_0^1 (|f(t)| + |f'(t)|)\, dt.$$

For $u \in C^\infty(\Omega)$ we apply this inequality to $u(t, \hat{x}_1)$ (where $\hat{x}_1 = (x_2, \dots, x_n)$) to obtain

$$\begin{aligned} |u(x)| &\le \int_{\xi_1}^{\xi_2} \big(|u(t, \hat{x}_1)| + |D_1 u(t, \hat{x}_1)|\big)\, dt \\ &\le \int_{\ell} \big(|u(t, \hat{x}_1)| + |D_1 u(t, \hat{x}_1)|\big)\, dt. \end{aligned}$$

Let Ω_1 be the orthogonal projection of Ω onto the hyperplane of coordinates $\hat{x}_1$, and let

$$u_1(\hat{x}_1) = \left(\int_{\ell} \big(|u(t, \hat{x}_1)| + |D_1 u(t, \hat{x}_1)|\big)\, dt\right)^{1/(n-1)}.$$

(Evidently $u_1(\hat{x}_1)$ is independent of x_1) We have

$$\|u_1\|_{1/(n-1),\Omega_1} = \int_{\Omega_1} |u_1(x)|^{n-1}\, d\hat{x}_1 \le \|u\|_{1,1,\Omega}.$$

Similarly, for $2 \le j \le n$ we can define u_j to be independent of x_j and to satisfy $|u(x)| \le \big(u_j(x)\big)^{1/(n-1)}$ and

$$\|u_j\|_{1/(n-1),\Omega_j} \le \|u\|_{1,1,\Omega}.$$

Since $|u(x)|^{n/(n-1)} \le \prod_{j=1}^{n} u_j(x)$, applying inequality (14) with $k = n - 1 = \lambda$ now gives

$$\int_{\Omega} |u(x)|^{n/(n-1)}\, dx \le \prod_{j=1}^{n} \int_{\Omega_j} |u_j|(\hat{x}_j))|^{n-1}\, d\hat{x}_j \le \|u\|_{1,1,\Omega}^{n/(n-1)}.$$

For the original domain Ω, this will imply that

$$\|u\|_{n/(n-1),\Omega} \le K\, \|u\|_{1,1,\Omega}$$

where the constant K depends on n and the cone C of the cone condition. These determine the number N of subdomains needed, and the size of the determinant of the linear transformation needed to transform the parallelepipeds for each subdomain into Q. The imbedding $W^{1,1}(\Omega) \to L^p(\Omega)$ for $1 \le p \le n/(n-1)$ now follows by L^p interpolation (Theorem 2.11) ∎

4.25 (Proof of Part I, Cases B and C of Theorem 4.12 for $p = 1$, $k > n - m$) Let $m \le n$. By the above lemma and previously proved parts of Cases B and C for $p > 1$, we have

$$W^{m,1}(\Omega) \to W^{m-1,p}(\Omega) \qquad \text{for } 1 \le p \le n/(n-1).$$

Since $k > n - m$, therefore $k \ge n - m + 1 > n - (m-1)p$ for any $p > 1$. Therefore $W^{m-1,p}(\Omega) \to L^q(\Omega_k)$ holds for

$$1 \le q \le p^* = \frac{kp}{n-(m-1)p} = \frac{kn/(n-1)}{n-(m-1)n/(n-1)} = \frac{k}{n-m}.$$

Combining these imbeddings we get $W^{m,1}(\Omega) \to L^p(\Omega)$, $1 \le q \le k/(n-m)$. For $p = 1$, $m = n$ the imbedding $W^{n,1}(\Omega) \to L^q(\Omega_k)$, $1 \le q \le \infty$, $1 \le k \le n$ was already proved under Case A. ∎

4.26 (Proof of Part I, Case C of Theorem 4.12 for $p = 1$, $k = n - m$) In this case we want to show $W^{m,1}(\Omega) \to L^1(\Omega_k)$. As in the proof in Paragraph 4.24 it is sufficient to establish the imbedding for a domain Ω that is a union of parallel translates of a unit cube with edges parallel to the coordinate axes. We can also assume that $0 \in \Omega$ and that

$$\Omega_k = \{x = (x', x'') \in \Omega : x' = 0\},$$

where $x' = (x_1, \dots x_m)$ and $x'' = (x_{m+1}, \dots, x_n)$. For $x \in \Omega$ let Ω_x be the intersection of Ω with the m-plane of variables x' passing through x. Ω_x contains an m-cube Q_x of edge 1 containing x, and so by Case A of Theorem 4.12 applied to this cube, we have for $u \in C^\infty(\Omega)$,

$$|u(x)| \le K \sum_{|\alpha| \le m} \int_{\Omega_x} |D^\alpha u(x', x'')| \, dx'.$$

Integrating x'' over Ω_k then gives

$$\int_{\Omega_k} |u(x)| \, dx'' \le K \sum_{|\alpha| \le m} \int_\Omega |D^\alpha u(x)| \, dx.$$

The proof of Part I of Theorem 4.12 is now complete. ∎

Imbeddings into Lipschitz Spaces

4.27 To prove Part II of Theorem 4.12, we now assume that the domain $\Omega \subset \mathbb{R}^n$ satisfies the strong local Lipschitz condition defined in Paragraph 4.9, and that $mp > n \geq (m-1)p$. We shall show that $W^{m,p}(\Omega) \to C^{0,\lambda}(\overline{\Omega})$ where:

(i) $0 < \lambda \leq m - (n/p)$ if $n > (m-1)p$, or

(ii) $0 < \lambda < 1$ if $n = (m-1)p$ and $p > 1$, or

(iii) $0 < \lambda \leq 1$ if $n = m - 1$ and $p = 1$.

In particular, therefore, $W^{m,p}(\Omega) \to C^0(\overline{\Omega})$. The imbedding constants may depend on m, p, n, and the parameters δ and M specified in the definition of the strong local Lipschitz condition. Since that condition implies the cone condition, we already know that $W^{m,p}(\Omega) \to C_B^0(\Omega)$, so if $u \in W^{m,p}(\Omega)$, then

$$\sup_{x\in\Omega} |u(x)| \leq K_1 \|u\|_{m,p,\Omega}.$$

It is therefore sufficient to establish further that for the appropriate λ,

$$\sup_{\substack{x,y\in\Omega \\ x\neq y}} \frac{|u(x)-u(y)|}{|x-y|^\lambda} \leq K_2 \|u\|_{m,p,\Omega}.$$

Since $mp > n \geq (m-1)p$, Cases B and C of Part I of Theorem 4.12 yields the imbedding $W^{m,p}(\Omega) \to W^{1,r}(\Omega)$ where:

(i) $r = np/(n-m+1)p$ and so $1 - (n/r) = m - (n/p)$ if $n > (m-1)p$, or

(ii) $p < r < \infty$ and so $0 < 1 - (n/r) < 1$ if $n > (m-1)p$, or

(iii) $r = \infty$ and so $1 - (n/r) = 1$ if $n = m - 1$ and $p = 1$.

It is therefore sufficient to establish the special case $m = 1$.

4.28 LEMMA Let Ω satisfy the strong local Lipschitz condition. If u belongs to $W^{1,p}(\Omega)$ where $n < p \leq \infty$, and if $0 \leq \lambda \leq 1 - (n/p)$, then

$$\sup_{\substack{x,y\in\Omega \\ x\neq y}} \frac{|u(x)-u(y)|}{|x-y|^\lambda} \leq K \|u\|_{1,p,\Omega}. \tag{15}$$

Proof. Suppose, for the moment, that Ω is a cube having unit edge length. For $0 < t < 1$ let Q_t denote a subset of Ω that is a closed cube having edge length t and faces parallel to those of Ω. If $x, y \in \Omega$ and $|x-y| = \sigma < 1$, then there is a fixed such cube Q_σ such that $x, y \in Q_\sigma$.

Let $u \in C^\infty(\Omega)$ If $z \in Q_\sigma$, then

$$u(x) - u(z) = \int_0^1 \frac{d}{dt} u\big((x + t(z-x)\big)\, dt,$$

so that

$$|u(x) - u(z)| \le \sigma\sqrt{n} \int_0^1 |\operatorname{grad} u((x + t(z - x))|\, dt.$$

It follows that

$$\begin{aligned}
\left| u(x) - \frac{1}{\sigma^n} \int_{Q_\sigma} u(z)\, dz \right| &= \left| \frac{1}{\sigma^n} \int_{Q_\sigma} \big(u(x) - u(z)\big)\, dz \right| \\
&\le \frac{\sqrt{n}}{\sigma^{n-1}} \int_{Q_\sigma} dz \int_0^1 |\operatorname{grad} u\big((x + t(z - x)\big)|\, dt \\
&= \frac{\sqrt{n}}{\sigma^{n-1}} \int_0^1 t^{-n}\, dt \int_{Q_{t\sigma}} |\operatorname{grad} u(\zeta)|\, d\zeta \\
&\le \frac{\sqrt{n}}{\sigma^{n-1}} \|\operatorname{grad} u\|_{0,p,\Omega} \int_0^1 \big(\operatorname{vol}(Q)_{t\sigma}\big)^{1/p'} t^{-n}\, dt \quad (16) \\
&\le K\sigma^{1-(n/p)} \|\operatorname{grad} u\|_{0,p,\Omega}\,,
\end{aligned}$$

where $K = K(n, p) = \sqrt{n} \int_0^1 t^{-n/p}\, dt < \infty$. A similar inequality holds with y in place of x and so

$$|u(x) - u(y)| \le 2K|x - y|^{1-(n/p)} \|\operatorname{grad} u\|_{0,p,\Omega}\,.$$

It follows that (15) holds for $0 < \lambda \le 1 - (n/p)$ for Ω a cube, and therefore via a nonsingular linear transformation, for Ω a parallelepiped.

Now suppose that Ω is an arbitrary domain satisfying the strong local Lipschitz condition. Let δ, M, Ω_δ, U_j and V_j be as specified in the definition of that condition in Paragraph 4.9. There exists a parallelepiped P of diameter δ whose dimensions depend only on δ and M such that to each j there corresponds a parallelepiped P_j congruent to P and having one vertex at the origin, such that for every $x \in V_j \cap \Omega$ we have $x + P_j \subset \Omega$. Furthermore, there exist constants δ_0 and δ_1 depending only on δ and P, with $\delta_0 \le \delta$, such that if $x, y \in V_j \cap \Omega$ and $|x - y| < \delta_0$, then there exists $z \in (x + P_j) \cap (y + P_j)$ with $|x - z| + |y - z| \le \delta_1 |x - y|$. If follows from applications of (15) to $x + P_j$ and $y + P_j$ that if $u \in C^\infty(\Omega)$, then

$$\begin{aligned}
|u(x) - u(y)| &\le |u(x) - u(z)| + |u(y) - u(z)| \\
&\le K|x - z|^\lambda \|u\|_{1,p,\Omega} + K|y - z|^\lambda \|u\|_{1,p,\Omega} \\
&\le K_1|x - y|^\lambda \|u\|_{1,p,\Omega}\,. \qquad (17)
\end{aligned}$$

Now let x, y be arbitrary points in Ω. If $|x - y| < \delta_0 \le \delta$ and $x, y \in \Omega_\delta$, then $x, y \in V_j$ for some j and (17) holds. If $|x - y| < \delta_0$, $x \in \Omega_\delta$, $y \in \Omega - \Omega_\delta$, then $x \in V_j$ for some j and (17) still follows by an applications of (15) to $x + P_j$ and $y + P_j$. If $|x - y| < \delta_0$, $x, y \in \Omega - \Omega_\delta$, then (17) follows from applications of (15) to $x + P'$ and $y + P'$ where P' is any parallelepiped congruent to p and having one vertex at the origin. Finally, if $|x - y| \geq \delta_0$, then

$$|u(x) - u(y)| \leq |u(x)| + |u(y)| \leq K_1 \|u\|_{1,p,\Omega} \leq K\delta_0^{-\lambda}|x - y|^\lambda \|u\|_{1,p,\Omega}.$$

Thus (15) holds for all $u \in C^\infty(\Omega)$ and, by Theorem 3.17, for all $u \in C_B^0(\Omega)$. ∎
This completes the proof of Part II of Theorem 4.12 and therefore of the whole theorem since, as remarked earlier, Part III follows from the fact that Parts I and II hold for $\Omega = \mathbb{R}^n$.

Sobolev's Inequality

4.29 (Seminorms) For $1 \leq p < \infty$ and for integers j, $0 \leq j \leq m$, we introduce functionals $|\cdot|_{j,p}$ on $W^{m,p}(\Omega)$ as follows:

$$|u|_{j,p} = |u|_{j,p,\Omega} = \left(\sum_{|\alpha|=j} |D^\alpha u(x)|^p \, dx \right)^{1/p}.$$

Clearly $|u|_{0,p} = \|u\|_{0,p} = \|u\|_p$ is the norm on $L^p(\Omega)$ and

$$\|u\|_{m,p} = \left(\sum_{j=0}^{m} |u|_{j,p}^p \right)^{1/p}.$$

If $j \geq 1$, we call $|\cdot|_{j,p}$ a *seminorm*. It has all the properties of a norm except that $|u|_{j,p} = 0$ need not imply $u = 0$ in $W^{m,p}(\Omega)$. For example, u may be a nonzero constant function if Ω has finite volume. Under certain circumstances which we begin to investigate in Paragraph 6.29, $|\cdot|_{m,p}$ is a norm on $W_0^{m,p}(\Omega)$ equivalent to the usual norm $\|\cdot\|_{m,p}$. In particular, this is so if Ω is bounded.

For now we will confine our attention to these seminorms as they apply to functions in $C_0^\infty(\mathbb{R}^n)$.

4.30 The Sobolev imbedding theorem tells us that $W_0^{m,p}(\mathbb{R}^n) \to L^q(\mathbb{R}^n)$ for certain finite values of q depending on m, p, and n; for such q there is a finite constant K such that for all $\phi \in C_0^\infty(\mathbb{R}^n)$ we have

$$\|\phi\|_q \leq K \|\phi\|_{m,p}.$$

We now ask whether such an inequality can hold with $|\cdot|_{m,p}$ in place of $\|\cdot\|_{m,p}$. That is, do there exist constants $K < \infty$ and $q \geq 1$ such that for all $\phi \in C_0^\infty(\mathbb{R}^n)$

$$\int_{\mathbb{R}^n} |\phi(x)|^q \, dx \leq K^q \left(\sum_{|\alpha|=m} \int_{\mathbb{R}^n} |D^\alpha \phi(x)|^p \, dx \right)^{q/p} ? \qquad (18)$$

If so, for any given $\phi \in C_0^\infty(\mathbb{R}^n)$, the inequality must also hold for all dilates $\phi_t(x) = \phi(tx)$, $0 < t < \infty$, as these functions also belong to $C_0^\infty(\mathbb{R}^n)$. Since $\|\phi_t\|_q = t^{-n/q}\|\phi\|_q$ and $\|D^\alpha\phi_t\|_p = t^{m-(n/p)}\|D^\alpha\phi\|_p$ if $|\alpha| = m$, we must have

$$\int_{\mathbb{R}^n} |\phi(x)|^q\,dx \le K^q t^{n+mq-(nq/p)} \left(\sum_{|\alpha|=m} \int_{\mathbb{R}^n} |D^\alpha\phi(x)|^p\,dx\right)^{q/p}$$

This is clearly not possible for all $t > 0$ unless the exponent of t on the right side is zero, that is, unless $q = p^* = np/(n-mp)$. Thus no inequality of the form (18) is possible unless $mp < n$ and $q = p^* = np/(n-mp)$. We now show that (18) does hold if these conditions are satisfied.

4.31 THEOREM (Sobolev's Inequality) When $mp < n$, there exists a finite constant K such that (18) holds for every $\phi \in C_0^\infty(\mathbb{R}^n)$:

$$\|\phi\|_{q,\mathbb{R}^n} \le K\,|\phi|_{m,p,\mathbb{R}^n} \tag{19}$$

if and only if $q = p^* = np/(n-mp)$. This is known as *Sobolev's inequality.*

Proof. The "only if" part was demonstrated above. For the "if" part note first that it is sufficient to establish the inequality for $m = 1$ as its validity for higher m (with $mp < n$) can be confirmed by induction on m. We leave the details to the reader.

Next, it suffices to prove the case $m = 1$, $p = 1$, that is

$$\int_{\mathbb{R}^n} |\phi(x)|^{n/(n-1)}\,dx \le K\left(\sum_{j=1}^n \int_{\mathbb{R}^n} |D_j\phi(x)|\,dx\right)^{n/(n-1)}, \tag{20}$$

for if $1 < p < n$ and $p^* = np/(n-p)$ we can apply (20) to $|\phi(x)|^s$ where $s = (n-1)p^*/n$ and obtain, using Hölder's inequality,

$$\begin{aligned}\int_{\mathbb{R}^n} |\phi(x)|^{p^*}\,dx &\le K\left(\sum_{j=1}^n s|\phi(x)|^{s-1}|D_j\phi(x)|\,dx\right)^{n/(n-1)}\\ &\le K_1\left(\sum_{j=1}^n \|\phi\|_{(s-1)p'}^{s-1}\left\|D_j\phi\right\|_p\right)^{n/(n-1)}.\end{aligned}$$

Since $(s-1)p' = p^*$ and $p^* - (s-1)n/(n-1) = n/(n-1)$, it follows by cancellation that

$$\|\phi\|_{p^*} \le K_2\,|\phi|_{1,p}.$$

It remains, therefore, to prove (20). Let $\phi \in C_0^\infty(\mathbb{R}^n)$ and for $x \in \mathbb{R}^n$ and $1 \le j \le n$ let $\hat{x}_j = (x_1, \dots, x_{j-1}, x_{j+1}, \dots, x_n)$. Let

$$u_j(\hat{x}_j) = \left(\sum_{i=1}^{n} \int_{-\infty}^{\infty} |D_j\phi(x)|\, dx_j \right)^{1/(n-1)},$$

which is evidently independent of x_j and satisfies

$$\left(\|u_j\|_{n-1,\mathbb{R}^{n-1}} \right)^{n-1} \le |u|_{1,1,\mathbb{R}^n}.$$

Since

$$\phi(x) = \int_{-\infty}^{x_1} D_1\phi(t, \hat{x}_1)\, dt$$

we have

$$|\phi(x)| \le \int_{-\infty}^{\infty} |D_1\phi(t, \hat{x}_1)|\, dt \le \left(u_1(\hat{x}_1)\right)^{n-1}.$$

Similarly, $|\phi(x)| \le \left(u_j(\hat{x}_j)\right)^{n-1}$. Applying the inequality (14) from Lemma 4.23 with $k = n - 1 = \lambda$ we obtain

$$\begin{aligned} \int_{\mathbb{R}^n} |\phi(x)|^{n/(n-1)}\, dx &\le \int_{\mathbb{R}^n} \prod_{j=1}^{n} u_j(\hat{x}_j)\, dx \\ &\le \left(\prod_{j=1}^{n} \int_{\mathbb{R}^{n-1}} |u_j(\hat{x}_j)|^{n-1}\, d\hat{x}_j \right)^{1/(n-1)} \le |u|_{1,1,\mathbb{R}^n}^{n/(n-1)}, \end{aligned}$$

which completes the proof of (20) and hence the theorem. ∎

4.32 (REMARK) For the case $m = 1$, $1 < p < n$, Talenti [T] and Aubin, as exposed in Section 2.6 of [Au], obtained the best constant for the equivalent form of Sobolev's inequality

$$\|\phi\|_{np/(n-p),\mathbb{R}^n} \le K \, \|\mathrm{grad}\, \phi\|_{p,\mathbb{R}^n} \tag{21}$$

by showing that the ratio

$$\frac{\|\phi\|_{np/(n-p)}}{\|\mathrm{grad}\, \phi\|_{1,p}}$$

is maximized if u is a radially symmetric function of the form

$$u(x) = \left(a + b|x|^{p/(p-1)}\right)^{1-(n/p)}$$

which, while not in $C_0^\infty(\mathbb{R}^n)$ is a limit of functions in that space. His method involved first showing that replacing an arbitrary function u vanishing at infinity with a radially symmetric, non-increasing, equimeasurable rearrangement of u decreased $\|\operatorname{grad} u\|_{p,\mathbb{R}^n}$ while, of course, leaving $\|u\|_{np/(n-p),\mathbb{R}^n}$ unchanged. Talenti's best constant for (21) is

$$K = \pi^{-1/2} n^{-1/p} \left(\frac{p-1}{n-p} \right)^{1/p'} \left(\frac{\Gamma(1+n/2)\Gamma(n)}{\Gamma(n/p)\Gamma(1+n-(n/p))} \right)^{1/n}.$$

Variations of Sobolev's Inequality

4.33 Mixed-norm L^p estimates of the type considered in Paragraphs 2.48–2.51 and used in the proof of Gagliardo's averaging lemma 4.23 can contribute to generalizations of Sobolev's inequality. We examine briefly two such generalizations:

(a) **anisotropic Sobolev inequalities**, in which different L^p norms are used for different partial derivatives on the right side of (19), and

(b) **reduced Sobolev inequalities**, in which the seminorm $|\phi|_{m,p,\mathbb{R}^n}$ on the right side of (19) is replaced with a similar seminorm involving only a subset of the partial derivatives of order m of ϕ.

Questions of this sort are discussed in [BIN1] and [BIN2]. We follow the treatment in [A3] and [A4] and most of the details will be omitted here.

4.34 (A First-Order Anisotropic Sobolev Inequality) If $p_j \ge 1$ for each j with $1 \le j \le n$ and $\phi \in C_0^\infty(\mathbb{R}^n)$, then an inequality of the form

$$\|\phi\|_q \le K \sum_{j=1}^{n} \|D_j \phi\|_{p_j} \tag{22}$$

is a (first-order) *anisotropic* Sobolev inequality because different L^p norms are used to estimate the derivatives of ϕ in different coordinate directions. A dilation argument involving $\phi(\lambda_1 x_1, \ldots, \lambda_n x_n)$ for $0 < \lambda_j < \infty$, $1 \le j \le n$ shows that no such anisotropic inequality is possible for finite q unless

$$\sum_{j=1}^{n} \frac{1}{p_j} > 1 \qquad \text{and} \qquad \frac{1}{q} = \frac{1}{n} \sum_{j=1}^{n} \frac{1}{p_j} - \frac{1}{n}.$$

If these conditions are satisfied, then (22) does hold. The proof is a generalization of that of Theorem 4.31 and uses the mixed-norm Hölder and permutation inequalities. (See [A3] for the details.)

4.35 (Higher-Order Anisotropic Sobolev Inequalities) The generalization of (22) to an mth order inequality by induction on m is somewhat more problematic.

The mth order isotropic inequality (19) follows from its special case $m = 1$ by simple induction. We can also obtain

$$\|\phi\|_q \le K \sum_{|\alpha|=m} \|D^\alpha \phi\|_{p_\alpha},$$

where

$$\frac{1}{q} = \frac{1}{n^m} \sum_{|\alpha|=m} \binom{m}{\alpha} \frac{1}{p_\alpha} - \frac{m}{n}, \qquad \binom{m}{\alpha} = \frac{m!}{\alpha_1!\alpha_2!\cdots\alpha_n!}$$

by induction from (22) under suitable restrictions on the exponents p_α, but the restriction

$$\frac{1}{n^m} \sum_{|\alpha|=m} \binom{m}{\alpha} \frac{1}{p_\alpha} > \frac{m}{n}$$

will not suffice in general for the induction even though $\sum_{|\alpha|=m} \binom{m}{\alpha} = n^m$. The conditions $mp_\alpha < n$ for each α with $|\alpha| = m$ will suffice, but are stronger than necessary.

For any multi-index β and $1 \le j \le n$, let

$$\beta[j] = (\beta_1, \dots, \beta_{j-1}, \beta_j + 1, \beta_{j+1}, \dots, \beta_n).$$

Evidently, $|\beta[j]| = |\beta| + 1$ and it can be verified that if the numbers p_α are defined for all α with $|\alpha| = m$, then

$$\sum_{|\beta|=m-1} \binom{m-1}{\beta} \sum_{j=1}^{n} \frac{1}{p_{\beta[j]}} = \sum_{|\alpha|=m} \binom{m}{\alpha} \frac{1}{p_\alpha}.$$

This provides the induction step necessary to verify the following theorem, for which the details can again be found in [A3].

4.36 THEOREM Let $p_\alpha \ge 1$ for all α with $|\alpha| = m$. Suppose that for every β with $|\beta| = m - 1$ we have

$$\sum_{j=1}^{n} \frac{1}{p_{\beta[j]}} > m.$$

Then there exists a constant K such that the inequality

$$\|\phi\|_q \le K \sum_{|\alpha|=m} \|D^\alpha \phi\|_{p_\alpha}$$

holds for all $\phi \in C_0^\infty(\mathbb{R}^n)$, where

$$\frac{1}{q} = \frac{1}{n^m} \sum_{|\alpha|=m} \binom{m}{\alpha} \frac{1}{p_\alpha} - \frac{m}{n}.$$

4.37 (Reduced Sobolev Inequalities) Another variation of Sobolev's inequality addresses the question of whether the number of derivatives estimated in the seminorm on the right side of (19) (or, equivalently, (18)) can be reduced without jeopardizing the validity of the inequality for all $\phi \in C_0^\infty(\mathbb{R}^n)$. If $m \geq 2$, the answer is yes; only those partial derivatives of order m that are "completely mixed" (in the sense that all m differentiations are taken with respect to different variables) need be included in the seminorm. Specifically, if we denote

$$\mathcal{M} = \mathcal{M}(n, m) = \{\alpha : |\alpha| = m, \quad \alpha_j = 0 \text{ or } \alpha_j = 1 \text{ for } 1 \leq j \leq n,$$

then the *reduced* Sobolev inequality

$$\|\phi\|_q \leq K \sum_{\alpha \in \mathcal{M}} \|D^\alpha \phi\|_p$$

holds for all $\phi \in C_0^\infty(\mathbb{R}^n)$, provided $mp < n$ and $q = np/(n - mp)$. Again the proof depends on mixed-norm estimates; it can be found in [A4] where the possibility of further reductions in the number of derivatives estimated on the right side of Sobolev's inequality is also considered. See also Section 13 in [BIN1].

$W^{m,p}(\Omega)$ as a Banach Algebra

4.38 Given u and v in $W^{m,p}(\Omega)$, where Ω is a domain in $\mathbb{R}^n$, one cannot in general expect that their pointwise product uv will belong to $W^{m,p}(\Omega)$. The imbedding theorem, however, shows that this is the case provided $mp > n$ and Ω satisfies the cone condition. (See [Sr] and [Mz2].)

4.39 THEOREM Let Ω be a domain in $\mathbb{R}^n$ satisfying the cone condition. If $mp > n$ or $p = 1$ and $m \geq n$, then there exists a constant K^* depending on m, p, n, and the cone C determining the cone condition for Ω, such that for $u, v \in W^{m,p}(\Omega)$ the product uv, defined pointwise a.e. in Ω, satisfies

$$\|uv\|_{m,p,\Omega} \leq K^* \|u\|_{m,p,\Omega} \|v\|_{m,p,\Omega}. \tag{23}$$

In particular, equipped with the equivalent norm $\|\cdot\|^*_{m,p,\Omega}$ defined by

$$\|u\|^*_{m,p,\Omega} = K^* \|u\|_{m,p,\Omega},$$

$W^{m,p}(\Omega)$ is a commutative Banach algebra with respect to pointwise multiplication in that

$$\|uv\|^*_{m,p,\Omega} \leq \|u\|^*_{m,p,\Omega} \|v\|^*_{m,p,\Omega}.$$

Proof. We assume $mp > n$; the case $p = 1$, $m = n$ is simpler. In order to establish (23) it is sufficient to show that if $|\alpha| \leq m$, then

$$\int_\Omega |D^\alpha[u(x)v(x)]|^p \leq K_\alpha \|u\|_{m,p,\Omega} \|v\|_{m,p,\Omega},$$

where $K_\alpha = K_\alpha(m, p, n, C)$. Let us assume for the moment that $u \in C^\infty(\Omega)$. By the Leibniz rule for distributional derivatives, that is,

$$D^\alpha(uv) = \sum_{\beta \le \alpha} \binom{\alpha}{\beta} D^\beta u D^{\alpha-\beta} v,$$

it is sufficient to show that for any $\beta \le \alpha$, $|\alpha| \le m$, we have

$$\int_\Omega |D^\beta u(x) D^{\alpha-\beta} v(x)|^p \, dx \le K_{\alpha,\beta} \|u\|^p_{m,p,\Omega} \|v\|^p_{m,p,\Omega},$$

where $K_{\alpha,\beta} = K_{\alpha,\beta}(m, p, n, C)$. By the imbedding theorem there exists, for any β with $|\beta| \le m$, a constant $K(\beta) = K(\beta, m, p, n, C)$ such that for any $w \in W^{m,p}(\Omega)$,

$$\int_\Omega |D^\beta w(x)|^r \, dx \le K(\beta) \|w\|^r_{m,p,\Omega}, \tag{24}$$

provided $(m - |\beta|)p \le n$ and $p \le r \le np/(n - [m - |\beta|]p)$ [or $p \le r < \infty$ if $(m - |\beta|)p = n$], or alternatively

$$|D^\beta w(x)| \le K(\beta) \|w\|_{m,p,\Omega} \qquad \text{a.e. in } \Omega$$

provided $(m - |\beta|)p > n$.

Let k be the largest integer such that $(m - k)p > n$. Since $mp > n$ we have $k \ge 0$. If $|\beta| \le k$, then $(m - |\beta|)p > n$, so

$$\begin{aligned}\int_\Omega |D^\beta u(x) D^{\alpha-\beta} v(x)|^p \, dx &\le K(\beta)^p \|u\|^p_{m,p,\Omega} \left\|D^{\alpha-\beta} v\right\|^p_{0,p,\Omega} \\ &\le K(\beta)^p \|u\|^p_{m,p,\Omega} \|v\|^p_{m,p,\Omega}.\end{aligned}$$

Similarly, if $|\alpha - \beta| \le k$, then

$$\int_\Omega |D^\beta u(x) D^{\alpha-\beta} v(x)|^p \, dx \le K(\alpha - \beta)^p \|u\|^p_{m,p,\Omega} \|v\|^p_{m,p,\Omega}.$$

Now if $|\beta| > k$ and $|\alpha - \beta| > k$, then, in fact, $|\beta| \ge k + 1$ and $|\alpha - \beta| \ge k + 1$ so that $n \ge (m - |\beta|)p$ and $n \ge (m - |\alpha - \beta|)p$. Moreover,

$$\frac{n - (m - |\beta|)p}{n} + \frac{n - (m - |\alpha - \beta|)p}{n} = 2 - \frac{(2m - |\alpha|)p}{n} \le 2 - \frac{mp}{n} < 1.$$

Hence there exist positive numbers r and r' with $(1/r) + (1/r') = 1$ such that

$$p \le rp < \frac{np}{n - (m - |\beta|)p}, \qquad p \le r'p < \frac{np}{n - (m - |\alpha - \beta|)p}.$$

Thus by Hölder's inequality and (24) we have

$$\int_\Omega |D^\beta u(x) D^{\alpha-\beta} v(x)|^p \, dx \le \left(\int_\Omega |D^\beta u(x)|^{rp} \, dx\right)^{1/r} \left(\int_\Omega |D^{\alpha-\beta} v(x)|^{r'p} \, dx\right)^{1/r'}$$
$$\le \big(K(\beta)\big)^{1/r} \big(K(\alpha-\beta)\big)^{1/r'} \|u\|^p_{m,p,\Omega} \|v\|^p_{m,p,\Omega} .$$

This completes the proof of (23) for $u \in C^\infty(\Omega)$, $v \in W^{m,p}(\Omega)$.

If $u \in W^{m,p}(\Omega)$ then by Theorem 3.17 there exists a sequence $\{u_j\}$ of $C^\infty(\Omega)$ functions converging to u in $W^{m,p}(\Omega)$. By the above argument, $\{u_j v\}$ is a Cauchy sequence in $W^{m,p}(\Omega)$ and so it converges to an element w of that space. Since $mp > n$, u and v may be assumed to be continuous and bounded on Ω. Thus

$$\begin{aligned}\|w - uv\|_{0,p,\Omega} &\le \left\|w - u_j v\right\|_{0,p,\Omega} + \left\|(u_j - u)v\right\|_{0,p,\Omega} \\ &\le \left\|w - u_j v\right\|_{0,p,\Omega} + \|v\|_{0,\infty,\Omega} \left\|u_j - u\right\|_{0,p,\Omega} \\ &\to 0 \qquad \text{as } j \to \infty.\end{aligned}$$

Hence $w = uv$ in $L^p(\Omega)$ and so $w = uv$ in the sense of distributions. Therefore, $w = uv$ in $W^{m,p}(\Omega)$ and

$$\|uv\|_{m,p,\Omega} = \|w\|_{m,p,\Omega} \le \limsup_{j\to\infty} \left\|u_j v\right\|_{m,p,\Omega} \le K^* \|u\|_{m,p,\Omega} \|v\|_{m,p,\Omega}$$

as was to be shown. ∎

We remark that the Banach algebra $W^{m,p}(\Omega)$ has an identity element if an only if Ω is bounded. That is, the function $e(x) = 1$ belongs to $W^{m,p}(\Omega)$ if and only if Ω has finite volume, but there are no unbounded domains of finite volume that satisfy the cone condition.

Optimality of the Imbedding Theorem

4.40 The imbeddings furnished by the Sobolev Imbedding Theorem 4.12 are "best possible" in the sense that no imbeddings of the types asserted there are possible for any domain for parameter values m, p, q, λ etc. not satisfying the restrictions imposed in the statement of the theorem. We present below a number of examples to illustrate this fact. In these examples it is the local behaviour of functions in $W^{m,p}(\Omega)$ rather than their behaviour near the boundary that prevents extending the parameter intervals for imbeddings.

There remains the possibility that a weaker version of Part I of the imbedding theorem may hold for certain domains not nice enough to satisfy the (weak) cone condition. We will examine some such possibilities later in this chapter.

4.41 EXAMPLE Let k be an integer such that $1 \le k \le n$ and suppose that $mp < n$ and $q > p^* = kp/(n - mp)$. We construct a function $u \in W^{m,p}(\Omega)$ such that $u \notin L^q(\Omega_k)$, where Ω_k is the intersection of Ω with a k-dimensional plane, thus showing that $W^{m,p}(\Omega)$ does not imbed into $L^q(\Omega_k)$.

Without loss of generality, we can assume that the origin belongs to Ω and that $\Omega_k = \{x \in \Omega : x_{k+1} = \cdots = x_n = 0\}$. For $R > 0$, let $B_R = \{x \in \mathbb{R}^n : |x| < R\}$. We fix R small enough that $\overline{B_{2R}} \subset \Omega$. Let $v(x) = |x|^\mu$; the value of μ will be set later. Evidently $v \in C^\infty(\mathbb{R}^n - \{0\})$. Let $u \in C^\infty(\mathbb{R}^n - \{0\})$ be a function satisfying $u(x) = v(x)$ in B_R and $u(x) = 0$ outside B_{2R}. The membership of u in $W^{m,p}(\Omega)$ depends only on the behaviour of v near the origin:

$$u \in W^{m,p}(\Omega) \quad \Longleftrightarrow \quad v \in W^{m,p}(B_R).$$

It is easily checked by induction on $|\alpha|$ that

$$D^\alpha v(x) = P_\alpha(x)|x|^{\mu - 2|\alpha|},$$

where $P_\alpha(x)$ is a polynomial homogeneous of degree $|\alpha|$ in the components of x. Thus $|D^\alpha v(x)| \le K_\alpha |x|^{\mu - |\alpha|}$ and, setting $\rho = |x|$,

$$\int_{B_R} |D^\alpha v(x)|^p \, dx \le K_n K_\alpha^p \int_0^R \rho^{(\mu - |\alpha|)p + n - 1} \, d\rho,$$

where K_n is the $(n-1)$-measure of the sphere of radius 1 in $\mathbb{R}^n$. Therefore $v \in W^{m,p}(B_R)$ and $u \in W^{m,p}(\Omega)$ provided $\mu > m - (n/p)$.

On the other hand, denoting $\tilde{x}_k = (x_1, \ldots, x_k)$ and $r = |\tilde{x}_k|$, we have

$$\int_{\Omega_k} |u(x)|^q \, d\tilde{x}_k \ge \int_{(B_R)_k} |v(x)|^q \, d\tilde{x}_k = K_k \int_0^R r^{\mu q + k - 1} \, dr.$$

Thus $u \notin L^q(\Omega_k)$ if $\mu \le -(k/q)$.

Since $q > kp/(n - mp)$ we can pick μ so that $m - (n/p) < \mu \le -(k/q)$, thus completing the specification of u. ∎

Note that $\mu < 0$, so u is unbounded near the origin. Hence no imbedding of the form $W^{m,p}(\Omega) \to C_B^0(\Omega)$ is possible if $mp < n$.

4.42 EXAMPLE Suppose $mp > n > (m-1)p$, and let $\lambda > m - (n/p)$. Fix μ so that $m - (n/p) < \mu < \lambda$. Then the function u constructed in Example 4.41 continues to belong to $W^{m,p}(\Omega)$. However, if $|x| < R$,

$$\frac{|u(x) - u(0)|}{|x - 0|^\lambda} = |x|^{\mu - \lambda} \to \infty \quad \text{as } |x| \to 0.$$

Thus $u \notin C^{0,\lambda}(\overline{\Omega})$, and the imbedding $W^{m,p}(\Omega) \to C^{0,\lambda}(\overline{\Omega})$ is not possible. ∎

4.43 EXAMPLE Suppose $p > 1$ and $mp = n$. We construct a function u in $W^{m,p}(\Omega)$ such that $u \notin L^\infty(\Omega)$. Hence the imbedding $W^{m,p}(\Omega) \to L^q(\Omega)$, valid for $p \le q < \infty$, cannot be extended to yield $W^{m,p}(\Omega) \to L^\infty(\Omega)$ or $W^{m,p}(\Omega) \to C^0(\overline{\Omega})$ unless $p = 1$ and $m = n$. (See, however, Theorem 8.27.)

Again we assume $0 \in \Omega$ and define $u(x)$ as in Example 4.41 except with a different function $v(x)$ defined by

$$v(x) = \log\bigl(\log(4R/|x|)\bigr).$$

Clearly v is not bounded near the origin, so $u \notin L^\infty(\Omega)$. It can be checked by induction on $|\alpha|$ that

$$D^\alpha v(x) = \sum_{j=1}^{|\alpha|} P_{\alpha,j}(x)|x|^{-2|\alpha|}\bigl(\log(4R/|x|)\bigr)^{-j},$$

where $P_{\alpha,j}(x)$ is a polynomial homogeneous of degree $|\alpha|$ in the components of x. Since $p = n/m$, we have

$$|D^\alpha v(x)|^p \le \sum_{j=1}^{|\alpha|} K_{\alpha,j}|x|^{-|\alpha|n/m}\bigl(\log(4R/|x|)\bigr)^{-jp},$$

so that, setting $\rho = |x|$,

$$\int_{B_R} |D^\alpha v(x)|^p\,dx \le K \sum_{j=1}^{|\alpha|} \int_0^R \bigl(\log(4R/\rho)\bigr)^{-jp} \rho^{-|\alpha|n/m+n-1}\,d\rho.$$

The right side of the above inequality is certainly finite if $|\alpha| < m$. If $|\alpha| = m$, we have, setting $\sigma = \log(4R/\rho)$,

$$\int_{B_R} |D^\alpha v(x)|^p\,dx \le K \sum_{j=1}^{|\alpha|} \int_{\log 4}^{\infty} \sigma^{-jp}\,d\sigma$$

which is finite since $p > 1$. Thus $v \in W^{m,p}(B_R)$ and $u \in W^{m,p}(\Omega)$. ∎

It is interesting that the same function v (and hence u) works for any choice of m and p with $mp = n$.

4.44 EXAMPLE Suppose $(m-1)p = n$ and $p > 1$. We construct u in $W^{m,p}(\Omega)$ such that $u \notin C^{0,1}(\overline{\Omega})$. Hence the imbedding $W^{m,p}(\Omega) \to C^{0,\lambda}(\overline{\Omega})$, valid for $0 < \lambda < 1$ whenever Ω satisfies the strong local Lipschitz condition, cannot be extended to yield $W^{m,p}(\Omega) \to C^{0,1}(\overline{\Omega})$ unless $p = 1$ and $m - 1 = n$. Here u is constructed as in the previous example except using

$$v(x) = |x| \log\bigl(\log(4R/|x|)\bigr).$$

Since $|v(x) - v(0)|/|x - 0| = \log\bigl(\log(4R/|x|)\bigr) \to \infty$ as $x \to 0$ it is clear that $v \notin C^{0,1}(\overline{B_R})$ and therefore $u \notin C^{0,1}(\overline{\Omega})$. The fact that $v \in W^{m,p}(B_R)$ and hence $u \in W^{m,p}(\Omega)$ is shown just as in the previous example. ∎

Nonimbedding Theorems for Irregular Domains

4.45 The above examples show that even for very regular domains there can exist no imbeddings of the types considered in Theorem 4.12 except those explicitly stated there. It remains to be seen whether any imbeddings of those types can exist for domains that do not satisfy the cone condition (or at least the weak cone condition). We will show below that Theorem 4.12 can be extended, with weakened conclusions, to certain types of irregular domains, but first we show that no extension is possible if the domain is "too irregular." This can happen if either the domain is unbounded and too narrow at infinity, or if it has a cusp of exponential sharpness on its boundary.

An unbounded domain $\Omega \subset \mathbb{R}^n$ may have a smooth boundary and still fail to satisfy the cone condition if it becomes narrow at infinity, that is, if

$$\lim_{\substack{|x|\to\infty \\ x\in\Omega}} \operatorname{dist}(x, \operatorname{bdry}\Omega) = 0.$$

The following theorem shows that Parts I and II of Theorem 4.12 fail completely for any unbounded Ω which has finite volume.

4.46 THEOREM Let Ω be an unbounded domain in $\mathbb{R}^n$ having finite volume, and let $q > p$. Then $W^{m,p}(\Omega)$ is not imbedded in $L^q(\Omega)$.

Proof. We construct a function $u(x)$ depending only on distance $\rho = |x|$ of x from the origin whose growth as ρ increases is rapid enough to prevent membership in $L^q(\Omega)$ but not so rapid as to prevent membership in $W^{m,p}(\Omega)$.

Without loss of generality we assume $\operatorname{vol}(\Omega) = 1$. Let $A(\rho)$ denote the surface area ($(n-1)$-measure) of the intersection of Ω with the surface $|x| = \rho$. Then

$$\int_0^\infty A(\rho)\, d\rho = 1.$$

Let $r_0 = 0$ and define r_k for $k = 1, 2, \ldots$ by

$$\int_{r_k}^\infty A(\rho)\, d\rho = \frac{1}{2^k} = \int_{r_{k-1}}^{r_k} A(\rho)\, d\rho.$$

Since Ω is unbounded, r_k increases to infinity with k. Let $\Delta r_k = r_{k+1} - r_k$ and fix ϵ such that $0 < \epsilon < [1/(mp)] - [1/(mq)]$. There must exist an increasing sequence $\{k_j\}_{j=1}^{\infty}$ such that $\Delta r_{k_j} \geq 2^{-\epsilon k_j}$, for otherwise $\Delta r_k < 2^{-\epsilon k}$ for all but possibly finitely many values of k and we would have $\sum_{k=0}^{\infty} \Delta r_k < \infty$, contradicting $\lim r_k = \infty$. For convenience we assume $k_1 \geq 1$ so $k_j \geq j$ for all j. Let $a_0 = 0$, $a_j = r_{k_j+1}$, and $b_j = r_{k_j}$. Note that $a_{j-1} \leq b_j < a_j$ and $a_j - b_j = \Delta r_{k_j} \geq 2^{-\epsilon k_j}$. Let f be an infinitely differentiable function on $\mathbb{R}$ having the properties:

(i) $0 \leq f(t) \leq 1$ for all t,

(ii) $f(t) = 0$ if $t \leq 0$ and $f(t) = 1$ if $t \geq 1$,

(iii) $|(d/dt)^\kappa f(t)| \leq M$ for all t if $1 \leq \kappa \leq m$.

If $x \in \Omega$ and $\rho = |x|$, set

$$u(x) = \begin{cases} 2^{k_{j-1}/q} & \text{for } a_{j-1} \leq \rho \leq b_j \\ 2^{k_{j-1}/q} + \left(2^{k_j/q} - 2^{k_{j-1}/q}\right) f\left(\dfrac{\rho - b_j}{a_j - b_j}\right) & \text{for } b_j \leq \rho \leq a_j. \end{cases}$$

Clearly $u \in C^\infty(\Omega)$. Denoting $\Omega_j = \{x \in \Omega : a_{j-1} \leq \rho \leq a_j\}$, we have

$$\begin{aligned} \int_{\Omega_j} |u(x)|^p \, dx &= \left(\int_{a_{j-1}}^{b_j} + \int_{b_j}^{a_j}\right) (u(x))^p A(\rho)\, d\rho \\ &\leq 2^{k_{j-1}p/q} \int_{a_{j-1}}^{\infty} A(\rho)\, d\rho + 2^{k_j p/q} \int_{b_j}^{a_j} A(\rho)\, d\rho \\ &= \frac{2^{-k_{j-1}(1-p/q)} + 2^{-k_j(1-p/q)}}{2} \leq \frac{1}{2^{(j-1)(1-p/q)}}. \end{aligned}$$

Since $p < q$, the above inequality forces

$$\int_\Omega |u(x)|^p \, dx = \sum_{j=1}^{\infty} \int_{\Omega_j} |u(x)|^p \, dx < \infty.$$

Also, if $1 \leq \kappa \leq m$, we have

$$\begin{aligned} \int_{\Omega_j} \left|\frac{d^\kappa u}{d\rho^\kappa}\right|^p dx &= \int_{b_j}^{a_j} \left|\frac{d^\kappa u}{d\rho^\kappa}\right|^p A(\rho)\, d\rho \\ &\leq M^p 2^{k_j p/q} (a_j - b_j)^{-\kappa p} \int_{b_j}^{a_j} A(\rho)\, d\rho \\ &= \frac{M^p 2^{-k_j(1-p/q-\epsilon\kappa p)}}{2} \leq \frac{M^p 2^{-Cj}}{2}, \end{aligned}$$

where $C = 1 - p/q - \epsilon\kappa p > 0$ because of the choice of ϵ. Hence $D^\alpha u \in L^p(\Omega)$ for $|\alpha| \le m$, that is, $u \in W^{m,p}(\Omega)$. However, $u \notin L^q(\Omega)$ because we have for each j,

$$\int_{\Omega_j} |u(x)|^q \, dx \ge 2^{k_{j-1}} \int_{a_{j-1}}^{a_j} A(\rho)\, d\rho$$
$$= 2^{k_{j-1}} \left(2^{-k_{j-1}-1} - 2^{-k_j-1}\right) \ge \frac{1}{4}.$$

Therefore $W^{m,p}(\Omega)$ cannot be imbedded in $L^q(\Omega)$. ∎

The conclusion of the above theorem can be extended to unbounded domains having infinite volume but satisfying

$$\limsup_{N\to\infty} \operatorname{vol}(\{x \in \Omega : N \le |x| \le N+1\}) = 0.$$

(See Theorem 6.41.)

4.47 Parts I and II of Theorem 4.12 also fail completely for domains with sufficiently sharp boundary cusps. If Ω is a domain in $\mathbb{R}^n$ and x_0 is a point on its boundary, let $B_r = B_r(x_0)$ denote the open ball of radius r and centre at x_0. Let $\Omega_r = B_r \cap \Omega$, let $S_r = (\operatorname{bdry} B_r) \cap \Omega$, and let $A(r, \Omega)$ be the surface area ($(n-1)$-measure) of S_r. We shall say that Ω has a *cusp of exponential sharpness* at its boundary point x_0 if for every real number k we have

$$\lim_{r\to 0+} \frac{A(r, \Omega)}{r^k} = 0. \tag{25}$$

4.48 THEOREM If Ω is a domain in $\mathbb{R}^n$ having a cusp of exponential sharpness at a point x_0 on its boundary, then $W^{m,p}(\Omega)$ is not imbedded in $L^q(\Omega)$ for any $q > p$.

Proof. We construct $u \in W^{m,p}(\Omega)$ which fails to belong to $L^q(\Omega)$ because it becomes unbounded too rapidly near x_0. Without loss of generality we may assume $x_0 = 0$, so that $r = |x|$. Let $\Omega^* = \{x/|x|^2 : x \in \Omega,\ |x| < 1\}$. Then Ω^* is unbounded and has finite volume by (25), and

$$A(r, \Omega^*) = r^{2(n-1)} A(1/r, \Omega).$$

Let t satisfy $p < t < q$. By Theorem 4.46 there exists a function $\tilde{v} \in C^m(0, \infty)$ such that

(i) $\tilde{v}(r) = 0$ if $0 < r \le 1$,

(ii) $\displaystyle\int_1^\infty |\tilde{v}^{(j)}|^t A(r, \Omega^*)\, dr < \infty$ if $0 \le j \le m$,

(iii) $\int_1^\infty |\tilde{v}(r)|^q A(r,\Omega^*)\,dr = \infty.$

[Specifically, $v(y) = \tilde{v}(|y|)$ defines $v \in W^{m,t}(\Omega^*)$ but $v \notin L^q(\Omega^*)$.] Let $x = y/|y|^2$ so that $\rho = |x| = 1/|y| = 1/r$. Set $\lambda = 2n/q$ and define

$$u(x) = \tilde{u}(\rho) = r^\lambda \tilde{v}(r) = |y|^\lambda v(y).$$

It follows for $|\alpha| = j \le m$ that

$$|D^\alpha u(x)| \le |\tilde{u}^{(j)}(\rho)| \le \sum_{i=1}^{j} c_{ij} r^{\lambda+j+i} \tilde{v}^{(i)}(r),$$

where the coefficients c_{ij} depend only on λ. Now $u(x)$ vanishes for $|x| \ge 1$ and so

$$\int_\Omega |u(x)|^q\,dx = \int_0^1 |\tilde{u}(\rho)|^q A(\rho,\Omega)\,d\rho = \int_1^\infty |\tilde{v}(r)|^q A(r,\Omega^*)\,dr = \infty.$$

On the other hand, if $0 \le |\alpha| = j \le m$, we have

$$\begin{aligned}\int_\Omega |D^\alpha u(x)|^p\,dx &\le \int_0^1 |\tilde{u}^{(j)}(\rho)|^p A(\rho,\Omega)\,d\rho \\ &\le K \sum_{i=0}^{j} \int_1^\infty |\tilde{v}^{(i)}(r)|^p r^{(\lambda+j+i)p-2n} A(r,\Omega^*)\,dr.\end{aligned}$$

If it happens that $(\lambda + 2m)p \le 2n$, then, since $p < t$ and $\mathrm{vol}(\Omega^*) < \infty$, all the integrals in the above sum are finite by Hölder's inequality, and $u \in W^{m,p}(\Omega)$. Otherwise let

$$k = ((\lambda + 2m)p - 2n)\frac{t}{t-p} + 2n.$$

By (25) there exists $a \le 1$ such that if $\rho \le a$, then $A(\rho,\Omega) \le \rho^k$. It follows that if $r \ge 1/a$, then

$$r^{k-2n} A(r,\Omega^*) \le r^{k-2}\rho^k = r^{-2}.$$

Thus

$$\begin{aligned}&\int_1^\infty |\tilde{v}^{(i)}(r)|^p r^{(\lambda+j+i)p-2n} A(r,\Omega^*)\,dr \\ &= \int_1^\infty |\tilde{v}^{(i)}(r)|^p r^{(k-2n)(t-p)/t} A(r,\Omega^*)\,dr \\ &\le \left(\int_1^\infty |\tilde{v}^{(i)}(r)|^t A(r,\Omega^*)\,dr\right)^{p/t} \left(\int_1^\infty r^{k-2n} A(r,\Omega^*)\,dr\right)^{(t-p)/t}\end{aligned}$$

which is finite. Hence $u \in W^{m,p}(\Omega)$ and the proof is complete. ∎

Imbedding Theorems for Domains with Cusps

4.49 Having proved that Theorem 4.12 fails completely for sufficiently irregular domains, we now propose to show that certain imbeddings of the types considered in that theorem do hold for less irregular domains that nevertheless fail to satisfy even the weak cone condition. Questions of this sort have been considered by many writers. The treatment here follows that in [A1].

We consider domains Ω in $\mathbb{R}^n$ whose boundaries consist only of $(n-1)$-dimensional surfaces, and it is assumed that Ω lies on only one side of its boundary. For such domains we shall say, somewhat loosely, that Ω has a *cusp* at point x_0 on its boundary if no finite open cone of positive volume contained in Ω can have its vertex at x_0. The failure of a domain to have any cusps does not, of course, imply that it satisfies the cone condition.

We consider a family of special domains in $\mathbb{R}^n$ that we call *standard cusps* and that have cusps of power sharpness (less sharp than exponential sharpness).

4.50 (Standard Cusps) If $1 \le k \le n-1$ and $\lambda > 1$, let the standard cusp $Q_{k,\lambda}$ be the set of points $x = (x_1, \dots, x_n)$ in $\mathbb{R}^n$ that satisfy the inequalities

$$x_1^2 + \cdots + x_k^2 < x_{k+1}^{2\lambda}, \quad x_{k+1} > 0, \dots, x_n > 0,$$
$$(x_1^2 + \cdots + x_k^2)^{1/\lambda} + x_{k+1}^2 + \cdots + x_n^2 < a^2, \tag{26}$$

where a is the radius of the ball of unit volume in $\mathbb{R}^n$. Note that $a < 1$. The cusp $Q_{k,\lambda}$ has axial plane spanned by the $x_k, \dots, x_n$ axes, and verticial plane (cusp plane) spanned by $x_{k+2}, \dots, x_n$. If $k = n-1$, the origin is the only vertex point of $Q_{k,\lambda}$. The outer boundary surface of $Q_{k,\lambda}$ corresponds to equality in (26) in order to simplify calculations later. A sphere or other suitable surface bounded and bounded away from the origin could be used instead.

Corresponding to the standard cusp $Q_{k,\lambda}$ we consider the associated *standard cone* $C_k = Q_{k,1}$ consisting of points $y = (y_1, \dots, y_n)$ in $\mathbb{R}^n$ that satisfy the inequalities

$$y_1^2 + \cdots + y_k^2 < y_{k+1}^2, \quad y_{k+1} > 0, \dots, y_n > 0,$$
$$y_1^2 + \cdots + y_n^2 < a^2.$$

Figure 3 illustrates the standard cusps $Q_{1,2}$ in $\mathbb{R}^2$, and $Q_{2,2}$ and $Q_{1,2}$ in $\mathbb{R}^3$, together with their associated standard cones. In $\mathbb{R}^3$ the cusp $Q_{2,2}$ has a single cusp point (vertex) at the origin, while $Q_{1,2}$ has a cusp line along the x_3-axis.

It is convenient to adopt a system of generalized "cylindrical" coordinates in $\mathbb{R}^n$, $(r_k, \phi_1, \dots, \phi_{k-1}, y_{k+1}, \dots, y_n)$, so that $r_k \ge 0$, $-\pi \le \phi_1 \le \pi$, $0 \le \phi_2, \dots,$

$\phi_{k-1} \leq \pi$, and

$$\begin{aligned}
y_1 &= r_k \sin\phi_1 \sin\phi_2 \cdots \sin\phi_{k-1} \\
y_2 &= r_k \cos\phi_1 \sin\phi_2 \cdots \sin\phi_{k-1} \\
y_3 &= \qquad r_k \cos\phi_2 \cdots \sin\phi_{k-1} \\
&\vdots \\
y_k &= \qquad\qquad r_k \cos\phi_{k-1}.
\end{aligned} \tag{27}$$

In terms of these coordinates, C_k is represented by

$$\begin{aligned}
0 \leq r_k < y_{k+1}, \quad & y_{k+1} > 0, \ldots, y_n > 0, \\
& r_k^2 + y_{k+1}^2 + \cdots + y_n^2 < a^2.
\end{aligned}$$

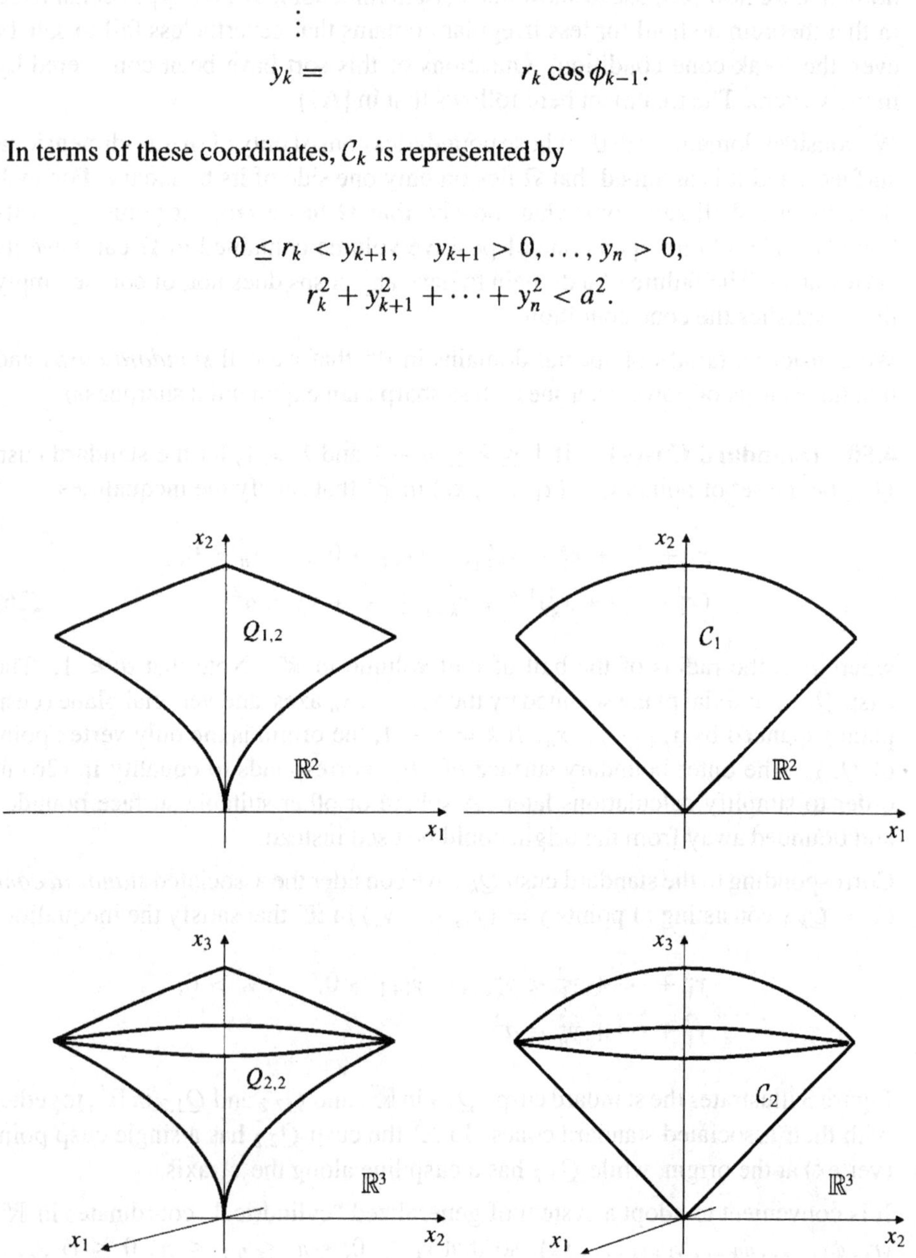

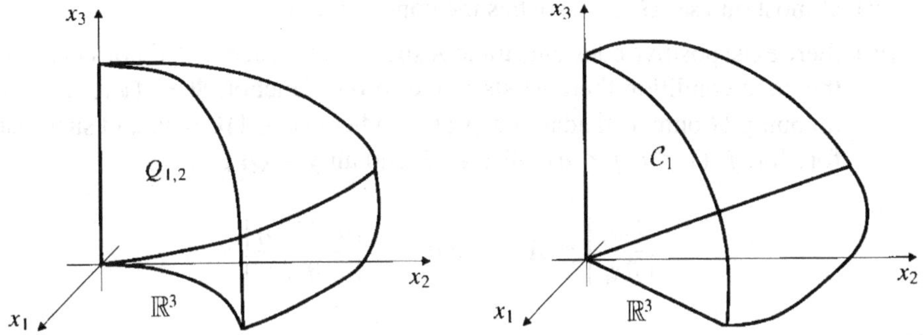

Fig. 3

The standard cusp $Q_{k,\lambda}$ may be transformed into the associated cone C_k by means of the one-to-one transformation

$$\begin{aligned}
x_1 &= r_k^\lambda \sin\phi_1 \sin\phi_2 \cdots \sin\phi_{k-1} \\
x_2 &= r_k^\lambda \cos\phi_1 \sin\phi_2 \cdots \sin\phi_{k-1} \\
x_3 &= \qquad\quad r_k^\lambda \cos\phi_2 \cdots \sin\phi_{k-1} \\
&\vdots \\
x_k &= \qquad\qquad\qquad\quad r_k^\lambda \cos\phi_{k-1} \\
x_{k+1} &= y_{k+1} \\
&\vdots \\
x_n &= y_n,
\end{aligned} \tag{28}$$

which has Jacobian determinant

$$\left| \frac{\partial(x_1, \ldots, x_n)}{\partial(y_1, \ldots, y_n)} \right| = \lambda r_k^{(\lambda-1)k}. \tag{29}$$

We now state three theorems extending imbeddings of the types considered in Theorem 4.12 (except the trace imbeddings) to domains with boundary irregularities comparable to standard cusps. The proofs of these theorems will be given later in this chapter.

4.51 THEOREM Let Ω be a domain in $\mathbb{R}^n$ having the following property: There exists a family Γ of open subsets of Ω such that

(i) $\Omega = \bigcup_{G\in\Gamma} G$,

(ii) Γ has the finite intersection property, that is, there exists a positive integer N such that any $N+1$ distinct sets in Γ have empty intersection,

(iii) at most one set $G \in \Gamma$ satisfies the cone condition,

(iv) there exist positive constants ν and A such that for each $G \in \Gamma$ not satisfying the cone condition there exists a one-to-one function $\Psi = (\psi_1, \dots, \psi_n)$ mapping G onto a standard cusp $Q_{k,\lambda}$, where $(\lambda - 1)k \le \nu$, and such that for all i, j, $(1 \le i, j \le n)$, all $x \in G$, and all $y \in Q_{k,\lambda}$,

$$\left|\frac{\partial \psi_j}{\partial x_i}\right| \le A \qquad \text{and} \qquad \left|\frac{\partial (\psi^{-1})_j}{\partial y_i}\right| \le A.$$

If $\nu > mp - n$, then

$$W^{m,p}(\Omega) \to L^q(\Omega), \qquad \text{for} \quad p \le q \le \frac{(\nu + n)p}{\nu + n - mp}.$$

If $\nu = mp - n$, then the same imbedding holds for $p \le q < \infty$, and for $q = \infty$ if $p = 1$.

If $\nu < mp - n$, then the imbedding holds for $p \le q \le \infty$.

4.52 THEOREM Let Ω be a domain in $\mathbb{R}^n$ having the following property: There exist positive constants $\nu < mp - n$ and A such that for each $x \in \Omega$ there exists an open set G with $x \in G \subset \Omega$ and a one-to-one mapping $\Psi = (\psi_1, \dots, \psi_n)$ mapping G onto a standard cusp $Q_{k,\lambda}$, where $(\lambda - 1)k \le \nu$, and such that for all i, j, $(1 \le i, j \le n)$, all $x \in G$, and all $y \in Q_{k,\lambda}$,

$$\left|\frac{\partial \psi_j}{\partial x_i}\right| \le A \qquad \text{and} \qquad \left|\frac{\partial (\psi^{-1})_j}{\partial y_i}\right| \le A.$$

Then

$$W^{m,p}(\Omega) \to C_B^0(\Omega).$$

More generally, if $\nu < (m - j)p - n$ where $0 \le j \le m - 1$, then

$$W^{m,p}(\Omega) \to C_B^j(\Omega).$$

4.53 THEOREM Let Ω be a domain in $\mathbb{R}^n$ having the following property: There exist positive constants ν, δ, and A such that for each pair of points $x, y \in \Omega$ with $|x - y| \le \delta$ there exists an open set G with $x, y \in G \subset \Omega$ and a one-to-one mapping $\Psi = (\psi_1, \dots, \psi_n)$ mapping G onto a standard cusp $Q_{k,\lambda}$, where $(\lambda - 1)k \le \nu$, and such that for all i, j, $(1 \le i, j \le n)$, all $x \in G$, and all $y \in Q_{k,\lambda}$,

$$\left|\frac{\partial \psi_j}{\partial x_i}\right| \le A \qquad \text{and} \qquad \left|\frac{\partial (\psi^{-1})_j}{\partial y_i}\right| \le A.$$

Suppose that $(m-j-1)p < \nu+n < (m-j)p$ for some integer $j, (0 \le j \le m-1)$. Then

$$W^{m,p}(\Omega) \to C^{j,\mu}(\overline{\Omega}) \qquad \text{for} \quad 0 < \mu \le m - j - \frac{n+\nu}{p}.$$

If $(m-j-1)p = \nu + n$, then the same imbedding holds for $0 < \mu < 1$. In either event we have $W^{m,p}(\Omega) \to C^j(\overline{\Omega})$.

4.54 REMARKS

1. In these theorems the role played by the parameter ν is equivalent to an increase in the dimension n in Theorem 4.12, where increasing n results in weaker imbedding results for given m and p. Since $\nu \ge (\lambda - 1)k$, the sharper the cusp, the greater the equivalent increase in dimension.
2. The reader may wish to construct examples similar to those of Paragraphs 4.41–4.44 to show that the three theorems above give the best possible imbeddings for the domains and types of spaces considered.

4.55 EXAMPLE To illustrate Theorem 4.51, consider the domain

$$\Omega = \{x = (x_1, x_2, x_3) \in \mathbb{R}^3 \ : \ x_2 > 0, \ x_2^2 < x_1 < 3x_2^2\}.$$

If $a = (4\pi/3)^{-1/3}$, the radius of the ball of unit volume in $\mathbb{R}^3$, it is readily verified that the transformation

$$y_1 = x_1 + 2x_2^2, \quad y_2 = x_2, \quad y_3 = x_3 - (k/a), \quad k = 0, \pm 1, \pm 2, \ldots$$

transforms a subdomain G_k of Ω onto the standard cusp $Q_{1,2} \subset \mathbb{R}^3$ in the manner required of the transformation Ψ in the statement of the theorem. Moreover, $\{G\}_{k=-\infty}^{\infty}$ has the finite intersection property and covers Ω up to a set satisfying the cone condition. Using $\nu = 1$, we conclude that $W^{m,p}(\Omega) \to L^q(\Omega)$ for $p \le q \le 4p/(4-mp)$ if $mp < 4$, or for $p \le q < \infty$ if $mp = 4$, or for $p \le q \le \infty$ if $mp > 4$.

Imbedding Inequalities Involving Weighted Norms

4.56 The technique of mapping a standard cusp onto its associated standard cone via (28) and (29) is central to the proof of Theorem 4.51. Such a transformation introduces into any integrals involved a weight factor in the form of the Jacobian determinant (29). Accordingly, we must obtain imbedding inequalities for such standard cones involving L^p-norms weighted by powers of distance from the axial plane of the cone. Such inequalities are also useful in extending the imbedding theorem 4.12 to more general Sobolev spaces involving weighted norms. Many authors have treated the subject of weighted Sobolev spaces. We mention, in particular, Kufner's monograph [Ku] which focuses on a different class of weights depending on distance from the boundary of Ω.

We begin with some one-dimensional inequalities for functions continuously differentiable on an open interval $(0, T)$ in $\mathbb{R}$.

4.57 LEMMA Let $\nu > 0$ and $u \in C^1(0, T)$. If $\int_0^T |u'(t)|t^\nu \, dt < \infty$, then $\lim_{t\to 0+} |u(t)|t^\nu = 0$.

Proof. Let $\epsilon > 0$ be given and fix s in $(0, T/2)$ small enough so that for any t, $0 < t < s$, we have

$$\int_t^s |u'(\tau)|\tau^\nu \, d\tau < \epsilon/3.$$

Now there exists δ in $(0, s)$ such that

$$\delta^\nu |u'(T/2)| < \epsilon/3 \qquad \text{and} \qquad (\delta/s)^\nu \int_s^{T/2} |u'(\tau)|\tau^\nu \, d\tau < \epsilon/3.$$

If $0 < t \le \delta$, then

$$|u(t)| \le |u(T/2)| + \int_t^{T/2} |u'(\tau)| \, d\tau$$

so that

$$t^\nu |u(t)| \le \delta^\nu |u(T/2)| + \int_t^s |u'(\tau)|\tau^\nu \, d\tau + (\delta/s)^\nu \int_s^{T/2} |u'(\tau)|\tau^\nu \, d\tau < \epsilon.$$

Hence $\lim_{t\to 0+} |u(t)|t^\nu = 0$. ∎

4.58 LEMMA Let $\nu > 0$, $p \ge 1$, and $u \in C^1(0, T)$. Then

$$\int_0^T |u(t)|^p t^{\nu-1} \, dt \le \frac{\nu+1}{\nu T} \int_0^T |u(t)|^p t^\nu \, dt + \frac{p}{\nu} \int_0^T |u(t)|^{p-1} |u'(t)| t^\nu \, dt. \quad (30)$$

Proof. We may assume without loss of generality that the right side of (30) is finite and that $p = 1$. Integration by parts gives

$$\int_0^T |u(t)| \left(\nu t^{\nu-1} - \frac{\nu+1}{T} t^\nu \right) dt = -\int_0^T \left(t^\nu - \frac{1}{T} t^{\nu+1} \right) \frac{d}{dt} |u(t)| \, dt,$$

the previous lemma assuring the vanishing of the integrated term at zero. Transposition and estimation of the term on the right now yields

$$\nu \int_0^T |u(t)| t^{\nu-1} \, dt \le \frac{\nu+1}{T} \int_0^T |u(t)| t^\nu \, dt + \int_0^T |u'(t)| t^\nu \, dt,$$

which is (30) for $p = 1$. ∎

4.59 LEMMA Let $\nu > 0$, $p \ge 1$, and $u \in C^1(0, T)$. Then

$$\sup_{0<t<T} |u(t)|^p \le \frac{2}{T} \int_0^T |u(t)|^p \, dt + p \int_0^T |u(t)|^{p-1} |u'(t)| \, dt, \tag{31}$$

$$\sup_{0<t<T} |u(t)|^p t^\nu \le \frac{\nu+3}{T} \int_0^T |u(t)|^p t^\nu \, dt + 2p \int_0^T |u(t)|^{p-1} |u'(t)| t^\nu \, dt. \tag{32}$$

Proof. Again the inequalities need only be proved for $p = 1$. If $0 < t \le T/2$, we obtain by integration by parts

$$\int_0^{T/2} \left| u\left(t + \frac{T}{2} - \tau\right) \right| d\tau = \frac{T}{2} |u(t)| - \int_0^{T/2} \tau \frac{d}{d\tau} \left| u\left(t + \frac{T}{2} - \tau\right) \right| d\tau$$

whence

$$|u(t)| \le \frac{2}{T} \int_0^T |u(\sigma)| \, d\sigma + \int_0^T |u'(\sigma)| \, d\sigma.$$

For $T/2 \le t < T$ the same inequality results from the partial integration of $\int_0^{T/2} |u(t + \tau - T/2)| \, d\tau$. This proves (31) for $p = 1$. Replacing $u(t)$ by $u(t)t^\nu$ in this inequality, we obtain

$$\begin{aligned}
\sup_{0<t<T} |u(t)| t^\nu &\le \frac{2}{T} \int_0^T |u(t)| t^\nu \, dt + \int_0^T \left(|u'(t)| t^\nu + \nu |u(t)| t^{\nu-1} \right) dt \\
&\le \frac{2}{T} \int_0^T |u(t)| t^\nu \, dt + \int_0^T |u'(t)| t^\nu \, dt \\
&\quad + \nu \left(\frac{\nu+1}{\nu T} \int_0^T |u(t)| t^\nu \, dt + \frac{1}{\nu} \int_0^T |u'(t)| t^\nu \, dt \right),
\end{aligned}$$

where (30) has been used to obtain the last inequality. This is the desired result (32) for $p = 1$. ∎

4.60 Now we return to $\mathbb{R}^n$ for $n \ge 2$. If $x \in \mathbb{R}^n$, we shall make use of the spherical polar coordinate representation

$$x = (\rho, \phi) = (\rho, \phi_1, \ldots, \phi_{n-1}),$$

where $\rho \ge 0$, $-\pi \le \phi_1 \le \pi$, $0 \le \phi_2, \ldots, \phi_{n-1} \le \pi$, and

$$\begin{aligned}
x_1 &= \rho \sin\phi_1 \sin\phi_2 \cdots \sin\phi_{n-1}, \\
x_2 &= \rho \cos\phi_1 \sin\phi_2 \cdots \sin\phi_{n-1}, \\
x_3 &= \rho \cos\phi_2 \cdots \sin\phi_{n-1}, \\
&\ \ \vdots \\
x_n &= \rho \cos\phi_{n-1}.
\end{aligned}$$

The volume element is

$$dx = dx_1\, dx_2 \cdots dx_n = \rho^{n-1} \prod_{j=1}^{n-1} \sin^{j-1} \phi_j \, d\rho \, d\phi,$$

where $d\phi = d\phi_1 \cdots d\phi_{n-1}$.

We define functions $r_k = r_k(x)$ for $1 \le k \le n$ as follows:

$$r_1(x) = \rho |\sin \phi_1| \prod_{j=2}^{n-1} \sin \phi_j,$$

$$r_k(x) = \rho \prod_{j=k}^{n-1} \sin \phi_j, \qquad k = 2, 3, \ldots, n-1,$$

$$r_n(x) = \rho.$$

For $1 \le k \le n-1$, $r_k(x)$ is the distance of x from the coordinate plane spanned by the axes $x_{k+1}, \ldots, x_n$; of course $r_n(x)$ is the distance of x from the origin. In connection with the use of product symbols of the form $P = \prod_{j=k}^{m} P_j$, we follow the convention that $P = 1$ if $m < k$.

Let C be an open, conical domain in $\mathbb{R}^n$ specified by the inequalities

$$0 < \rho < a, \quad -\beta_1 < \phi_1 < \beta_1, \quad 0 \le \phi_j < \beta_j, \quad (2 \le j \le n-1), \tag{33}$$

where $0 < \beta_i \le \pi$. (Inequalities "$<$" in (33) corresponding to any $\beta_i = \pi$ are replaced by "$\le$." If all $\beta_i = \pi$, the first inequality is replaced with $0 \le \rho < a$.) Note that any standard cone C_k (introduced in section 4.50) is of the form (33) for some choice of the parameters β_i, $1 \le i \le n-1$.

4.61 LEMMA Let C be as specified by (33) and let $p \ge 1$. Suppose that either $m = k = 1$, or $2 \le m \le n$ and $1 \le k \le n$. Let $1 - k < \nu_1 \le \nu \le \nu_2 < \infty$. Then there exists a constant $K = K(m, k, n, p, \nu 1, \nu 2, \beta_1, \ldots, \beta_{n-1})$ independent of ν and a, such that for every $u \in C^1(C)$ we have

$$\int_C |u(x)|^p [r_k(x)]^\nu [r_m(x)]^{-1}\, dx$$
$$\le K \int_C |u(x)|^{p-1} \left(\frac{1}{a} |u(x)| + |\operatorname{grad} u(x)| \right) [r_k(x)]^\nu \, dx. \tag{34}$$

Proof. Once again it is sufficient to establish (34) for $p = 1$. Let C_+ be the set $\{x = (\rho, \phi) : \phi_1 \ge 0\}$ and C_- the set $\{x = (\rho, \phi) : \phi_1 \le 0\}$. Then $C = C_+ \cup C_-$. We prove (34) only for C_+ (which, however, we continue to call C); a similar proof holds for C_-, so that (34) holds for the given C. Accordingly, assume $C = C_+$.

For $k \le m$ we may write (34) in the form (taking $p = 1$)

$$\int_C |u| \prod_{j=2}^{k-1} \sin^{j-1} \phi_j \prod_{j=k}^{m-1} \sin^{\nu+j-1} \phi_j \prod_{j=m}^{n-1} \sin^{\nu+j-2} \phi_j \rho^{\nu+n-2}\, d\rho\, d\phi$$

$$\le K \int_C \left(\frac{1}{a}|u| + |\operatorname{grad} u| \right) \prod_{j=2}^{k-1} \sin^{j-1} \phi_j \prod_{j=k}^{n-1} \sin^{\nu+j-1} \phi_j \rho^{\nu+n-1}\, d\rho\, d\phi.$$

For $k > m \ge 2$ we may write (34) in the form

$$\int_C |u| \prod_{j=2}^{m-1} \sin^{j-1} \phi_j \prod_{j=m}^{k-1} \sin^{j-2} \phi_j \prod_{j=k}^{n-1} \sin^{\nu+j-2} \phi_j \rho^{\nu+n-2}\, d\rho\, d\phi$$

$$\le K \int_C \left(\frac{1}{a}|u| + |\operatorname{grad} u| \right) \prod_{j=2}^{k-1} \sin^{j-1} \phi_j \prod_{j=k}^{n-1} \sin^{\nu+j-1} \phi_j \rho^{\nu+n-1}\, d\rho\, d\phi.$$

By virtue of the restrictions placed on ν, m, and k in the statement of the lemma, each of the two inequalities above is a special case of

$$\begin{aligned} &\int_C |u| \prod_{j=1}^{i-1} \sin^{\mu_j} \phi_j \prod_{j=i}^{n-1} \sin^{\mu_j - 1} \phi_j \rho^{\nu+n-2}\, d\rho\, d\phi \\ &\qquad \le K \int_C \left(\frac{1}{a}|u| + |\operatorname{grad} u| \right) \prod_{j=1}^{n-1} \sin^{\mu_j} \phi_j \rho^{\nu+n-1}\, d\rho\, d\phi, \end{aligned} \tag{35}$$

where $1 \le i \le n$, $\mu_j \ge 0$, and $\mu_j > 0$ if $j \ge i$. We prove (35) by backwards induction on i. For $i = n$, (35) is obtained by applying Lemma 4.58 to u considered as a function of ρ on $(0, a)$, and then integrating the remaining variables with the appropriate weights. Assume, therefore, that (35) has been proved for $i = k + 1$ where $1 \le k \le n - 1$. We prove it must also hold for $i = k$.

If $\beta_k < \pi$, we have

$$\sin \phi_k \le \phi_k \le K_1 \sin \phi_k, \qquad 0 \le \phi_k \le \beta_k, \tag{36}$$

where $K_1 = K_1(\beta_k)$. By Lemma 4.58, and since

$$\left| \frac{\partial u}{\partial \phi_k} \right| \le \rho |\operatorname{grad} u| \prod_{j=k+1}^{n-1} \sin \phi_j,$$

we have

$$\int_0^{\beta_k} |u(\rho, \phi)| \sin^{\mu_k - 1} \phi_k\, d\phi_k$$

$$\leq \int_0^{\beta_k} |u|\phi_k^{\mu_k - 1}\, d\phi_k$$

$$\leq K_2 \int_0^{\beta_k} \left(|u| + |\operatorname{grad} u|\rho \prod_{j=k+1}^{n-1} \sin\phi_j \right) \phi_k^{\mu_k}\, d\phi_k$$

$$\leq K_3 \int_0^{\beta_k} \left(|u| + |\operatorname{grad} u|\rho \prod_{j=k+1}^{n-1} \sin\phi_j \right) \sin^{\mu_k}\phi_k\, d\phi_k. \tag{37}$$

Note that K_2, and hence K_3, depends on β_k but may be chosen independent of μ_k, and hence of ν, under the conditions of the lemma. If $\beta_k = \pi$, we obtain (37) by writing $\int_0^\pi = \int_0^{\pi/2} + \int_{\pi/2}^\pi$ and using the inequalities

$$\begin{aligned} \sin\phi_k \leq \phi_k \leq (\pi/2)\sin\phi_k \quad & \text{if} \quad 0 \leq \phi_k \leq \pi/2 \\ \sin\phi_k \leq \pi - \phi_k \leq (\pi/2)\sin\phi_k \quad & \text{if} \quad \pi/2 \leq \phi_k \leq \pi. \end{aligned} \tag{38}$$

We now obtain, using (37) and the induction hypothesis,

$$\int_C |u| \prod_{j=1}^{k-1} \sin^{\mu_j}\phi_j \prod_{j=k}^{n-1} \sin^{\mu_j - 1}\phi_j\, \rho^{\nu+n-2}\, d\rho\, d\phi$$

$$\leq \int_0^a \rho^{\nu+n-2}\, d\rho \prod_{j=1}^{k-1} \int_0^{\beta_j} \sin^{\mu_j}\phi_j\, d\phi_j$$

$$\times \prod_{j=k+1}^{n-1} \int_0^{\beta_j} \sin^{\mu_j - 1}\phi_j\, d\phi_j \times \int_0^{\beta_k} |u| \sin^{\mu_k - 1}\phi_k\, d\phi_k$$

$$\leq K_3 \int_C |\operatorname{grad} u| \prod_{j=1}^{n-1} \sin^{\mu_j}\phi_j \rho^{\nu+n-1}\, d\rho\, d\phi$$

$$+ K_3 \int_C |u| \prod_{j=1}^{k} \sin^{\mu_j}\phi_j \prod_{j=k+1}^{n-1} \sin^{\mu_j - 1}\phi_j \rho^{\nu+n-2}\, d\rho\, d\phi$$

$$\leq K \int_C \left(\frac{1}{a}|u| + |\operatorname{grad} u| \right) \prod_{j=1}^{n-1} \sin^{\mu_j}\phi_j \rho^{\nu+n-1}\, d\rho\, d\phi.$$

This completes the induction establishing (35) and hence the lemma. ∎

The following lemma provides a weighted imbedding inequality for the L^q-norm of a function defined on a conical domain of the type (33) in terms of the $W^{m,p}$-norm, both norms being weighted with a power of distance r_k from a coordinate $(n-k)$-plane. It provides the core of the proof of Theorem 4.51.

4.62 LEMMA Let C be as specified by (33) and let $p \geq 1$ and $1 \leq k \leq n$. Suppose that $\max\{1-k, p-n\} < \nu_1 < \nu_2 < \infty$. Then there exists a constant

$K = K(k, n, p, \nu_1, \nu_2, \beta_1, \ldots, \beta_{n-1})$, independent of a, such that for every ν satisfying $\nu_1 \le \nu \le \nu_2$ and every function $u \in C^1(C) \cap C(\overline{C})$ we have

$$\begin{aligned}\left(\int_C |u(x)|^q [r_k(x)]^\nu \, dx\right)^{1/q} \\ \le K \left(\int_C \left(\frac{1}{a^p}|u(x)|^p + |\operatorname{grad} u(x)|^p\right) [r_k(x)]^\nu \, dx\right)^{1/p},\end{aligned} \tag{39}$$

where $q = (\nu + n)p/(\nu + n - p)$.

Proof. Let $\delta = (\nu + n - 1)p/(\nu + n - p)$, let $s = (\nu + n - 1)/\nu$, and let $s' = (\nu + n - 1)/(n - 1)$. We have by Hölder's inequality and Lemma 4.61 (the case $m = k$)

$$\begin{aligned}\int_C |u(x)|^q [r_k(x)]^\nu \, dx &\le \left(\int_C |u|^\delta r_k^{\nu-1} \, dx\right)^{1/s} \left(\int_C |u|^{n\delta/(n-1)} r_k^{n\nu/(n-1)} \, dx\right)^{1/s'} \\ &\le K_1 \left(\int_C |u|^{\delta-1} \left(\frac{1}{a}|u| + |\operatorname{grad} u|\right) r_k^\nu \, dx\right)^{1/s} \\ &\quad \times \left(\int_C |u|^{n\delta/(n-1)} r_k^{n\nu/(n-1)} \, dx\right)^{1/s'}.\end{aligned} \tag{40}$$

In order to estimate the last integral above we adopt the notation

$$\rho^* = (\phi_1, \ldots, \phi_{n-1}), \qquad \phi_j^* = (\rho, \phi_1, \ldots, \hat{\phi}_j, \phi_{j+1}, \ldots, \phi_{n-1}), \quad 1 \le j \le n-1,$$

where the caret denotes omission of a component. Let

$$\begin{aligned}C_0^* &= \{\rho^* : (\rho, \rho^*) \in C \text{ for } 0 < \rho < a\} \\ C_j^* &= \{\phi_j^* : (\rho, \phi) \in C \text{ for } 0 < \phi_j < \beta_j\}.\end{aligned}$$

C_0^* and C_j^*, $(1 \le j \le n-1)$, are domains in $\mathbb{R}^{n-1}$. We define functions F_0 on C_0^* and F_j on C_j^* as follows:

$$\begin{aligned}(F_0(\rho^*))^{n-1} &= \sup_{0<\rho<a} \left(|u|^\delta \rho^{\nu+n-1}\right) \prod_{i=k}^{n-1} \sin^\nu \phi_i \prod_{i=2}^{n-1} \sin^{i-1} \phi_i, \\ (F_j(\phi_j^*))^{n-1} &= \left(\sup_{0<\phi_j<\beta_j} \left(|u|^\delta \sin^{\nu+n-1} \phi_j\right)\right) \rho^{\nu+n-2} \\ &\quad \times \prod_{i=k}^{n-1} \sin^\nu \phi_i \prod_{i=2}^{j-1} \sin^{i-1} \phi_i \prod_{i=j+1}^{n-1} \sin^{i-2} \phi_i.\end{aligned}$$

Then we have

$$|u|^{n\delta/(n-1)} r_k^{n\nu/(n-1)} \rho^{n-1} \prod_{i=2}^{n-1} \sin^{i-1} \phi_i \le F_0(\rho^*) \prod_{j=1}^{n-1} F_j(\phi_j^*).$$

Applying the combinatorial lemma 4.23 with $k = n - 1 = \lambda$ we obtain

$$\begin{aligned}
&\int_C |u|^{n\delta/(n-1)} r_k^{n\nu/(n-1)} \, dx \\
&\le \int_C F_0(\rho^*) \prod_{j=1}^{n-1} F_j(\phi_j^*) \, d\rho \, d\phi \\
&\le \left(\int_{C_0^*} \left(F_0(\rho^*)\right)^{n-1} d\phi \prod_{j=1}^{n-1} \int_{C_j^*} \left(F_j(\phi_j^*)\right)^{n-1} d\rho \, d\hat{\phi}_j \right)^{1/(n-1)}. \qquad (41)
\end{aligned}$$

Now by Lemma 4.59, and since $|\partial u/\partial \rho| \le |\operatorname{grad} u|$,

$$\sup_{0<\rho<a} |u|^\delta \rho^{\nu+n-1} \le K_2 \int_0^a |u|^{\delta-1} \left(\frac{1}{a}|u| + |\operatorname{grad} u| \right) \rho^{\nu+n-1} \, d\rho,$$

where K_2 is independent of ν for $1 - n < \nu_1 \le \nu \le \nu_2 < \infty$. It follows that

$$\int_{C_0^*} \left(F_0(\rho^*)\right)^{n-1} d\phi \le K_2 \int_C |u|^{\delta-1} \left(\frac{1}{a}|u| + |\operatorname{grad} u| \right) r_k^\nu \, dx. \qquad (42)$$

Similarly, by making use of (36) or (38) as in Lemma 4.61, we obtain from Lemma 4.59

$$\begin{aligned}
&\sup_{0<\phi_j<\beta_j} |u|^\delta \sin^{\nu+j-1} \phi_j \\
&\le K_{2,j} \int_0^{\beta_j} |u|^{\delta-1} \left(|u| + \left| \frac{\partial u}{\partial \phi_j} \right| \right) \sin^{\nu+j-1} \phi_j \, d\phi_j \\
&\le K_{2,j} \int_0^{\beta_j} |u|^{\delta-1} \left(|u| + |\operatorname{grad} u| \rho \prod_{i=j+1}^{n-1} \sin \phi_i \right) \sin^{\nu+j-1} \phi_j \, d\phi_j,
\end{aligned}$$

since $|\partial u/\phi_j| \le \rho \prod_{i=j+1}^{n-1} \sin \phi_i$. Hence

$$\begin{aligned}
&\int_{C_j^*} \left(F_j(\phi_j^*)\right)^{n-1} d\rho \, d\hat{\phi}_j \\
&\le K_{2,j} \int_C |\operatorname{grad} u| |u|^{\delta-1} r_k^\nu \, dx + K_{2,j} \int_C |u|^\delta r_k^\nu r_{j+1}^{-1} \, dx \\
&\le K_{3,j} \int_C |u|^{\delta-1} \left(\frac{1}{a}|u| + |\operatorname{grad} u| \right) r_k^\nu \, dx, \qquad (43)
\end{aligned}$$

where we have used Lemma 4.61 again to obtain the last inequality. Note that the constants $K_{2,j}$ and $K_{3,j}$ can be chosen independent of ν for the values of ν allowed. Substitution of (42) and (43) into (41) and then into (40) leads to

$$\int_C |u|^q r_k^\nu \, dx \le K_4 \left(\int_C |u|^{\delta-1} \left(\frac{1}{a}|u| + |\operatorname{grad} u| \right) r_k^\nu \, dx \right)^{1/s+n/((n-1)s')}$$

$$\le K_4 \left(\left[\int_C |u|^q r_k^\nu \, dx \right]^{(p-1)/p} \right.$$

$$\left. \times \left[2^{p-1} \int_C \left(\frac{1}{a^p}|u|^p + |\operatorname{grad} u|^p \right) r_k^\nu \, dx \right]^{1/p} \right)^{(\nu+n)/(\nu+n-1)} .$$

Since $(\nu + n - 1)/(\nu + n) - (p - 1)/p = 1/q$, inequality (39) follows by cancellation for, since u is bounded on C and $\nu > 1 - n$, $\int_C |u|^q r_k^\nu \, dx$ is finite. ∎

4.63 REMARKS

1. The assumption that $u \in C(\overline{C})$ was made only to ensure that the above cancellation was justified. In fact, the lemma holds for any $u \in C^1(C)$.
2. If $1 - k < \nu_1 < \nu_2 < \infty$ and $\nu_1 \le \nu \le \nu_2$, where $p \ge \nu + n$, then (39) holds for any q satisfying $1 \le q < \infty$. It is sufficient to prove this for large q. If $q \ge (\nu + n)/(\nu + n - 1)$, then $q = (\nu + n)s/(\nu + n - s)$ for some s satisfying $1 \le s < p$. Thus

$$\left(\int_C |u|^q r_k^\nu \, dx \right)^{s/q} \le K \int_C \left(\frac{1}{a^s}|u|^s + |\operatorname{grad} u|^s \right) r_k^\nu \, dx$$

$$\le K \left(2^{(p-2)/s} \int_C \left(\frac{1}{a^p}|u|^p + |\operatorname{grad} u|^p \right) r_k^\nu \, dx \right)^{s/p} \left(\int_C r_k^\nu \, dx \right)^{(p-s)/p},$$

which yields (39) since the last factor is finite.
3. If $\nu = m$, a positive integer, then (39) can be obtained very simply as follows. Let $y = (x, z) = (x_1, \ldots, x_n, z_1, \ldots, z_m)$ denote a point in $\mathbb{R}^{n+m}$ and define $u^*(y) = u(x)$ for $x \in C$. If

$$C^* = \left\{ y \in \mathbb{R}^{n+m} : y = (x, z), \, x \in C, \, 0 < z_j < r_k(x), \, 1 \le j \le m \right\},$$

then C^* satisfies the cone condition in $\mathbb{R}^{n+m}$, whence by Theorem 4.12 we have, putting $q = (n + m)p/(n + m - p)$,

$$\left(\int_C |u|^q r_k^m \, dx \right)^{1/q} = \left(\int_{C^*} |u^*(y)|^q \, dy \right)^{1/q}$$

$$\le K \left(\int_{C^*} \left(\frac{1}{a^p}|u^*(y)|^p + |\operatorname{grad} u^*(y)|^p \right) dy \right)^{1/p}$$

$$= K \left(\int_C \left(\frac{1}{a^p}|u|^p + |\operatorname{grad} u|^p \right) r_k^m \, dx \right)^{1/p}$$

since $|\operatorname{grad} u^*(y)| = |\operatorname{grad} u(x)|$, u^* being independent of z.

4. Suppose that $u \in C_0^\infty(\mathbb{R}^n)$, or, more generally, that

$$\int_{\mathbb{R}^n} |u(x)|^p \, [r_k(x)]^\nu \, dx < \infty$$

with ν as in the above lemma. If we take $\beta_i = \pi$, $1 \le i \le n-1$, and let $a \to \infty$ in (39), we obtain

$$\left(\int_{\mathbb{R}^n} |u(x)|^q \, [r_k(x)]^\nu \, dx \right)^{1/q} \le K \left(\int_{\mathbb{R}^n} |\operatorname{grad} u(x)|^p \, [r_k(x)]^\nu \, dx \right)^{1/p}.$$

This generalizes (the case $m = 1$ of) Sobolev's inequality, Theorem 4.31.

As final preparations for the proofs of Theorems 4.51–4.53 we need to obtain weighted analogs of the L^∞ and Hölder imbedding inequalities provided by Theorem 4.12. It is convenient here to deal with arbitrary domains satisfying the cone condition rather than the special case C considered in the lemmas above. The following elementary result will be needed.

4.64 LEMMA Let $z \in \mathbb{R}^k$ and let Ω be a domain of finite volume in $\mathbb{R}^k$. If $0 \le \nu < k$, then

$$\int_\Omega |x - z|^{-\nu} \, dx \le \frac{K}{k - \nu} (\operatorname{vol}(\Omega))^{1 - \nu/k},$$

where the constant K depends on ν and k, but not on z or Ω.

Proof. Let B be the ball in $\mathbb{R}^k$ having centre z and the same volume as Ω. It is easily seen that the left side of the above inequality does not exceed $\int_B |x - z|^{-\nu} \, dx$, and that the inequality holds for $\Omega = B$. ∎

4.65 LEMMA Let $\Omega \subset \mathbb{R}^n$ satisfy the cone condition. Let $1 \le k \le n$ and let P be an $(n-k)$-dimensional plane in $\mathbb{R}^n$. Denote by $r(x)$ the distance from x to P. If $0 \le \nu < p - n$, then for all $u \in C^1(\Omega)$ we have

$$\sup_{x \in \Omega} |u(x)| \le K \left(\int_\Omega \left(|u(x)|^p + |\operatorname{grad} u|^p \right) [r(x)]^\nu \, dx \right)^{1/p}, \tag{44}$$

where the constant K may depend on ν, n, p, k, and the cone C determining the cone condition for Ω, but not on u.

Proof. Throughout this proof A_i and K_i will denote various constants depending on one or more of the parameters on which K is allowed to depend above. It is sufficient to prove that if C is a finite cone contained in Ω having vertex at, say, the origin, then

$$|u(0)| \le K\left(\int_C \left(|u(x)|^p + |\operatorname{grad} u|^p\right)[r(x)]^\nu \, dx\right)^{1/p}. \tag{45}$$

For $0 \le j \le n$, let A_j denote the supremum of the Lebesgue j-dimensional measure of the projection of C onto $\mathbb{R}^j$, taken over all j-dimensional subspaces $\mathbb{R}^j$ of $\mathbb{R}^n$. Writing $x = (x', x'')$ where $x' = (x_1, \dots, x_{n-k})$ and $x'' = (x_{n-k+1}, \dots, x_n)$, we may assume, without loss of generality, that P is orthogonal to the coordinate axes corresponding to the components of x''. Define

$$S = \{x' \in \mathbb{R}^{n-k} : (x', x'') \in C \text{ for some } x'' \in \mathbb{R}^k\},$$
$$R(x') = \{x'' \in \mathbb{R}^k : (x', x'') \in C\} \quad \text{for each } x' \in S.$$

For $0 \le t \le 1$ we denote by C_t the cone $\{tx : x \in C\}$ so that $C_t \subset C$ and $C_t = C$ if $t = 1$. For C_t we define the quantities $A_{t,j}$, S_t, and $R_t(x')$ analogously to the similar quantities defined for C. Clearly $A_{t,j} = t^j A_j$. If $x \in C$, we have

$$u(x) = u(0) + \int_0^1 \frac{d}{dt} u(tx)\, dt,$$

so that

$$|u(0)| \le |u(x)| + |x| \int_0^1 |\operatorname{grad} u(tx)|\, dt.$$

Setting $V = \operatorname{vol}(C)$ and $a = \sup_{x \in C} |x|$, and integrating the above inequality over C, we obtain

$$\begin{aligned} V|u(0)| &\le \int_C |u(x)|\, dx + a \int_C \int_0^1 |\operatorname{grad} u(tx)|\, dt\, dx \\ &= \int_C |u(x)|\, dx + a \int_0^1 t^{-n}\, dt \int_{C_t} |\operatorname{grad} u(x)|\, dx. \end{aligned} \tag{46}$$

Let z denote the orthogonal projection of x onto P. Then $r(x) = |x'' - z''|$. Since $0 \le \nu < p - n$, we have $p > 1$, and so by the previous lemma

$$\begin{aligned} \int_{C_t} [r(x)]^{-\nu/(p-1)}\, dx &= \int_{S_t} dx' \int_{R_t(x')} |x'' - z''|^{-\nu/(p-1)}\, dx'' \\ &\le K_1 \int_{S_t} [A_{t,k}]^{1-\nu/(k(p-1))}\, dx' \\ &\le K_1 [A_{t,k}]^{1-\nu/(k(p-1))} [A_{t,n-k}] = K_2 t^{n-\nu/(p-1)}. \end{aligned}$$

It follows that

$$\begin{aligned}\int_{C_t} |\operatorname{grad} u(x)|\, dx &\le \left(\int_{C_t} |\operatorname{grad} u(x)|^p [r(x)]^\nu\, dx\right)^{1/p} \left(\int_{C_t} [r(x)]^{-\nu/(p-1)}\, dx\right)^{1/p'} \\ &\le K_3 t^{n-(\nu+n)/p} \left(\int_{C_t} |\operatorname{grad} u(x)|^p [r(x)]^\nu\, dx\right)^{1/p}. \qquad (47)\end{aligned}$$

Hence, since $\nu < p - n$,

$$\int_0^1 t^{-n}\, dt \int_{C_t} |\operatorname{grad} u(x)|\, dx \le K_4 \left(\int_C |\operatorname{grad} u(x)|^p [r(x)]^\nu\, dx\right)^{1/p}. \qquad (48)$$

Similarly,

$$\begin{aligned}\int_C |u(x)|\, dx &\le \left(\int_C |u(x)|^p [r(x)]^\nu\, dx\right)^{1/p} \left(\int_C [r(x)]^{-\nu/(p-1)}\, dx\right)^{1/p'} \\ &\le K_5 \left(\int_C |u(x)|^p [r(x)]^\nu\, dx\right)^{1/p}. \qquad (49)\end{aligned}$$

Inequality (45) now follows from (46), (48), and (49). ∎

4.66 LEMMA Suppose all the conditions of the previous lemma are satisfied and, in addition, Ω satisfies the strong local Lipschitz condition. Then for all $u \in C^1(\Omega)$ we have

$$\sup_{\substack{x,y\in\Omega \\ x\ne y}} \frac{|u(x)-u(y)|}{|x-y|^\mu} \le K \left(\int_\Omega \left(|u(x)|^p + |\operatorname{grad} u(x)|^p\right)[r(x)]^\nu\, dx\right)^{1/p}, \qquad (50)$$

where $\mu = 1 - (\nu + n)/p$ satisfies $0 < \mu < 1$, and K is independent of u.

Proof. The proof is the same as that given for inequality (15) in Lemma 4.28 except that the inequality

$$\int_{\Omega_{t\sigma}} |\operatorname{grad} u(z)|\, dz \le K_1 t^{n-(\nu+n)/p} \left(|\operatorname{grad} u(z)|^p [r(z)]^\nu\, dz\right)^{1/p} \qquad (51)$$

is used in (16) in place of the special case $\nu = 0$ actually used there. Inequality (51) is obtained in the same way as (47) above. ∎

Proofs of Theorems 4.51–4.53

4.67 LEMMA Let $\bar{\nu} \geq 0$. If $\bar{\nu} > p - n$, let $1 \leq q \leq (\bar{\nu} + n)/(\bar{\nu} + n - p)$; otherwise, let $1 \leq q < \infty$. There exists a constant $K = K(n, p, \bar{\nu})$ such that for every standard cusp $Q_{k,\lambda}$ (see Paragraph 4.50) for which $(\lambda - 1)k \equiv \nu \leq \bar{\nu}$, and every $u \in C^1(Q_{k,\lambda})$, we have

$$\|u\|_{0,q,Q_{k,\lambda}} \leq K \, \|u\|_{1,p,Q_{k,\lambda}} . \tag{52}$$

Proof. Since each $Q_{k,\lambda}$ has the segment property, it suffices to prove (52) for $u \in C^1(\overline{Q_{k,\lambda}})$. We first do so for given k and λ and then show that K may be chosen to be independent of these parameters.

First suppose $\bar{\nu} > p - n$. It suffices to prove (52) for

$$q = (\bar{\nu} + n)/(\bar{\nu} + n - p).$$

For $u \in C^1(\overline{Q_{k,\lambda}})$ define $\tilde{u}(y) = u(x)$, where y is related to x by (27) and (28). Thus $\tilde{u} \in C^1(C_k) \cap C(\overline{C_k})$, where C_k is the standard cone associated with $Q_{k,\lambda}$. By Lemma 4.62, and since $q \leq (\nu + n)p/(\nu + n - p)$, we have

$$\begin{aligned} \|u\|_{0,q,Q_{k,\lambda}} &= \left(\lambda \int_{C_k} |\tilde{u}(y)|^q [r_k(y)]^\nu \, dy\right)^{1/q} \\ &\leq K_1 \left(\int_{C_k} \left(|\tilde{u}(y)|^p + |\operatorname{grad} \tilde{u}(y)|^p\right)[r_k(y)]^\nu \, dy\right)^{1/q} . \end{aligned} \tag{53}$$

Now $x_j = r_k^{\lambda-1} y_j$ if $1 \leq j \leq k$ and $x_j = y_j$ if $k + 1 \leq j \leq n$. Since $r_k^2 = y_1^2 + \cdots + y_k^2$ we have

$$\frac{\partial x_j}{\partial y_i} = \begin{cases} \delta_{ij} r_k^{\lambda-1} + (\lambda - 1) r_k^{\lambda-3} y_i y_j & \text{if } 1 \leq i, j \leq k \\ \delta_{ij} & \text{otherwise,} \end{cases}$$

where $\delta_{ii} = 1$ and $\delta_{ij} = 0$ if $i \neq j$. Since $r_k(y) \leq 1$ on C_k it follows that

$$|\operatorname{grad} \tilde{u}(y)| \leq K_2 |\operatorname{grad} u(x)|.$$

Hence (52) follows from (53) in this case. For $\bar{\nu} \leq p - n$ and arbitrary q the proof is similar, being based on Remark 2 of Paragraph 4.63.

In order to show that the constant K in (52) can be chosen independent of k and λ provided $\nu = (\lambda - 1)k \leq \bar{\nu}$, we note that it is sufficient to prove that there is a constant $\tilde{K}$ such that for any such k, λ and all $v \in C^1(C_k) \cap C(\overline{C_k})$ we have

$$\begin{aligned} &\left(\int_{C_k} |v(y)|^q [r_k(y)]^\nu \, dy\right)^{1/q} \\ &\qquad \leq \tilde{K} \left(\int_{C_k} \left(|v(y)|^p + |\operatorname{grad} v(y)|^p\right)[r_k(y)]^\nu \, dy\right)^{1/p} . \end{aligned} \tag{54}$$

In fact, it is sufficient to establish (54) with $\tilde{K}$ depending on k as we can then use the maximum of $\tilde{K}(k)$ over the finitely many values of k allowed. We distinguish three cases.

Case I $\bar{\nu} < p - n$, $1 \le q < \infty$. By Lemma 4.65 we have for $0 \le \nu \le \bar{\nu}$,

$$\sup_{y \in C_k} |v(y)| \le K(\nu) \left(\int_{C_k} (|v(y)|^p + |\operatorname{grad} v(y)|^p)[r_k(y)]^\nu \, dy \right)^{1/p}. \tag{55}$$

Since the integral on the right decreases as ν increases, we have $K(\nu) \le K(\bar{\nu})$ and (54) now follows from (55) and the boundedness of C_k.

Case II $\bar{\nu} > p - n$. Again it is sufficient to deal with $q = (\bar{\nu}+n)p/(\bar{\nu}+n-p)$. From Lemma 4.62 we obtain

$$\left(\int_{C_k} |v|^s r_k^\nu \, dy \right)^{1/s} \le K_1 \left(\int_{C_k} (|v|^p + |\operatorname{grad} v|^p) r_k^\nu \, dy \right)^{1/p}, \tag{56}$$

where $s = (\nu+n)p/(\nu+n-p) \ge q$ and K_1 is independent of ν for $p-n < \nu_0 \le \bar{\nu}$. By Hölder's inequality, and since $r_k(y) \le 1$ on C_k, we have

$$\left(\int_{C_k} |v|^q r_k^\nu \, dy \right)^{1/q} \le \left(\int_{C_k} |v|^s r_k^\nu \, dy \right)^{1/s} (\operatorname{vol}(C_k))^{(s-q)/sq}$$

so that if $\nu_0 \le \nu \le \bar{\nu}$, then (54) follows from (56).

If $p - n < 0$, we can take $\nu_0 = 0$ and be done. Otherwise, $p \ge n \ge 2$. Fixing $\nu_0 = (\bar{\nu} - n + p)/2$, we can find ν_1 such that $0 \le \nu_1 \le p - n$ (or $\nu_1 = 0$ if $p = n$) such that for $\nu_1 \le \nu \le \nu_0$ we have

$$1 \le t = \frac{(\nu + n)(\bar{\nu} + n)p}{(\nu + n)(\bar{\nu} + n) + (\bar{\nu} - \nu)p} \le \frac{p}{1 + \epsilon_0},$$

where $\epsilon_0 > 0$ and depends only on $\bar{\nu}$, n, and p. Because of the latter inequality we may also assume $t - n < \nu_1$. Since $(\nu + n)t/(\nu + n - t) = q$ we have, again by Lemma 4.62 and Hölder's inequality,

$$\begin{aligned}\left(\int_{C_k} |v|^q r_k^\nu \, dy \right)^{1/q} &\le K_2 \left(\int_{C_k} (|v|^t + |\operatorname{grad} v|^t) r_k^\nu \, dy \right)^{1/t} \\ &2^{(p-t)/pt} K_2 \left(\int_{C_k} (|v|^p + |\operatorname{grad} v|^p) r_k^\nu \, dy \right)^{1/p} (\operatorname{vol}(C_k))^{(p-t)/pt},\end{aligned} \tag{57}$$

where K_2 is independent of ν for $\nu_1 \le \nu \le \nu_0$.

In the case $\nu_1 > 0$ we can obtain a similar (uniform) estimate for $0 \le \nu \le \nu_1$ by the method of Case I. Combining this with (56) and (57), we prove (54) for this case.

Case III $\bar{\nu} = p - n$, $1 \le q < \infty$. Fix $s \ge \max\{q, n/(n-1)\}$ and let $t = (\nu+n)s/(\nu+n+s)$, so $s = (\nu+n)t/(\nu+n-t)$. Then $1 \le t \le ps/(p+s) < p$ for $0 \le \nu \le \bar{\nu}$. Hence we can select $\nu_1 \ge 0$ such that $t - n < \nu_1 < p - n$. The rest of the proof is similar to Case II. This completes the proof of the lemma. ∎

4.68 (Proof of Theorem 4.51) It is sufficient to prove only the special case $m = 1$, for the general case then follows by induction on m. Let q satisfy $p \le q \le (\nu+n)p/(\nu+n-p)$ if $\nu+n > p$, or $p \le q < \infty$ otherwise. Clearly $q < np/(n-p)$ if $n > p$ so in either case we have by Theorem 4.12

$$\|u\|_{0,q,G} \le K_1 \|u\|_{1,p,G}$$

for every $u \in C^1(\Omega)$ and that element G of Γ that satisfies the cone condition (if such a G exists). If $G \in \Gamma$ does not satisfy the cone condition, and if $\Psi : G \to Q_{k,\lambda}$, where $(\lambda-1)k \le \nu$, is the 1-smooth mapping specified in the statement of the theorem. Then by Theorem 3.41 and Lemma 4.67

$$\|u\|_{0,q,G} \le K_2 \left\|u \circ \Psi^{-1}\right\|_{0,q,Q_{k,\lambda}} \le K_3 \left\|u \circ \Psi^{-1}\right\|_{1,p,Q_{k,\lambda}} \le K_4 \|u\|_{1,p,G},$$

where K_4 is independent of G. Thus, since $q/p \ge 1$,

$$\|u\|_{0,q,\Omega}^q \le \sum_{G\in\Gamma} \|u\|_{0,q,G}^q \le K_5 \sum_{G\in\Gamma} \left(\|u\|_{1,p,G}^p\right)^{q/p}$$

$$\le K_5 \left(\sum_{G\in\Gamma} \|u\|_{1,p,G}^p\right)^{q/p} \le K_5 N^{q/p} \|u\|_{1,p,\Omega}^q,$$

where we have used the finite intersection property of Γ to obtain the final inequality. The required imbedding inequality now follows by completion.

If $\nu < mp - n$, we require that $W^{m,p}(\Omega) \to L^q(\Omega)$ also holds for $q = \infty$. This is a consequence of Theorem 4.52 proved below. ∎

4.69 LEMMA Let $0 \le \bar{\nu} < mp - n$. Then there exists a constant $K = K(m, p, n, \bar{\nu})$ such that if $Q_{k,\lambda}$ is any standard cusp domain for which $(\lambda-1)k = \nu \le \bar{\nu}$ and if $u \in C^m(Q_{k,\lambda})$, then

$$\sup_{x\in Q_{k,\lambda}} |u(x)| \le K \|u\|_{m,p,Q_{k,\lambda}}. \tag{58}$$

Proof. Again it is sufficient to prove the lemma for the case $m = 1$. If u belongs to $C^1(Q_{k,\lambda})$ where $(\lambda-1)k = \nu \le \bar{\nu}$, then we have by Lemma 4.65 and via the method of the second paragraph of the proof of Lemma 4.67,

$$\begin{aligned}\sup_{x\in Q_{k,\lambda}}|u(x)| &= \sup_{y\in C_k}|\tilde{u}(y)|\\ &\le K_1\left(\int_{C_k}(|\tilde{u}(y)|^p+|\operatorname{grad}\tilde{u}(y)|^p)[r_k(y)]^\nu\,dy\right)^{1/p}\\ &\le K_2\left(\int_{Q_{k,\lambda}}(|u(x)|^p+|\operatorname{grad}u(x)|^p)\,dx\right)^{1/p}. \qquad (59)\end{aligned}$$

Since $r_k(y)\le 1$ for $y\in C_k$ it is evident that K_1, and hence K_2, can be chosen independent of k and λ provided $0\le\nu=(\lambda-1)k\le\bar{\nu}$. ∎

4.70 (Proof of Theorem 4.52) It is sufficient to prove that

$$W^{m,p}(\Omega)\to C_B^0(\Omega).$$

Let $u\in C^\infty(\Omega)$. If $x\in\Omega$, then $x\in G\subset\Omega$ for some domain G for which there exists a 1-smooth transformation $\Psi: G\to Q_{k,\lambda}$, $(\lambda-1)k\le\nu$, as specified in the statement of the theorem. Thus

$$\begin{aligned}|u(x)|\le\sup_{x\in G}|u(x)| &= \sup_{y\in Q_{k,\lambda}}|u\circ\Psi^{-1}(y)|\\ &\le K_1\left\|u\circ\Psi^{-1}\right\|_{m,p,Q_{k,\lambda}}\le K_2\|u\|_{m,p,G}\\ &\le K_2\|u\|_{m,p,\Omega}, \qquad (60)\end{aligned}$$

where K_1 and K_2 are independent of G. The rest of the proof is similar to the second paragraph of the proof in Paragraph 4.16. ∎

4.71 (Proof of Theorem 4.53) As in Lemma 4.28 it is sufficient to prove that

$$W^{1,p}(\Omega)\to C^{0,\mu}(\overline{\Omega})\qquad\text{if}\quad 0<\mu\le 1-\frac{n+\nu}{p},$$

that is, that

$$\sup_{\substack{x,y\in\Omega\\ x\ne y}}\frac{|u(x)-u(y)|}{|x-y|^\mu}\le K\|u\|_{1,p,\Omega} \qquad (61)$$

holds when $\nu+n<p$ and $0<\mu\le 1-(\nu+n)/p$. For $x,y\in\Omega$ satisfying $|x-y|\ge\delta$, (61) holds by virtue of (60). If $|x-y|<\delta$, then there exists $G\subset\Omega$ with $x,y\in G$, and a 1-smooth transformation Ψ from G onto a standard cusp $Q_{k,\lambda}$ with $(\lambda-1)k\le\nu$, satisfying the conditions of the theorem. Inequality (61) can then be derived from Lemma 4.66 by the same method used in the proof of Lemma 4.69. The details are left to the reader. ∎

5

INTERPOLATION, EXTENSION, AND APPROXIMATION THEOREMS

Interpolation on Order of Smoothness

5.1 We consider the problem of determining upper bounds for L^p norms of derivatives $D^\beta u$, $0 < |\beta| < m$, of functions in $W^{m,p}(\Omega)$ in terms of the L^p norms of u and its partial derivatives of order m. Such estimates are conveniently expressed in terms of the seminorms $|\cdot|_{j,p}$ defined in Paragraph 4.29. Theorem 5.2 below provides such an estimate for the seminorm $|u|_{j,p}$ in terms of $|u|_{m,p}$ and $\|u\|_p$, as well as some elementary consequences of this estimate. Such estimates arose in the work of of Ehrling [E], Nirenberg [Nr1, Nr2], Gagliardo [Ga1, Ga2], and Browder [Br1, Br2], and were frequently proved under the assumption that Ω satisfies the uniform cone condition, at least if Ω is unbounded. However, we will prove Theorem 5.2 assuming only the cone condition. In fact, even the weak cone condition is sufficient for the proof, as is shown in [AF1].

5.2 THEOREM Let Ω be a domain in $\mathbb{R}^n$ satisfying the cone condition. For each $\epsilon_0 > 0$ there exist finite constants K and K', each depending on n, m, p, ϵ_0 and the dimensions of the cone C providing the cone condition for Ω such that if $0 < \epsilon \le \epsilon_0$, $0 \le j \le m$, and $u \in W^{m,p}(\Omega)$, then

$$|u|_{j,p} \le K\left(\epsilon\, |u|_{m,p} + \epsilon^{-j/(m-j)}\, \|u\|_p\right), \tag{1}$$

$$\|u\|_{j,p} \le K'\left(\epsilon\, \|u\|_{m,p} + \epsilon^{-j/(m-j)}\, \|u\|_p\right), \tag{2}$$

$$\|u\|_{j,p} \le 2K'\, \|u\|_{m,p}^{j/m}\, \|u\|_p^{(m-j)/m}\,. \tag{3}$$

5.3 Inequality (2) follows from repeated applications of (1), and (3) by setting $\epsilon_0 = 1$ in (2) and choosing ϵ in (2) so that the two terms on the right side are equal. Furthermore, (1) holds when $\epsilon < \epsilon_0$ if it holds for $\epsilon < \epsilon_1$ for any specific positive ϵ_1; to see this just replace ϵ by $\epsilon\epsilon_1/\epsilon_0$ and suitably adjust K. Thus we need only prove (1), and that for just one value of ϵ_0.

We carry out the proof in three lemmas. The first develops a one-dimensional version for the case $m = 2$, $j = 1$. The second establishes (1) for $m = 2$, $j = 1$ for general Ω satisfying the cone condition. The third shows that (1) is valid for general $m \geq 2$ and $1 \leq j \leq m - 1$ whenever the case $m = 2$, $j = 1$ is known to hold.

5.4 LEMMA If $\rho > 0$, $1 \leq p < \infty$, $K_p = 2^{p-1}9^p$, and $g \in C^2([0, \rho])$, then

$$|g'(0)|^p \leq \frac{K_p}{\rho}\left(\rho^p \int_0^\rho |g''(t)|^p\,dt + \rho^{-p}\int_0^\rho |g(t)|^p\,dt\right). \tag{4}$$

Proof. Let $f \in C^2([0, 1])$, let $x \in [0, 1/3]$, and let $y \in [2/3, 1]$. By the mean-value theorem there exists $z \in (x, y)$ such that

$$|f'(z)| = \left|\frac{f(y) - f(x)}{y - x}\right| \leq 3|f(x)| + 3|f(y)|.$$

Thus

$$|f'(0)| = \left|f'(z) - \int_0^z f''(t)\,dt\right|$$

$$\leq 3|f(x)| + 3|f(y)| + \int_0^1 |f''(t)|\,dt.$$

Integration of x over $[0, 1/3]$ and y over $[2/3, 1]$ yields

$$\frac{1}{9}|f'(0)| \leq \int_0^{1/3} |f(x)|\,dx + \int_{2/3}^1 |f(y)|\,dy + \frac{1}{9}\int_0^1 |f''(t)|\,dt.$$

For $p \geq 1$ we therefore have (using Hölder's inequality if $p > 1$)

$$|f'(0)|^p \leq K_p\left(\int_0^1 |f''(t)|^p\,dt + \int_0^1 |f(t)|^p\,dt\right).$$

where $K_p = 2^{p-1}9^p$.

Inequality (4) now follows by substituting $f(t) = g(\rho t)$. ∎

5.5 LEMMA If $1 \leq p < \infty$ and the domain $\Omega \subset \mathbb{R}^n$ satisfies the cone condition, then there exists a constant K depending on n, p, and the height ρ_0 and aperture angle κ of the cone C providing the cone condition for Ω such that for all ϵ, $0 < \epsilon \le \rho_0$ and all $u \in W^{2,p}(\Omega)$ we have

$$|u|_{1,p} \le K\big(\epsilon\,|u|_{2,p} + \epsilon^{-1}\,\|u\|_p\big). \tag{5}$$

Proof. Let $\Sigma = \{\sigma \in \mathbb{R}^n : |\sigma| = 1\}$ be the unit sphere in $\mathbb{R}^n$ with volume element $d\sigma$ and $(n-1)$-volume $K_0 = K_0(n) = \int_\Sigma d\sigma$. If $x \in \Omega$ let σ_x be the unit vector in the direction of the axis of a cone $C_x \subset \Omega$ congruent to C and having vertex at x, and let $\Sigma_x = \{\sigma \in \Sigma : \angle(\sigma, \sigma_x) \le \kappa/2\}$.
Let $u \in C^\infty(\Omega)$. If $x \in \Omega$, $\sigma \in \Sigma_x$, and $0 < \rho \le \rho_0$, then

$$|\sigma \cdot \operatorname{grad} u(x)|^p \le \frac{K_p}{\rho}\, I(\rho, p, u, x, \sigma),$$

where

$$I(\rho, p, u, x, \sigma) = \rho^p \int_0^\rho |D_t^2 u(x + t\sigma)|^p\, dt + \rho^{-p} \int_0^\rho |u(x + t\sigma)|^p\, dt.$$

There exists a constant $K_1 = K_1(n, p, \kappa)$ such that

$$\int_\Sigma |\sigma \cdot \operatorname{grad} u(x)|^p\, d\sigma \ge \int_{\Sigma_x} |\sigma \cdot \operatorname{grad} u(x)|^p\, d\sigma \ge K_1 |\operatorname{grad} u(x)|^p.$$

Accordingly,

$$\int_\Omega |\operatorname{grad} u(x)|^p\, dx \le \frac{K_p}{K_1 \rho} \int_\Sigma d\sigma \int_\Omega I(\rho, p, u, x, \sigma)\, dx.$$

In order to estimate the inner integral on the right, regard u and its derivatives as extended to all of $\mathbb{R}^n$ so as to be identically zero outside Ω. For simplicity, we suppose $\sigma = e_n = (0, \ldots, 0, 1)$ and write $x = (x', x_n)$ with $x' \in \mathbb{R}^{n-1}$. We have

$$\begin{aligned}
&\int_\Omega I(\rho, p, u, x, e_n)\, dx \\
&= \int_{\mathbb{R}^{n-1}} dx' \int_{-\infty}^{\infty} dx_n \int_0^\rho \big(\rho^p |D_n^2 u(x', x_n + t)|^p + \rho^{-p} |u(x', x_n + t)|^p\big)\, dt \\
&= \int_{\mathbb{R}^{n-1}} dx' \int_0^\rho dt \int_{-\infty}^{\infty} \big(\rho^p |D_n^2 u(x)|^p + \rho^{-p} |u(x)|^p\big)\, dx_n \\
&\le \rho \int_\Omega \big(\rho^p |D_n^2 u(x)|^p + \rho^{-p} |u(x)|^p\big)\, dx,
\end{aligned}$$

In general, for $\sigma \in \Sigma$

$$\int_\Omega I(\rho, p, u, x, \sigma)\, dx \le \rho \int_\Omega \big(\rho^p\, |u|_{2,p}^p + \rho^{-p}\, \|u\|_p^p\big)\, dx,$$

and since $|D_j(u)| \le |\text{grad}\, u|$ and the measure of Σ is K_0,

$$|u|_{1,p}^p \le \frac{nK_pK_0}{K_1}\left(\rho^p\,|u|_{2,p}^p + \rho^{-p}\,\|u\|_p^p\right).$$

Inequality (5) now follows by taking pth roots, replacing ρ with ϵ, and noting that $C^\infty(\Omega)$ is dense in $W^{2,p}(\Omega)$. ∎

5.6 LEMMA Let $m \ge 2$, let $0 < \delta_0 < \infty$, and let $\epsilon_0 = \min\{\delta_0, \delta_0^2, \dots, \delta_0^{m-1}\}$. Suppose that for given p, $1 \le p < \infty$, and given $\Omega \subset \mathbb{R}^n$ there exists a constant $K = K(\delta_0, p, \Omega)$ such that for every δ satisfying $0 < \delta \le \delta_0$ and every $u \in W^{2,p}(\Omega)$, we have

$$|u|_{1,p} \le K\delta\,|u|_{2,p} + K\delta^{-1}\,|u|_{0,p}\,. \tag{6}$$

Then there exists a constant $K = K(\epsilon_0, m, p, \Omega)$ such that for every ϵ satisfying $0 < \epsilon \le \epsilon_0$, every integer j satisfying $0 \le j \le m-1$, and every $u \in W^{m,p}(\Omega)$, we have

$$|u|_{j,p} \le K\epsilon\,|u|_{m,p} + K\epsilon^{-j/(m-j)}\,|u|_{0,p}\,. \tag{7}$$

Proof. Since (7) is trivial for $j = 0$, we consider only the case $1 \le j \le m-1$. The proof is accomplished by a double induction on m and j. The constants $K_1, K_2, \dots$ appearing in the argument may depend on δ_0 (or ϵ_0), m, p, and Ω. First we prove (7) for $j = m-1$ by induction on m, so that (6) is the special case $m = 2$. Assume, therefore, that for some k, $2 \le k \le m-1$,

$$|u|_{k-1,p} \le K_1\delta\,|u|_{k,p} + K_1\delta^{-(k-1)}\,|u|_{0,p} \tag{8}$$

holds for all δ, $0 < \delta \le \delta_0$, and all $u \in W^{k,p}(\Omega)$. If $u \in W^{k+1,p}(\Omega)$, we prove (8) with $k+1$ replacing k (and a different constant K_1). If $|\alpha| = k-1$ we obtain from (6)

$$|D^\alpha u|_{1,p} \le K_2\delta\,|D^\alpha u|_{2,p} + K_2\delta^{-1}\,|D^\alpha u|_{0,p}\,.$$

Combining this inequality with (8) we obtain, for $0 < \eta \le \delta_0$,

$$\begin{aligned}|u|_{k,p} &\le K_3 \sum_{|\alpha|=k-1} |D^\alpha u|_{1,p} \\ &\le K_4\delta\,|u|_{k+1,p} + K_4\delta^{-1}\,|u|_{k-1,p} \\ &\le K_4\delta\,|u|_{k+1,p} + K_4K_1\delta^{-1}\eta\,|u|_{k,p} + K_4K_1\delta^{-1}\eta^{1-k}\,|u|_{0,p}\,.\end{aligned}$$

We may assume without prejudice that $2K_1K_4 \ge 1$. Therefore, we may take $\eta = \delta/(2K_1K_4)$ and so obtain

$$\begin{aligned}|u|_{k,p} &\le 2K_4\delta\,|u|_{k+1,p} + \left(\delta/(2K_1K_4)\right)^{-k}\,|u|_{0,p} \\ &\le K_5\delta\,|u|_{k+1,p} + K_5\delta^{-k}\,|u|_{0,p}\,.\end{aligned}$$

This completes the induction establishing (8) for $0 < \delta \le \delta_0$ and hence (7) for $j = m - 1$ and $0 < \epsilon \le \delta_0$.

We now prove by downward induction on j that

$$|u|_{j,p} \le K_6 \delta^{m-j} |u|_{m,p} + K_6 \delta^{-j} |u|_{0,p} \tag{9}$$

holds for $1 \le j \le m - 1$ and $0 < \delta \le \delta_0$. Note that (8) with $k = m$ is the special case $j = m - 1$ of (9). Assume, therefore, that (9) holds for some j, $2 \le j \le m - 1$. We prove that it also holds with j replaced by $j - 1$ (and a different constant K_6). From (8) and (9) we obtain

$$\begin{aligned} |u|_{j-1,p} &\le K_7 \delta |u|_{j,p} + K_7 \delta^{1-j} |u|_{0,p} \\ &\le K_7 \delta \big(K_6 \delta^{m-j} |u|_{m,p} + K_6 \delta^{-j} |u|_{0,p} \big) + K_7 \delta^{1-j} |u|_{0,p} \\ &\le K_8 \delta^{m-(j-1)} |u|_{m,p} + K_8 \delta^{1-j} |u|_{0,p} . \end{aligned}$$

Thus (9) holds, and (7) follows by setting $\delta = \epsilon^{1/(m-j)}$ in (7) and noting that $\epsilon \le \epsilon_0$ if $\delta \le \delta_0$. ∎

This completes the proof of Theorem 5.2

5.7 REMARK Careful consideration of the proofs of the previous two lemmas shows that if the height of the cone providing the cone condition for Ω is infinite, then inequalities (5) and (7) (and therefore (1) and (2)) hold for all $\epsilon > 0$, the corresponding constants K being independent of ϵ. This is the case, for example, if $\Omega = \mathbb{R}^n$ or a half-space like $\mathbb{R}^n_+$.

Interpolation on Degree of Summability

The following two interpolation theorems provide sharp estimates for L^q norms of functions in $W^{m,p}(\Omega)$. Some of these estimates follow from Theorem 4.12 while others have traditionally been obtained for regular domains from imbeddings of Sobolev spaces of fractional order. (See Chapter 7.) We obtain them here assuming only that the domain satisfies the cone condition. Again, the weak cone condition would do as well; see [AF1].

5.8 THEOREM Let Ω be a domain in $\mathbb{R}^n$ satisfying the cone condition. If $mp > n$, let $p \le q \le \infty$; if $mp = n$, let $p \le q < \infty$; if $mp < n$, let $p \le q \le p^* = np/(n - mp)$. Then there exists a constant K depending on m, n, p, q and the dimensions of the cone C providing the cone condition for Ω, such that for all $u \in W^{m,p}(\Omega)$,

$$\|u\|_q \le K \, \|u\|_{m,p}^{\theta} \, \|u\|_p^{1-\theta} , \tag{10}$$

where $\theta = (n/mp) - (n/mq)$.

Proof. The case $mp < n,\ p \le q \le p^*$ follows directly from Theorems 2.11 and 4.12:

$$\|u\|_q \le \|u\|_{p^*}^{\theta}\, \|u\|_p^{1-\theta} \le K\, \|u\|_{m,p}^{\theta}\, \|u\|_p^{1-\theta},$$

where $1/q = (\theta/p^*)+(1-\theta)/p$ from which it follows that $\theta = (n/mp)-(n/mq)$. For the cases $mp = n,\ p \le q < \infty$, and $mp > n,\ p \le q \le \infty$ we use the local bound obtained in Lemma 4.15. If $0 < r \le \rho$ (the height of the cone C), then

$$|u(x)| \le K_1 \left(\sum_{|\alpha| \le m-1} r^{|\alpha|-n} \chi_r * |D^\alpha u|(x) + \sum_{|\alpha|=m} (\chi_r \omega_m) * |D^\alpha u|(x) \right), \tag{11}$$

where χ_r is the characteristic function of the ball of radius r centred at the origin in $\mathbb{R}^n$, and $\omega_m(x) = |x|^{m-n}$. We estimate the L^q norms of both terms on the right side of (11) using Young's inequality from Corollary 2.25. If $(1/p)+(1/s) = 1+(1/q)$, then

$$\|\chi_r * |D^\alpha u|\|_q \le \|\chi_r\|_s \|D^\alpha u\|_p = K_2 r^{n-(n/p)+(n/q)} \|D^\alpha u\|_p$$
$$\|(\chi_r \omega_m) * |D^\alpha u|\|_q \le \|\chi_r \omega_m\|_s \|D^\alpha u\|_p = K_3 r^{m-(n/p)+(n/q)} \|D^\alpha u\|_p .$$

(Note that $m - (n/p) + (n/q) > 0$ if q satisfies the above restrictions.) Hence

$$\|u\|_q \le K_4 \left(\sum_{j=0}^{m-1} r^{j-(n/p)+(n/q)} |u|_{j,p} + r^{m-(n/p)+(n/q)} |u|_{m,p} \right).$$

By Theorem 5.2,

$$|u|_{j,p} \le K_5 \big(r^{m-j} |u|_{m,p} + r^{-j} \|u\|_p\big),$$

so

$$\|u\|_q \le K_6 \big(r^{m-(n/p)+(n/q)} \|u\|_{m,p} + r^{-(n/p)+(n/q)} \|u\|_p\big).$$

Adjusting K_6 if necessary, we can assume this inequality holds for all $r \le 1$. Choosing r to make the two terms on the right side equal, we obtain (10). ∎

A special case of the above Theorem asserts that if $mp > n$, then

$$\|u\|_\infty \le K\, \|u\|_{m,p}^{n/mp}\, \|u\|_p^{1-(n/mp)}. \tag{12}$$

A similar inequality with $\|u\|_p$ replaced by a more general $\|u\|_q$ is sometimes useful.

5.9 THEOREM Let Ω be a domain in $\mathbb{R}^n$ satisfying the cone condition. Let $p > 1$ and $mp > n$. Suppose that either $1 \le q \le p$ or both $q > p$ and $mp - p < n$. Then there exists a constant K depending on m, n, p, q and the dimensions of the cone C providing the cone condition for Ω, such that for all $u \in W^{m,p}(\Omega)$,

$$\|u\|_\infty \le K\,\|u\|_{m,p}^{\theta}\,\|u\|_q^{1-\theta},$$

where $\theta = np/[np + (mp - n)q]$.

Proof. It is sufficient to show that the inequality

$$|u(x)| \le K\,\|u\|_{m,p}^{\theta}\,\|u\|_q^{1-\theta}, \qquad \theta = np/[np + (mp - n)q] \tag{13}$$

holds for all $x \in \Omega$ and all $u \in W^{m,p}(\Omega) \cap C^\infty(\Omega)$.

First we observe that (13) is a straightforward consequence of Theorems 5.8 and 2.11 if $1 \le q \le p$; since (12) holds we can substitute

$$\|u\|_p \le \|u\|_q^{q/p}\,\|u\|_\infty^{1-(q/p)}$$

and obtain (13) by cancellation.

Now suppose $q > p$, and, for the moment, that $m = 1$ and $p > n$. We reuse the local bound (11); in this case it says

$$|u(x)| \le K_1\big(r^{-n}\chi_r * |u|(x) + \sum_{|\alpha|=1}(\chi_r\omega_1) * |D^\alpha u|(x)\big),$$

for $0 < r \le \rho$, the height of the cone C. By Hölder's inequality,

$$\chi_r * |u|(x) \le K_2 r^{n-(n/q)}\,\|u\|_q,$$

and, for $|\alpha| = 1$,

$$(\chi_r\omega_1) * |D^\alpha u|(x) \le K_3 r^{1-(n/p)}\,\|D^\alpha u\|_p. \tag{14}$$

Since $\|u\|_q \le K_5\,\|u\|_{1,p}$ (by Part I Case A of Theorem 4.12), and since inequality (14) may be assumed to hold for all r such that $0 < r^{1-(n/p)+(n/q)} \le K_5$ provided K_4 is suitably adjusted, we can choose r to make the two upper bounds above equal. This choice yields (13) with $m = 1$.

For general m, we have $W^{m,p}(\Omega) \to W^{1,r}(\Omega)$, where $r = np/(n - mp + p)$ satisfies $n < r < \infty$ since $(m-1)p < n < mp$. Hence, if $u \in W^{m,p}(\Omega) \cap C^\infty(\Omega)$, we have

$$|u(x)| \le K_6\,\|u\|_{1,r}^{\theta}\,\|u\|_q^{1-\theta} \le K_7\,\|u\|_{m,p}^{\theta}\,\|u\|_q^{1-\theta},$$

where $\theta = nr/[nr + (r - n)q] = np/[np + (mp - n)q]$. ∎

The following theorem makes use of the above result to provide an alternate direct proof of Part I Case C of the Sobolev imbedding theorem 4.12 as well as a hybrid imbedding inequality that will prove useful for establishing compactness of some of these imbeddings in the next chapter.

5.10 THEOREM Let Ω be a domain in $\mathbb{R}^n$ satisfying the cone condition. Let m and k be positive integers and let $p > 1$. Suppose that $mp < n$ and $n - mp < k \le n$. Let ν be the largest integer less than mp, so that $n - \nu \le k$. Let Ω_k be the intersection of Ω with a k-dimensional plane in $\mathbb{R}^n$. Then there exists a constant K such that the inequality

$$\|u\|_{0,kq/n,\Omega_k} \le K \|u\|_{0,q,\Omega}^{1-\theta} \|u\|_{m,p,\Omega}^{\theta} \tag{15}$$

holds for all $u \in W^{m,p}(\Omega)$, where

$$q = p^* = \frac{np}{n - mp} \qquad \text{and} \qquad \theta = \frac{\nu p}{\nu p + (mp - \nu)q}.$$

Note that $0 < \theta < 1$.

Proof. Again it is sufficient to establish the inequality for functions in $W^{m,p}(\Omega) \cap C^\infty(\Omega)$. Without loss of generality we assume that H is a coordinate k-plane $\mathbb{R}^k$ in $\mathbb{R}^n$, and, as we did in Lemma 4.24, that Ω is a union of coordinate cubes of fixed edge length, say 2.

Let $\mu = \binom{k}{n-\nu}$, and let E^i, $1 \le i \le \mu$, denote the various coordinate planes in $\mathbb{R}^k$ having dimension $n - \nu$. Let Ω^i be the projection of Ω_k onto E^i, and for each $x \in \Omega^i$ let Ω^i_x denote the intersection of Ω with the ν-dimensional plane through x perpendicular to E^i. Then Ω^i_x contains a ν dimensional cube of unit edge length having a vertex at x, so it satisfies a cone condition with parameters independent of i and x. By Theorem 5.9

$$\|u\|_{0,\infty,\Omega^i_x} \le K_1 \|u\|_{0,q,\Omega^i_x}^{1-\theta} \|u\|_{m,p,\Omega^i_x}^{\theta}.$$

Let $s = (n - \nu)p/(n - mp)$, and let dx^i and dx^i_* denote the volume elements in E^i and its orthogonal complement (in $\mathbb{R}^n$) respectively. Since

$$s(1-\theta) = \frac{q(mp-\nu)}{mp} \qquad \text{and} \qquad s\theta = \frac{\nu}{m},$$

we have

$$\begin{aligned}
&\int_{\Omega^i} \sup_{y \in \Omega^i_x} |u(y)|^s \, dx^i \\
&\le K_1 \int_{\Omega^i} \left[\int_{\Omega^i_x} |u(x)|^q \, dx^i_* \right]^{(mp-\nu)/mp} \left[\int_{\Omega^i_x} \sum_{|\alpha| \le m} |D^\alpha u(x)|^p \, dx^i_* \right]^{\nu/mp} \\
&\le K_1 \|u\|_{0,q,\Omega}^{s(1-\theta)} \|u\|_{m,p,\Omega}^{s\theta},
\end{aligned}$$

the last line being an application of Hölder's inequality.

Let dx^k denote the k-dimensional volume element in H. We apply the averaging Lemma 4.23 to the family of μ subspaces E^i of $\mathbb{R}^k$. The parameter λ for this application of the lemma is $\lambda = \binom{k-1}{n-\nu-1} = (n-\nu)\mu/k$. Since $(kq/n)(\lambda/\mu) = s$, we obtain

$$\begin{aligned}\|u\|_{0,kq/n,\Omega_k}^{kq/n} &\le K_2 \int_{\Omega_k} \prod_{i=1}^{\mu} \sup_{y\in\Omega_x^i} |u(y)|^{kq/\mu n}\, dx^k \\ &\le K_2 \prod_{i=1}^{\mu} \left[\int_{\Omega^i} \sup_{y\in\Omega_x^i} |u(y)|^s\, dx^i \right]^{1/\lambda} \\ &\le K_3 \prod_{i=1}^{\mu} \|u\|_{0,q,\Omega}^{kq(1-\theta)/\mu n} \|u\|_{m,p,\Omega}^{kq\theta/\mu n},\end{aligned}$$

so that

$$\|u\|_{0,kq/n,\Omega_k} \le K\, \|u\|_{0,q,\Omega}^{1-\theta} \|u\|_{m,p,\Omega}^{\theta}$$

as required. ∎

5.11 REMARK If we take $k = n$ in inequality (15), then the imbedding $W^{m,p}(\Omega) \to L^q(\Omega)$ follows for $q = np/(n-mp)$ by cancellation. The corresponding imbedding inequality $\|u\|_{0,q,\Omega} \le K\,\|u\|_{m,p,\Omega}$ can then be used to further estimate the right side of (15), yielding the trace imbedding $W^{m,p}(\Omega) \to L^r(\Omega_k)$ for $r = kp/(n-mp)$.

Interpolation Involving Compact Subdomains

Sometimes it is useful to have bounds for intermediate derivatives $D^\beta u$, of a function $u \in W^{m,p}(\Omega)$, where $1 \le |\beta| \le m-1$, in terms of the seminorm $|u|_{m,p,\Omega}$ and the L^p-norm of u over a compact subdomain $\Omega' \Subset \Omega$. Such inequalities are typically not possible unless Ω is bounded, but for bounded Ω they can be established under the assumption that Ω satisfies either the segment condition or the cone condition. (A bounded domain Ω satisfying the cone condition can be decomposed into a finite union of subdomains each of which satisfies the strong local Lipschitz condition, and therefore the segment condition. See Lemma 4.22.) We will prove the following hybrid interpolation theorem. (See Agmon [Ag].)

5.12 THEOREM Let Ω be a bounded domain in $\mathbb{R}^n$ satisfying the segment condition. Let $0 < \epsilon_0 < \infty$, let $1 \le p < \infty$, and let j and m be integers with $0 \le j \le m-1$. There exists a constant $K = K(\epsilon_0, m, p, \Omega)$ and for each ϵ satisfying $0 < \epsilon \le \epsilon_0$ a domain $\Omega_\epsilon \Subset \Omega$ such that for every $u \in W^{m,p}(\Omega)$

$$|u|_{j,p,\Omega} \le K\epsilon\, |u|_{m,p,\Omega} + K\epsilon^{-j/(m-j)} \|u\|_{p,\Omega_\epsilon}. \tag{16}$$

Note that this theorem implies Theorem 5.2 extends to bounded domains satisfying the segment condition.

As in the proof of Theorem 5.12, we begin with a one-dimensional inequality.

5.13 LEMMA Let $1 \le p < \infty$ and let $0 < l_1 < l_2 < \infty$. Then there exists a constant $K = K(p, l_1, l_2)$ and, for every $\epsilon > 0$, a number $\delta = \delta(\epsilon, l_1, l_2)$ satisfying $0 < 2\delta < l_1$ such that if (a, b) is a finite open interval in $\mathbb{R}$ whose length $b - a$ satisfies $l_1 \le b - a \le l_2$, and $g \in C^1(a, b)$, then

$$\int_a^b |g(t)|^p \, dt \le K\epsilon \int_a^b |g'(t)|^p \, dt + K \int_{a+\delta}^{b-\delta} |g(t)|^p \, dt. \tag{17}$$

Proof. If $f \in C^1(0, 1)$, $0 < t < 1$, and $1/3 < \tau < 2/3$, then

$$|f(s)| = \left| f(\tau) + \int_\tau^s f'(\xi)\, d\xi \right| \le |f(\tau| + \int_0^1 |f'(\xi)|\, d\xi.$$

Integrating τ over $(1/3, 2/3)$, applying Hölder's inequality if $p > 1$, and finally integrating s over $(0, 1)$ gives

$$\int_0^1 |f(s)|^p \le K_p \int_{1/3}^{2/3} |f(s)|^p \, ds + K_p \int_0^1 |f'(s)|^p \, ds,$$

where $K_p = 3 \cdot 2^{p-1}$. Now substitute $f(s) = g\big(a + s(b - a)\big) = g(t)$ to obtain

$$\int_a^b |g(t)|^p \le K_p (b-a)^p \int_a^b |g'(t)|^p \, dt + K_p \int_{(2a+b)/3}^{(a+2b)/3} |g(t)|^p \, dt.$$

For given $\epsilon > 0$ pick a positive integer k such that $k^{-p} \le \epsilon$. Let $a_j = a + (b-a)j/k$ for $j = 0, 1, \ldots, k$ and pick δ so that $0 < \delta \le (b - a)/3k$. Then

$$\begin{aligned}
\int_a^b |g(t)|^p \, dt &= \sum_{j=1}^k \int_{a_{j-1}}^{a_j} |g(t)|^p \, dt \\
&\le K_p \sum_{j=1}^k \left[\left(\frac{b-a}{k} \right)^p \int_{a_{j-1}}^{a_j} |g'(t)|^p \, dt + \int_{a_{j-1}+\delta}^{a_j-\delta} |g(t)|^p \, dt \right] \\
&\le K_p \max\{1, (b-a)^p\} \left[\epsilon \int_a^b |g'(t)|^p \, dt + \int_{a+\delta}^{b-\delta} |g(t)|^p \, dt \right]
\end{aligned}$$

which is the desired inequality (17). ∎

5.14 LEMMA Let Ω be a bounded domain in $\mathbb{R}^n$ that satisfies the segment condition. Then there exists a constant $K = K(p, \Omega)$ and, for any positive number ϵ, a domain $\Omega_\epsilon \Subset \Omega$, such that

$$|u|_{0,p,\Omega} \le K\epsilon\, |u|_{1,p,\Omega} + K\, |u|_{0,p,\Omega_\epsilon} \tag{18}$$

holds for every $u \in W^{1,p}(\Omega)$.

Proof. Since Ω is bounded, and its boundary is therefore compact, the open cover $\{U_j\}$ of bdry Ω and corresponding set $\{y_j\}$ of nonzero vectors referred to in the definition of the segment condition (Paragraph 3.21) are both finite sets. Therefore open sets $V_j \Subset U_j$ can be found such that bdry $\Omega \subset \bigcup_j V_j$ and even, for sufficiently small δ, $\Omega_\delta = \{x \in \Omega : \operatorname{dist}(x, \operatorname{bdry}\Omega) < \delta\} \subset \bigcup_j V_j$. Thus $\Omega = \bigcup_j (V_j \cap \Omega) \cup \tilde{\Omega}$, where $\tilde{\Omega} \Subset \Omega$. It is thus sufficient to prove that for each j

$$|u|_{0,p,V_j\cap\Omega^p} \le K_1\epsilon^p\, |u|^p_{1,p,\Omega} + K_1\, |u|^p_{0,p,\Omega_{\epsilon,j}}$$

for some $\Omega_{\epsilon,j} \Subset \Omega$. For simplicity, we now drop the subscripts j.

Consider the sets Q, Q_η, $0 \le \eta < 1$, defined by

$$Q = \{x + ty : x \in U \cap \Omega, 0 < t < 1\},$$
$$Q_\eta = \{x + ty : x \in V \cap \Omega, \eta < t < 1\}.$$

If $\eta > 0$, then $Q_\eta \Subset Q$, and by the segment condition, $Q \subset \Omega$. Any line ℓ parallel to y and passing through a point in $V \cap \Omega$ intersects Q_0 in one or more intervals each having length between $|y|$ and diam Ω. By 5.13 there exists $\eta > 0$ and a constant K_1 such that for every $u \in C^\infty(\Omega)$ and any such line ℓ

$$\int_{\ell\cap Q_0} |u(x)|^p\, ds \le K_1\epsilon^p \int_{\ell\cap Q_0} |D_y u(x)|^p\, ds + K_1 \int_{\ell\cap Q_\eta} |u(x)|^p\, ds,$$

D_y denoting differentiation in the direction of y and ds being the length element in that direction. We integrate this inequality over the projection of Q_0 on a hyperplane perpendicular to y and so obtain

$$\begin{aligned} |u|^p_{0,p,V\cap\Omega} \le |u|^p_{0,p,Q_0} &\le K_1\epsilon^p\, |u|^p_{1,p,Q_0} + K_1\, |u|^p_{0,p,Q_\eta} \\ &\le K_1\epsilon^p\, |u|^p_{1,p,\Omega} + K_1\, |u|^p_{0,p,Q_\epsilon}, \end{aligned}$$

where $\Omega_\epsilon = \Omega_\eta \Subset \Omega$. By density, this inequality holds for every $u \in W^{1,p}(\Omega)$.

5.15 (Completion of the Proof of Theorem 5.12) We apply Lemma 5.14 to derivatives $D^\beta u$, $|\beta| = m - 1$ to obtain

$$|u|_{m-1,p,\Omega} \le K\epsilon\, |u|_{m,p,\Omega} + K_1\, |u|_{m-1,p,\Omega_\epsilon}, \tag{19}$$

where $\Omega_\epsilon \Subset \Omega$. Since $\overline{\Omega_\epsilon}$ is a compact subset of Ω, there exists a constant $\delta > 0$ such that $\operatorname{dist}(\overline{\Omega_\epsilon}, \operatorname{bdry}\Omega) > \delta$. The union Ω' of open balls of radius δ about points in $\overline{\Omega_\epsilon}$ clearly satisfies the cone condition and also $\Omega' \Subset \Omega$. We can use Ω' in place of Ω_ϵ in (19), and so we can assume Ω_ϵ satisfies the cone condition. By Theorem 5.2, for given $\epsilon_0 > 0$ the inequality

$$|u|_{m-1,p,\Omega_\epsilon} \le K_2\epsilon\, |u|_{m,p,\Omega_\epsilon} + K_2\epsilon^{-(m-1)}\, |u|_{0,p,\Omega_\epsilon}.$$

Combining this with inequality (19) we obtain the case $j = m-1$ of (16). The rest of the proof is by downward induction on j. Assuming that (16) holds for some j satisfying $1 \le j \le m-1$, and replacing ϵ with ϵ^{m-j} (with consequent alterations to K and Ω_ϵ), we obtain

$$|u|_{j,p,\Omega} \le K_3\epsilon^{m-j}\, |u|_{m,p,\Omega} + K_3\epsilon^{-j}\, |u|_{0,p,\Omega_{\epsilon,1}}.$$

Also, by the case already proved,

$$|u|_{j-1,p,\Omega} \le K_4\epsilon\, |u|_{j,p,\Omega} + K_4\epsilon^{-(j-1)}\, |u|_{0,p,\Omega_{\epsilon,2}}.$$

Combining these we get

$$|u|_{j-1,p,\Omega} \le K_5\epsilon^{m-(j-1)}\, |u|_{m,p,\Omega} + K_5\epsilon^{-(j-1)}\, |u|_{0,p,\Omega_\epsilon},$$

where $K_5 = K_4(K_3+1)$ and $\Omega_\epsilon = \Omega_{\epsilon,1} \cup \Omega_{\epsilon,2}$. Replacing ϵ by $\epsilon^{1/(m-j+1)}$ we complete the induction. ∎

5.16 REMARK The conclusion of Theorem 5.12 is also valid for bounded domains satisfying the cone condition. Although the cone condition does not imply the segment condition, the decomposition of a domain Ω satisfying the cone condition into a finite union of subdomains each of which is a union of parallel translates of a parallelepiped (see Lemma 4.22) can be refined, for bounded Ω, so that each of the subdomains satisfies a strong local Lipschitz condition and therefore also the segment condition.

Extension Theorems

5.17 (Extension Operators) Let Ω be a domain in $\mathbb{R}^n$. For given m and p a linear operator E mapping $W^{m,p}(\Omega)$ into $W^{m,p}(\mathbb{R}^n)$ is called a *simple* (m,p)-*extension operator for* Ω if there exists a constant $K = K(m,p)$ such that for every $u \in W^{m,p}(\Omega)$ the following conditions hold:

(i) $Eu(x) = u(x)$ a.e. in Ω,

(ii) $\|Eu\|_{m,p,\mathbb{R}^n} \le K\, \|u\|_{m,p,\Omega}$.

E is called a *strong m-extension operator for* Ω if E is a linear operator mapping functions defined a.e. in Ω to functions defined a.e. in $\mathbb{R}^n$ and if, for every p, $1 \le p < \infty$, and every integer k, $0 \le k \le m$, the restriction of E to $W^{k,p}(\Omega)$ is a simple (k, p)-extension operator for Ω.

Finally, E is called a *total extension operator for* Ω if E is a strong m-extension operator for Ω for every m. Such a total extension operator necessarily extends functions in $C^m(\overline{\Omega})$ to lie in $C^m(\mathbb{R}^n)$.

5.18 The existence of even a simple (m, p)-extension operator for Ω guarantees that $W^{m,p}(\Omega)$ inherits many properties possessed by $W^{m,p}(\mathbb{R}^n)$. For instance, if an imbedding $W^{m,p}(\mathbb{R}^n) \to L^q(\mathbb{R}^n)$ is known to hold, so that

$$\|u\|_{q,\mathbb{R}^n} \le K_1 \|u\|_{m,p,\mathbb{R}^n},$$

then the imbedding $W^{m,p}(\Omega) \to L^q(\Omega)$ must also hold, for if $u \in W^{m,p}(\Omega)$, then

$$\|u\|_{0,q,\Omega} \le \|Eu\|_{0,q,\mathbb{R}^n} \le K_1 \|Eu\|_{m,p,\mathbb{R}^n} \le K_1 K \|u\|_{m,p,\Omega}.$$

The reason we did not use this technique to prove the Sobolev imbedding theorem 4.12 is that extension theorems cannot be obtained for some domains satisfying such weak conditions as the cone condition or even the weak cone condition.

We will construct extension operators of each of the three types defined above. First we will use successive reflections in smooth boundaries to construct strong and total extension operators for half spaces, and strong extension operators for domains with suitably smooth boundaries. The method is attributed to Whitney [W] and later Hestenes [He] and Seeley [Se]. Stein [St] obtained a total extension operator under the minimal assumption that Ω satisfies the strong local Lipschitz condition. He used integral averaging instead of reflections. We will give only an outline of his proof here, leaving the interested reader to consult [St] for the details. See also [Ry]. The third construction, due to Calderón [Ca1] involves the use of the Calderón-Zygmund theory of singular integrals. It is less transparent than the reflection or averaging methods, and only works when $1 < p < \infty$, but requires only that the domain Ω satisfies the uniform cone condition. Unlike the other methods, it has the property that if the trivial extension $\tilde{u}$ belongs to $W^{m,p}(\mathbb{R}^n)$, then $\tilde{u}$ is the extension produced by the method. By Theorem 5.29 below, this happens if and only if $u \in W_0^{m,p}(\Omega)$. The paper [Jn] provides an extension method that works under a geometric hypothesis that is necessary and sufficient in $\mathbb{R}^2$, and is nearly optimal in higher dimensions.

Except for very simple domains all of our constructions require the use of partitions of unity subordinate to open covers of bdry Ω chosen in such a way that the functions in the partition have uniformly bounded derivatives.

To illustrate the reflection technique we begin by constructing a strong m-extension operator and a total extension operator for a half-space. Then we extend these to apply to domains that satisfy the uniform C^m-regularity condition and also have a bounded boundary.

5.19 THEOREM Let Ω be the half-space $\mathbb{R}^n_+ = \{x \in \mathbb{R}^n : x_n > 0\}$. Then there exists a strong m-extension operator E for Ω. Moreover, for every multi-index α satisfying $|\alpha| \le m$ there exists a strong $(m - |\alpha|)$-extension operator E_α for Ω, such that

$$D^\alpha Eu(x) = E_\alpha D^\alpha u(x).$$

Proof. For functions u defined a.e. on $\mathbb{R}^n_+$ we define Eu and $E_\alpha u$, $|\alpha| \le m$ a.e. on $\mathbb{R}^n$ via

$$Eu(x) = \begin{cases} u(x) & \text{if } x_n > 0 \\ \sum_{j=1}^{m+1} \lambda_j u(x_1, \dots, x_{n-1}, -jx_n) & \text{if } x_n < 0, \end{cases}$$

$$E_\alpha u(x) = \begin{cases} u(x) & \text{if } x_n > 0 \\ \sum_{j=1}^{m+1} (-j)^{\alpha_n} \lambda_j u(x_1, \dots, x_{n-1}, -jx_n) & \text{if } x_n < 0, \end{cases}$$

where the coefficients $\lambda_1, \dots, \lambda_{m+1}$ are the unique solutions of the $(m+1) \times (m+1)$ system of linear equations

$$\sum_{j=1}^{m+1} (-j)^k \lambda_j = 1, \qquad k = 0, \dots, m.$$

If $u \in C^m(\overline{\mathbb{R}^n_+})$, it is readily checked that $Eu \in C^m(\mathbb{R}^n)$ and

$$D^\alpha Eu(x) = E_\alpha D^\alpha u(x), \qquad |\alpha| \le m.$$

Thus

$$\begin{aligned} &\int_{\mathbb{R}^n} |D^\alpha Eu(x)|^p \, dx \\ &= \int_{\mathbb{R}^n_+} |D^\alpha u(x)|^p \, dx + \int_{\mathbb{R}^n_-} \left| \sum_{j=1}^{m+1} (-j)^{\alpha_n} \lambda_j u(x_1, \dots, x_{n-1}, -jx_n) \right|^p dx \\ &\le K(m, p, \alpha) \int_{\mathbb{R}^n_+} |D^\alpha u(x)|^p \, dx. \end{aligned}$$

By Theorem 3.22, the above inequality extends to functions $u \in W^{k,p}(\mathbb{R}^n_+)$, $m \ge k \ge |\alpha|$. Hence, E is a strong m-extension operator for $\mathbb{R}^n_+$. Since $D^\beta E_\alpha u(x) = E_{\alpha+\beta} u(x)$, a similar calculations shows that E_α is a strong $(m-|\alpha|)$-extension. ∎

The reflection technique used in the above proof can be modified to yield a total extension operator. The proof, due to Seeley [Se], is based on the following lemma.

5.20 LEMMA There exists a sequence $\{a_k\}_{k=0}^{\infty}$ such that for every nonnegative integer n we have

$$\sum_{k=0}^{\infty} 2^{nk} a_k = (-1)^n, \tag{20}$$

and

$$\sum_{k=0}^{\infty} 2^{nk} |a_k| < \infty. \tag{21}$$

Proof. For fixed N, let $a_{k,N}$, $k = 0, 1, \dots, N$ be the solution of the system of linear equations

$$\sum_{k=0}^{N} 2^{nk} a_{k,N} = (-1)^n, \qquad n = 0, 1, \dots, N. \tag{22}$$

In terms of the Vandermonde determinant

$$V(x_0, x_1, \dots, x_N) = \begin{vmatrix} 1 & 1 & \cdots & 1 \\ x_0 & x_1 & \cdots & x_N \\ x_0^2 & x_1^2 & \cdots & x_N^2 \\ \vdots & \vdots & & \vdots \\ x_0^N & x_1^N & \cdots & x_N^N \end{vmatrix} = \prod_{\substack{i,j=0 \\ i<j}}^{N} (x_j - x_i),$$

$a_{k,N}$ as given by Cramer's rule is

$$\begin{aligned} a_{k,N} &= \frac{V(1, 2, \dots, 2^{k-1}, -1, 2^{k+1}, \dots, 2^N)}{V(1, 2, \dots, 2^N)} \\ &= \left[\prod_{\substack{i,j=0 \\ i,j \neq k \\ i<j}} (2^j - 2^i) \prod_{i=0}^{k-1} (-1 - 2^i) \prod_{j=k+1}^{N} (2^j + 1) \right] \cdot \left[\prod_{\substack{i,j=0 \\ i<j}}^{N} (2^j - 2^i) \right]^{-1} \\ &= A_k B_{k,N} \end{aligned}$$

where

$$A_k = \prod_{i=1}^{k-1} \frac{1 + 2^i}{2^i - 2^k}, \qquad B_{k,N} = \prod_{j=k+1}^{N} \frac{1 + 2^j}{2^j - 2^k},$$

it being understood that $\prod_{i=l}^{m} P_i = 1$ if $l > m$. Now

$$|A_k| \le \prod_{i=1}^{k-1} \frac{2^{i+1}}{2^{k-1}} \le 2^{(5k-k^2)/2}.$$

Also

$$\log B_{k,N} = \sum_{j=k+1}^{N} \log\left(1 + \frac{1+2^k}{2^j - 2^k}\right)$$

$$< \sum_{j=k+1}^{N} \frac{1+2^k}{2^j - 2^k} < (1+2^k) \sum_{j=k+1}^{N} \frac{1}{2^{j-1}} < 4,$$

where we have used the inequality $\log(1+x) < x$ valid for $x > 0$. It follows that the increasing sequence $\{B_{k,N}\}_{N=0}^{\infty}$ converges to a limit $B_k \le e^4$. Let $a_k = A_k B_k$, so that

$$|a_k| \le e^4 \cdot 2^{(5k-k^2)/2}.$$

Then for any n

$$\sum_{k=0}^{\infty} 2^{nk} |a_k| \le e^4 \sum_{k=0}^{\infty} 2^{(2nk+5k-k^2)/2} < \infty.$$

Letting $n \to \infty$ in (22) completes the proof. ∎

5.21 THEOREM Let Ω be a half-space in $\mathbb{R}^n$. Then there exists a total extension operator E for Ω.

Proof. The restrictions to $\mathbb{R}^n_+$ of functions $\phi \in C_0^\infty(\mathbb{R}^n)$ being dense in $W^{m,p}(\mathbb{R}^n_+)$ for any m and p, we need only define the extension operator for such functions. Let f be a real-valued function, infinitely differentiable on $[0, \infty)$ and satisfying $f(t) = 1$ if $0 \le t \le 1/2$ and $f(t) = 0$ if $t \ge 1$, If $\phi \in C_0^\infty(\mathbb{R}^n)$, let

$$E\phi(x) = E\phi(x', x_n) = \begin{cases} \phi(x) & \text{if } x_n \ge 0, \\ \sum_{k=0}^{\infty} a_k f(-2^k x_n)\phi(x', -2^k x_n) & \text{if } x_n < 0, \end{cases}$$

where $\{a_k\}$ is the sequence constructed in the previous lemma. $E\phi$ is well-defined on $\mathbb{R}^n$ since the sum above has only finitely many nonvanishing terms for any particular $x \in \mathbb{R}^n_- = \{x \in \mathbb{R}^n : x_n < 0\}$. Moreover, $E\phi$ has compact support and belongs to $C^\infty(\overline{\mathbb{R}^n_+}) \cap C^\infty(\overline{\mathbb{R}^n_-})$. If $x \in \mathbb{R}^n_-$, we have

$$D^\alpha E\phi(x) = \sum_{k=0}^{\infty} \sum_{j=0}^{\alpha_n} \binom{\alpha_n}{j} (-2^k)^{\alpha_n} f^{(\alpha_n - j)}(-2^k x_n) D_n^j D^{\alpha'} \phi(x', -2^k x_n)$$

$$= \sum_{k=0}^{\infty} \psi_k(x).$$

Since $\psi_k(x) = 0$ when $-x_n > 1/2^{k-1}$ it follows from (21) that the above series converges absolutely and uniformly as $x_n \to 0-$. Hence by (20)

$$\begin{aligned}\lim_{x_n \to 0-} D^\alpha E\phi(x) &= \sum_{k=0}^{\infty} (-2^k)^{\alpha_n} a_k D^\alpha \phi(x', 0+) \\ &= D^\alpha \phi(x', 0+) = \lim_{x_n \to 0+} D^\alpha E\phi(x) = D^\alpha E\phi(0).\end{aligned}$$

Thus $E\phi \in C_0^\infty(\mathbb{R}^n)$. Moreover, if $|\alpha| \le m$,

$$|\psi_k(x)|^p \le K_1^p |a_k|^p 2^{kmp} \sum_{|\beta| \le m} |D^\beta \phi(x', -2^k x_n)|^p,$$

where K_1 depends only on m, p, n, and f. Thus

$$\begin{aligned}\|\psi_k\|_{0,p,\mathbb{R}^n_-} &\le K_1 |a_k| 2^{km} \left(\sum_{|\beta| \le m} \int_{\mathbb{R}^n_-} |D^\beta \phi(x', -2^k x_n)|^p \, dx \right)^{1/p} \\ &= K_1 |a_k| 2^{km} \left(\frac{1}{2^k} \sum_{|\beta| \le m} \int_{\mathbb{R}^n_+} |D^\beta \phi(y)|^p \, dy \right)^{1/p} \\ &\le K_1 |a_k| 2^{km} \, \|\phi\|_{m,p,\mathbb{R}^n_+} .\end{aligned}$$

It follows from (21) that

$$\|D^\alpha E\phi\|_{0,p,\mathbb{R}^n_-} \le K_1 \, \|\phi\|_{m,p,\mathbb{R}^n_+} \sum_{k=0}^{\infty} 2^{km} \, |a_k| \le K_2 \, \|\phi\|_{m,p,\mathbb{R}^n_+} .$$

Combining this with a similar (trivial) estimate for $\|D^\alpha E\phi\|_{0,p,\mathbb{R}^n_+}$, we obtain

$$\|E\phi\|_{m,p,\mathbb{R}^n} \le K_3 \, \|\phi\|_{m,p,\mathbb{R}^n_+}$$

with $K_3 = K_3(m, p, n)$. This completes the proof. ∎

5.22 THEOREM Let Ω be a domain in $\mathbb{R}^n$ satisfying the uniform C^m-regularity condition and having a bounded boundary. Then there exists a strong m-extension operator E for Ω. Moreover, if α and γ are multi-indices with $|\gamma| \le |\alpha| \le m$, then there exists a linear operator $E_{\alpha\gamma}$ continuous from $W^{j,p}(\Omega)$ into $W^{j,p}(\mathbb{R}^n)$ for $1 \le j \le m - |\alpha|$, $1 \le p < \infty$, such that if $u \in W^{|\alpha|,p}(\Omega)$, then

$$D^\alpha (Eu)(x) = \sum_{|\gamma| \le |\alpha|} E_{\alpha\gamma} D^\gamma u(x). \tag{23}$$

Proof. Since Ω is uniformly C^m-regular and has a bounded boundary the open cover $\{U_j\}$ of bdry Ω and the corresponding m-smooth maps Φ_j from U_j onto B referred to in Paragraph 4.10 are finite collections, say $1 \le j \le N$. Let $Q = \{y = (y', y_n) \in \mathbb{R}^n : |y'| < 1/2, |y_n| < \sqrt{3}/2\}$. Then

$$\{y \in \mathbb{R}^n : |y| < 1/2\} \subset Q \subset B = \{y \in \mathbb{R}^n : |y| < 1\}.$$

By condition (ii) of Paragraph 4.10 the open sets $V_j = \Psi_j(Q)$, $1 \le j \le N$, form an open cover of $\Omega_\delta = \{x \in \Omega : \operatorname{dist}(x, \operatorname{bdry}\Omega) < \delta\}$ for some $\delta > 0$. There exists an open set $V_0 \subset \Omega$, bounded away from bdry Ω, such that $\Omega \subset \bigcup_{j=0}^N V_j$. By Theorem 3.15 we can find infinitely differentiable functions $\omega_0, \omega_1, \dots, \omega_N$ such that the support of ω_j is a subset of V_j and $\sum_{j=0}^N \omega_j(x) = 1$ for all $x \in \Omega$. (Note that the support of ω_0 need not be compact if Ω is unbounded.)

Since Ω is uniformly C^m-regular it satisfies the segment condition and so restrictions to Ω of functions in $C_0^\infty(\mathbb{R}^n)$ are dense in $W^{k,p}(\Omega)$. If $\phi \in C_0^\infty(\mathbb{R}^n)$, then for $x \in \Omega$, $\phi(x) = \sum_{j=0}^N \phi_j(x)$, where $\phi_j = \omega_j \cdot \phi$.

For $j \ge 1$ and $y \in B$ let $\psi_j(y) = \phi_j(\Psi_j(y))$. Then $\psi_j \in C_0^\infty(Q)$. We extend ψ_j to be identically zero outside Q. With E and E_α defined as in Theorem 5.19, we have $E\psi_j \in C_0^m(Q)$, $E\psi_j = \psi_j$ on $Q_+ = \{y \in Q : y_n > 0\}$, and

$$\|E\psi_j\|_{k,p,Q} \le K_1 \|\psi_j\|_{k,p,Q_+}, \qquad 0 \le k \le m,$$

where K_1 depends on k, m, and p. If $\theta_j(x) = E\psi_j(\Phi_j(x))$, then $\theta_j \in C_0^\infty(V_j)$ and $\theta_j(x) = \phi_j(x)$ if $x \in \Omega$. It may be checked by induction that if $|\alpha| \le m$, then

$$D^\alpha \theta_j(x) = \sum_{|\beta| \le |\alpha|} \sum_{|\gamma| \le |\alpha|} a_{j;\alpha\beta}(x) \big[E_\beta\big(b_{j;\beta\gamma} \cdot (D^\gamma \phi_j \circ \Psi_j)\big)\big]\big(\Phi_j(x)\big),$$

where $a_{j;\alpha\beta} \in C^{m-|\alpha|}(\overline{U_j})$ and $b_{j;\beta\gamma} \in C^{m-|\beta|}(\overline{B})$ depend on the transformations Φ_j and $\Psi_j = \Phi_j^{-1}$ and satisfy

$$\sum_{|\beta| \le |\alpha|} a_{j;\alpha\beta}(x) b_{j;\beta\gamma}\big(\Phi_j(x)\big) = \begin{cases} 1 & \text{if } \gamma = \alpha \\ 0 & \text{otherwise.} \end{cases}$$

By Theorem 3.41 we have for $k \le m$,

$$\|\theta_j\|_{k,p,\mathbb{R}^n} \le K_2 \|E\psi_j\|_{k,p,Q} \le K_1 K_2 \|\psi_j\|_{k,p,Q_+} \le K_3 \|\psi_j\|_{k,p,\Omega},$$

where K_3 may be chosen to be independent of j. The operator $\tilde{E}$ defined by

$$\tilde{E}\phi(x) = \phi_0(x) + \sum_{j=1}^N \theta_j(x)$$

clearly satisfies $\tilde{E}\phi(x) = \phi(x)$ if $x \in \Omega$, and

$$\left\|\tilde{E}\phi\right\|_{k,p,\mathbb{R}^n} \le \|\phi_0\|_{k,p,\Omega} + K_3 \sum_{j=1}^{N} \left\|\phi_j\right\|_{k,p,\Omega} \le K_4(1+NK_3)\,\|\phi\|_{k,p,\Omega}, \tag{24}$$

where

$$K_4 = \max_{0\le j\le N}\ \max_{|\alpha|\le m}\ \sup\,|D^\alpha \omega_j(x)| < \infty.$$

Thus $\tilde{E}$ is a strong m-extension operator for Ω. Also

$$D^\alpha \tilde{E}\phi(x) = \sum_{|\gamma|\le|\alpha|} (E_{\alpha\gamma} D^\gamma \phi)(x),$$

where

$$E_{\alpha\gamma} v(x) = \sum_{j=1}^{N} \sum_{|\beta|\le|\alpha|} a_{j;\alpha\beta}(x)\big[E_\beta\big(b_{j;\beta\gamma}\cdot(v\cdot\omega_j)\circ\Psi_j\big)\big]\big(\Phi_j(x)\big)$$

if $\alpha \ne \gamma$, and

$$E_{\alpha\alpha} v(x) = (v\cdot\omega_0)(x) + \sum_{j=1}^{N} \sum_{|\beta|\le|\alpha|} a_{j;\alpha\beta}(x)\big[E_\beta\big(b_{j;\beta\gamma}\cdot(v\cdot\omega_j)\circ\Psi_j\big)\big]\big(\Phi_j(x)\big).$$

We note that if $x \in \Omega$, then $E_{\alpha\gamma}v(x) = 0$ for $\alpha \ne \gamma$ and $E_{\alpha\alpha}v(x) = v(x)$. Clearly $E_{\alpha\gamma}$ is a linear operator. By the differentiability properties of $a_{j;\alpha\beta}$ and $b_{j;\beta\gamma}$, $E_{\alpha\gamma}$ is continuous on $W^{j,p}(\Omega)$ into $W^{j,p}(\mathbb{R}^n)$ for $1 \le j \le m - |\alpha|$. This completes the proof. ∎

5.23 REMARKS

1. If Ω is uniformly C^m-regular for all m, and has a bounded boundary, then we can use the total extension operator of Theorem 5.21 in place of that of Theorem 5.19 in the above proof to obtain a total extension operator for Ω.
2. The restriction that bdry Ω be bounded was imposed in Theorem 5.22 so that the open cover $\{V_j\}$ would be finite. This finiteness was used in two places in the proof, first in asserting the existence of the constant K_4, and secondly in obtaining the last inequality in (24). This latter use is, however, not essential for the proof because (24) could still be obtained from the finite intersection property (condition (i) in Paragraph 4.10) even if the cover $\{V_j\}$ were not finite. Theorem 5.22 extends to any suitably regular domain for which there exists a partition of unity $\{\omega_j\}$ subordinate to $\{V_j\}$ with $D^\alpha\omega_j$ bounded on $\mathbb{R}^n$ uniformly in j for any given α. The reader may find it

interesting to construct, by the above techniques, extension operators for domains not covered by the above theorems, for example, quadrants, strips, rectangular boxes, and smooth images of these.

3. The previous remark also applies to the Calderón Extension Theorem 5.28 given below. Although it is proved by methods quite different from the reflection methods used above, the proof still makes use of a partition of unity in the same way as does that of Theorem 5.22. Accordingly, the above considerations also apply to it. The theorem is proved under a strengthened uniform cone condition that reduces to the uniform cone condition of Paragraph 4.8 if Ω has a bounded boundary.

Clearly subsuming the extension theorems obtained above is the following theorem of Stein [St].

5.24 THEOREM (The Stein Extension Theorem) If Ω is a domain in $\mathbb{R}^n$ satisfying the strong local Lipschitz condition, then there exists a total extension operator for Ω.

We will provide here only an outline of the proof. The details can be found in Chapter 6 of [St].

5.25 (Outline of the Proof of the Stein Extension Theorem) .

1. Let $\Omega_e = \mathbb{R}^n - \overline{\Omega}$ be the open exterior of Ω. The function $\delta(x) = \operatorname{dist}(x, \overline{\Omega})$ is Lipschitz continuous on Ω_e since

$$|\delta(x) - \delta(y)| \le |x - y| \qquad \text{for} \quad x, y \in \Omega_e,$$

but might not be smooth there. However, there exists a function Δ in $C^\infty(\Omega_e)$ and positive constants c_1, c_2, and C_α for all multiindices α such that for all $x \in \Omega_e$,

$$c_1\delta(x) \le \Delta(x) \le c_2\delta(x), \qquad \text{and}$$
$$|D^\alpha \Delta(x)| \le C_\alpha \big(\delta(x)\big)^{1-|\alpha|}.$$

2. There exists a continuous function ϕ on $[1, \infty)$ for which
   (a) $\lim_{t\to\infty} t^k \phi(t) = 0$ for $k = 0, 1, 2, \ldots$,
   (b) $\displaystyle\int_1^\infty \phi(t)\,dt = 1$
   (c) $\displaystyle\int_1^\infty t^k \phi(t) = 0$ for $k = 1, 2, \ldots$.

   In fact, $\phi(t) = \dfrac{e}{\pi t}\operatorname{Im}\left(e^{-w(t-1)^{1/4}}\right)$, where $w = e^{-i\pi/4}$, is such a function.

3. For the special case $\Omega = \{(x, y) : x \in \mathbb{R}^{n-1}, y \in \mathbb{R}, y > f(x)$ where f satisfies a Lipschitz condition $|\phi(x) - \phi(x')| \le M|x - x'|$, there exists a constant c such that if $(x, y) \in \Omega_e$, then $\phi(x) - y \le c\Delta(x, y)$.
4. For Ω as specified in 3, $\Delta^*(x, y) = 2c\Delta(x, y)$, and $u \in C_0^\infty(\mathbb{R}^n)$, the operator E defined by

$$E(u)(x, y) = \begin{cases} u(x, y) & \text{if } y > f(x) \\ \int_1^\infty u\big(x, y + t\Delta^*(x, y)\big)\phi(t)\,dt & \text{if } y < f(x) \end{cases}$$

satisfies, for every $m \ge 0$ and $1 \le p \le \infty$,

$$\|E(u)\|_{m,p,\mathbb{R}^n} \le K\,\|u\|_{m,p,\Omega}\,, \tag{25}$$

where $K = K(m, p, n, M)$. Since Ω satisfies the strong local Lipschitz condition it also satisfies the segment condition and so, by Theorem 3.22 the restrictions to Ω of functions in $C_0^\infty(\mathbb{R}^n)$ are dense in $W^{m,p}(\Omega)$ and so (25) holds for all $u \in W^{m,p}(\Omega)$. Thus Stein's theorem holds for this Ω.
5. The case of general Ω satisfying the strong local Lipschitz condition now follows via a partition of unity subordinate to an open cover of bdry Ω by open sets in each of which (a rotated version of) the special case 4 can be applied. ∎

5.26 The proof of the Calderón extension theorem is based on a special case, suitable for our purposes, of a well-known inequality of Calderón and Zygmund [CZ] for convolutions involving kernels with nonintegrable singularities. The proof of this inequality is rather lengthy and can be found in many sources (e.g. Stein and Weiss [SW]). It will be omitted here. Neither the inequality nor the extension theorem itself will be required hereafter in this monograph.

Let $B_R = \{x \in \mathbb{R}^n : |x| \le R\}$, let $\Sigma_R = \{x \in \mathbb{R}^n : |x| = R\}$, and let $d\sigma_R$ be the area element (Lebesgue $(n-1)$-volume element) on Σ_R. A function g is said to be *homogeneous of degree* μ on $B_R - \{0\}$ if $g(tx) = t^\mu g(x)$ for all $x \in B_R - \{0\}$ and $0 < t \le 1$.

5.27 THEOREM (The Calderón Zygmund Inequality) Let

$$g(x) = G(x)|x|^{-n},$$

where

(i) G is bounded on $\mathbb{R}^n - \{0\}$ and has compact support,
(ii) G is homogeneous of degree 0 on $B_R - \{0\}$ for some $R > 0$, and
(iii) $\int_{\Sigma_R} G(x)\,d\sigma_R = 0$.

If $1 < p < \infty$ and $u \in L^p(\mathbb{R}^n)$, then the principal-value convolution integral

$$u * g(x) = \lim_{\epsilon \to 0+} \int_{\mathbb{R}^n - B_\epsilon} u(x-y)g(y)\,dy$$

exists for almost all $x \in \mathbb{R}^n$, and there exists a constant $K = K(G, p)$ such that for all such u

$$\|u * g\|_p \le K\,\|u\|_p\,.$$

Conversely, if G satisfies (i) and (ii) and if $u * g$ exists for all $u \in C_0^\infty(\mathbb{R}^n)$, then G satisfies (iii).

5.28 THEOREM (The Calderón Extension Theorem) Let Ω be a domain in $\mathbb{R}^n$ satisfying the uniform cone condition (Paragraph 4.8) modified as follows:

(i) the open cover $\{U_j\}$ of bdry Ω is required to be finite, and

(ii) the sets U_j are not required to be bounded.

Then for any $m \in \{1, 2, \ldots\}$ and any p satisfying $1 < p < \infty$, there exists a simple (m, p)-extension operator $E = E(m, p)$ for Ω.

Proof. Let $\{U_1, \ldots, U_N\}$ be the open cover of bdry Ω given by the uniform cone condition, and let U_0 be an open subset of Ω bounded away from bdry Ω such that $\Omega \subset \bigcup_{j=0}^{N} U_j$. (Such a U_0 exists by condition (ii) of Paragraph 4.8.) Let $\omega_0, \omega_1, \ldots, \omega_N$ be a C^∞ partition of unity for Ω with $\operatorname{supp}(\omega)_j \subset U_j$. For $1 \le j \le N$ we shall define operators E_j so that if $u \in W^{m,p}(\Omega)$, then $E_j u \in W^{m,p}(\mathbb{R}^n)$ and satisfies

$$E_j u = u \qquad \text{in} \quad U_j \cap \Omega,$$
$$\|E_j u\|_{m,p,\mathbb{R}^n} \le K_{m,p,j}\,\|u\|_{m,p,\Omega}\,.$$

The desired extension operator is then clearly given by

$$Eu = \omega_0 u + \sum_{j=1}^{N} \omega_j E_j u.$$

We shall write $x \in \mathbb{R}^n$ in the polar coordinate form $x = \rho\sigma$ where $\rho \ge 0$ and σ is a unit vector. Let C_j, the the cone associated with U_j in the description of the uniform cone condition, have vertex at 0. Let ϕ_j be a nontrivial function defined in $\mathbb{R}^n - \{0\}$ satisfying

(i) $\phi_j(x) \ge 0$ for all $x \ne 0$,

(ii) $\operatorname{supp}(\phi_j) \subset -C_j \cup \{0\}$,

(iii) $\phi_j \in C^\infty(\mathbb{R}^n - \{0\})$, and

(iv) for some $\epsilon > 0$, ϕ_j is homogeneous of degree $m-n$ in $B_\epsilon - \{0\}$.

Now $\rho^{n-1}\phi_j$ is homogeneous of degree $m-1 \geq 0$ on $B_\epsilon - \{0\}$ and so the function $\psi_j(x) = (\partial/\partial\rho)^m\big(\rho^{n-1}\phi_j(x)\big)$ vanishes on $B_\epsilon - \{0\}$. Hence ψ_j, extended to be zero at $x = 0$, belongs to $C_0^\infty(-C_j)$. Define

$$E_j u(y) = K_j\left((-1)^m \int_\Sigma \int_0^\infty \phi_j(\rho\sigma)\rho^{n-1}\left(\frac{\partial}{\partial\rho}\right)^m \tilde{u}(y-\rho\sigma)\,d\rho\,d\sigma \right.$$
$$\left. - \int_\Sigma \int_0^\infty \psi_j(\rho\sigma)\tilde{u}(y-\rho\sigma)\,d\rho\,d\sigma\right) \tag{26}$$

where $\tilde{u}$ is the zero extension of u outside Ω and where the constant K_j will be determined shortly. If $y \in U_j \cap \Omega$, then, assuming for the moment that $u \in C^\infty(\Omega)$, we have, for $\rho\sigma \in \operatorname{supp}(\phi_j)$, by condition (iii) of Paragraph 4.8, that $\tilde{u}(y-\rho\sigma) = u(y-\rho\sigma)$ is infinitely differentiable. Now integration by parts m times yields

$$\begin{aligned}
&(-1)^m \int_0^\infty \rho^{n-1}\phi_j(\rho\sigma)\left(\frac{\partial}{\partial\rho}\right)^m u(y-\rho\sigma)\,d\rho \\
&= \sum_{k=0}^{m-1}(-1)^{m-k}\left(\frac{\partial}{\partial\rho}\right)^k \big(\rho^{n-1}\phi_j(\rho\sigma)\big)\left(\frac{\partial}{\partial\rho}\right)^{m-k-1} u(y-\rho\sigma)\Bigg|_{\rho=0}^{\rho=\infty} \\
&\quad + \int_0^\infty \left(\frac{\partial}{\partial\rho}\right)^m \big(\rho^{n-1}\phi_j(\rho\sigma)\big)u(y-\rho\sigma)\,d\rho \\
&= \left(\frac{\partial}{\partial\rho}\right)^{m-1}\big(\rho^{n-1}\phi_j(\rho\sigma)\big)\Big|_{\rho=0} u(y) + \int_0^\infty \psi_j(\rho\sigma)u(y-\rho\sigma)\,d\rho.
\end{aligned}$$

Hence

$$E_j u(y) = K_j u(y)\int_\Sigma \left(\frac{\partial}{\partial\rho}\right)^{m-1}\big(\rho^{n-1}\phi_j(\rho\sigma)\big)\Big|_{\rho=0} d\sigma.$$

Since $(\partial/\partial\rho)^{m-1}\big(\rho^{n-1}\phi_j(\rho\sigma)\big)$ is homogeneous of degree zero near 0, the above integral does not vanish if ϕ_j is not identically zero. Hence K_j can be chosen so that $E_j u(y) = u(y)$ for $y \in U_j \cap \Omega$ and all $u \in C^\infty(\Omega)$. Since $C^\infty(\Omega)$ is dense in $W^{m,p}(\Omega)$ we have $E_j u(y) = u(y)$ a.e. in $U_j \cap \Omega$ for every $u \in W^{m,p}(\Omega)$. The same argument shows that if $\tilde{u} \in W^{m,p}(\mathbb{R}^n)$, then $E_j u(y) = \tilde{u}(y)$ a.e. in $\mathbb{R}^n$.

It remains, therefore, to show that

$$\left\|D^\alpha E_j u\right\|_{0,p,\mathbb{R}^n} \leq K_\alpha \|u\|_{m,p,\Omega}$$

holds for any α with $|\alpha| \leq m$ and all $u \in C^\infty(\Omega) \cap W^{m,p}(\Omega)$. The last integral in (26) is of the form $\theta_j * \tilde{u}(y)$, where $\theta_j(x) = \psi_j(x)|x|^{1-n}$. Since $\theta_j \in L^1(\mathbb{R}^n)$ and has compact support, we obtain via Young's inequality for convolution (Corollary 2.25),

$$\left\|D^\alpha(\theta_j * \tilde{u})\right\|_{0,p,\mathbb{R}^n} = \left\|\theta_j * (\widetilde{D^\alpha u})\right\|_{0,p,\mathbb{R}^n} \le \left\|\theta_j\right\|_{0,1,\mathbb{R}^n} \|D^\alpha u\|_{0,p,\Omega}.$$

It now remains to be shown that the first integral in (26) defines a bounded map from $W^{m,p}(\Omega)$ into $W^{m,p}(\mathbb{R}^n)$. Since $(\partial/\partial\rho)^m = \sum_{|\alpha|=m}(m!/\alpha!)\sigma^\alpha D^\alpha$ we obtain

$$\begin{aligned}\int_\Sigma \int_0^\infty \phi_j(\rho\sigma)\rho^{n-1}\left(\frac{\partial}{\partial\rho}\right)^m \tilde{u}(y-\rho\sigma)\,d\rho\,d\sigma &= \sum_{|\alpha|=m}\frac{m!}{\alpha!}\int_{\mathbb{R}^n}\phi_j(x)\widetilde{D_x^\alpha u}(y-x)\sigma^\alpha\,dx \\ &= \sum_{|\alpha|=m}\xi_\alpha * \widetilde{D^\alpha u},\end{aligned}$$

where $\xi_\alpha = (-1)^{|\alpha|}(m!/\alpha!)\sigma^\alpha\phi_j$ is homogeneous of degree $m-n$ in $B_\epsilon - \{0\}$ and belongs to $C^\infty(\mathbb{R}^n - \{0\})$. It is now clearly sufficient to show that for any β satisfying $|\beta| \le m$

$$\left\|D^\beta(\xi_\alpha * v)\right\|_{0,p,\mathbb{R}^n} \le K_{\alpha,\beta}\,\|v\|_{0,p,\mathbb{R}^n}. \tag{27}$$

If $|\beta| \le m-1$, then $D^\beta\xi_\alpha$ is homogeneous of degree not exceeding $1-n$ in $B_\epsilon - \{0\}$ and so belongs to $L^1(\mathbb{R}^n)$. Inequality (27) now follows by Young's inequality for convolution. Thus we need consider only the case $|\beta| = m$, in which we write $D^\beta = (\partial/\partial x_i)D^\gamma$ for some i, $1 \le i \le n$, and some γ with $|\gamma| = m-1$. Suppose, for the moment, that $v \in C_0^\infty(\mathbb{R}^n)$. Then we may write

$$\begin{aligned}D^\beta(\xi_\alpha * v)(x) = (D^\gamma\xi_\alpha) * \left[\left(\frac{\partial}{\partial x_i}\right)v\right](x) &= \int_{\mathbb{R}^n} D_i v(x-y)D^\gamma\xi_\alpha(y)\,dy \\ &= \lim_{\delta\to 0+}\int_{\mathbb{R}^n - B_\delta} D_i v(x-y)D^\gamma\xi_\alpha(y)\,dy.\end{aligned}$$

We now integrate by parts in the last integral to free v and obtain $D^\beta\xi_\alpha$ under the integral. The integrated term is a surface integral over the spherical boundary Σ_δ of B_δ of the product of $v(x-\cdot)$ and a function homogeneous of degree $1-n$ near zero. This surface integral must therefore tend to $Kv(x)$ as $\delta \to 0+$, for some constant K. Noting that $D_i v(x-y) = -(\partial/\partial y_i)v(x-y)$, we now have

$$D^\beta(\xi_\alpha * v)(x) = \lim_{\delta\to 0+}\int_{\mathbb{R}^n} v(x-y)D^\beta\xi_\alpha(y)\,dy + Kv(x).$$

Now $D^\beta\xi_\alpha$ is homogeneous of degree $-n$ near the origin and so, by the last assertion of Theorem 5.27, $D^\beta\xi_\alpha$ satisfies all the conditions for the singular kernel

g of that theorem. Since $1 < p < \infty$, we have for any $v \in L^p(\Omega)$ (regarded as being identically zero outside Ω)

$$\left\| D^\beta \xi_\alpha * v \right\|_{0,p,\mathbb{R}^n} \le K_{\alpha,\beta} \|v\|_{0,p,\mathbb{R}^n}.$$

This completes the proof. ∎

As observed in the proof of the above theorem, the Calderón extension of a function $u \in W^{m,p}(\Omega)$ coincides with the zero extension $\tilde{u}$ of u if $\tilde{u}$ belongs to $W^{m,p}(\mathbb{R}^n)$. The following theorem (which could have been proved in Chapter 3) shows that in this case u must belong to $W_0^{m,p}(\Omega)$.

5.29 THEOREM (Characterization of $W_0^{m,p}(\Omega)$ by Exterior Extension) Let Ω have the segment property. Then a function u on Ω belongs to $W_0^{m,p}(\Omega)$ if and only if the zero extension $\tilde{u}$ of u belongs to $W^{m,p}(R^n)$.

Proof. Lemma 3.27 shows, with no hypotheses on Ω, that if $u \in W_0^{m,p}(\Omega)$, then $\tilde{u} \in W^{m,p}(R^n)$.

Conversely, suppose that Ω has the segment property and that $\tilde{u} \in W^{m,p}(\mathbb{R}^n)$. Proceed as in the proof of Theorem 3.22, first multiplying u by a suitable smooth cutoff function f_ϵ to approximate u in $W^{m,p}(\Omega)$ by a function in that space with a bounded support. Replace u by that approximation; then $\tilde{u}$ is replaced by $f_\epsilon \tilde{u}$, and so still belongs to $W^{m,p}(\mathbb{R}^n)$. Now split this u into finitely-many pieces u_j, where $0 \le i \le k$, with u_j supported in a set V_j and the union of the sets V_j covering the support of u. In the context of that theorem, u_0 already belongs to $W_0^{m,p}(\Omega)$.

For the other values of j, use a translate $u_{j,t}$ of $\widetilde{u_j}$ mapping x to $\widetilde{u_j}(x - ty)$ rather than to $\widetilde{u_j}(x + ty)$ as we did in the proof of Theorem 3.22. For small enough positive values of t, using $x - ty$ shifts the support of $\widetilde{u_j}$ strictly inside the domain Ω. Then $u_{j,t}$ belongs to $W^{m,p}(R^n)$ since $\widetilde{u_j}$ does. Since $u_{j,t}$ vanishes outside a compact subset of Ω, the restriction of $u_{j,t}$ to Ω belongs to $W_0^{m,p}(\Omega)$. As $t \to 0+$, these restrictions converge to u_j in $W^{m,p}(\Omega)$. Thus each piece u_j belongs to $W_0^{m,p}(\Omega)$, and so does u. ∎

5.30 There is a close connection between the existence of extension operators and imbeddings into spaces of Hölder continuous functions. For example, it is shown in [Ko] that the imbedding $W^{m,p}(\Omega) \to C^{0,1-(n/p)}(\overline{\Omega})$ implies the existence of a simple $(1, q)$-extension operator for Ω provided $q > p$.

A short survey of extension theorems for Sobolev spaces can be found in [Bu2].

An Approximation Theorem

5.31 (The Approximation Property) The following question is involved in the matter of interpolation of Sobolev spaces on order of smoothness that will play a central role in the development of Besov spaces and Sobolev spaces of fractional order in Chapter 7:

> If $0 < k < m$ does there exist a constant C such that for every $u \in W^{k,p}(\Omega)$ and every sufficiently small ϵ there exists $u_\epsilon \in W^{m,p}(\Omega)$ satisfying
>
> $$\|u - u_\epsilon\|_p \le C\epsilon^k \|u\|_{k,p}, \qquad \text{and} \qquad \|u_\epsilon\|_{m,p} \le C\epsilon^{k-m} \|u\|_{k,p}?$$

If the answer is "yes," we will say that the domain Ω has the *approximation property*. Combined with the interpolation Theorem 5.2, this property will show that $W^{k,p}(\Omega)$ is suitably intermediate between $L^p(\Omega)$ and $W^{m,p}(\Omega)$ for purposes of interpolation. In Theorem 5.33 we prove that $\mathbb{R}^n$ itself has the approximation property. It will therefore follow that any domain Ω admitting a total extension operator will have the approximation property for any choice of k and m with $0 < k < m$. In particular, therefore, a domain satisfying the strong local Lipschitz condition has the approximation property.

There are domains with the approximation property that do not satisfy the strong local Lipschitz condition. The approximation property does not prevent a domain from lying on both sides of a boundary hypersurface. In [AF4] the authors obtain the property under the assumption that Ω satisfies the "smooth cone condition," which is essentially a cone condition with the added restriction that the cone must vary smoothly from point to point. Our proof of Theorem 5.33 is a simplified version of the proof in [AF4].

We begin by stating an elementary lemma.

5.32 LEMMA If $u \in L^p(\mathbb{R}^n)$ and $B_\epsilon(x)$ is the ball of radius ϵ about x, then

$$\int_{\mathbb{R}^n} \left(\int_{B_\epsilon(x)} |u(y)|\, dy \right)^p dx \le K_n^p \epsilon^{np} \|u\|_{p,\mathbb{R}^n}^p,$$

where K_n is the volume of the unit ball $B_1(0)$.

Proof. The proof is immediate using Hölder's inequality and a change or order of integration. ∎

5.33 THEOREM (An Approximation Theorem for $\mathbb{R}^n$) If $0 < k < m$, there exists a constant C such that for $u \in W^{k,p}(\mathbb{R}^n)$ and $0 < \epsilon \le 1$ there exists u_ϵ in $C^\infty(\mathbb{R}^n)$ such that the following seminorm inequalities hold:

$$\|u - u_\epsilon\|_p \le C\epsilon^k |u|_{k,p}, \qquad \text{and}$$

$$|u_\epsilon|_{j,p} \le \begin{cases} C\|u\|_{k,p} & \text{if } j \le k-1 \\ C\epsilon^{k-j} |u|_{k,p} & \text{if } j \ge k. \end{cases}$$

In particular, $\mathbb{R}^n$ has the approximation property.

Proof. It is sufficient to establish the inequalities for $u \in C_0^\infty(\mathbb{R}^n)$ which is dense in $W^{k,p}(\mathbb{R}^n)$. We apply Taylor's formula

$$f(1) = \sum_{j=0}^{k-1} \frac{1}{j!} f^{(j)}(0) + \frac{1}{(k-1)!} \int_0^1 (1-t)^{k-1} f^{(k)}(t)\, dt$$

to the function $f(t) = u\big(tx + (1-t)y\big)$ to obtain

$$\begin{aligned} u(x) = &\sum_{|\alpha| \le k-1} \frac{1}{\alpha!} D^\alpha u(y)(x-y)^\alpha \\ &+ \sum_{|\alpha|=k} \frac{k}{\alpha!}(x-y)^\alpha \int_0^1 (1-t)^{k-1} D^\alpha u\big(tx + (1-t)y\big)\, dt. \end{aligned}$$

Now let $\phi \in C_0^\infty\big(B_1(0)\big)$ satisfy $0 \le \phi(x) \le K_0$ for all x and $\int_{\mathbb{R}^n} \phi(x)\, dx = 1$. We multiply the above Taylor formula by $\epsilon^{-n}\phi\big((x-y)/\epsilon\big)$ and integrate y over $\mathbb{R}^n$ to obtain $u(x) = u_\epsilon(x) + R(x)$ where

$$\begin{aligned} u_\epsilon(x) &= \epsilon^{-n} \sum_{|\alpha| \le k-1} \frac{1}{\alpha!} \int_{\mathbb{R}^n} \phi\left(\frac{x-y}{\epsilon}\right)(x-y)^\alpha D^\alpha u(y)\, dy \\ R(x) &= \epsilon^{-n} \sum_{|\alpha|=k} \frac{k}{\alpha!} \int_{\mathbb{R}^n} \phi\left(\frac{x-y}{\epsilon}\right)(x-y)^\alpha\, dy \\ &\qquad \times \int_0^1 (1-t)^{k-1} D^\alpha u\big(tx + (1-t)y\big)\, dt. \end{aligned}$$

We can estimate $|u(x) - u_\epsilon(x)| = |R(x)|$ by reversing the order of the double integral, substituting $z = tx + (1-t)y$ (so that $z - x = (1-t)(y-x)$ and $dz = (1-t)^n\, dy$), and reversing the order of integration again:

$$\begin{aligned} |u(x) - u_\epsilon(x)| &\le K_0 \sum_{|\alpha|=k} \frac{k}{\alpha!} \epsilon^{-n} \int_0^1 (1-t)^{-1-n}\, dt \int_{B_{\epsilon(1-t)}(x)} |x-z|^k |D^\alpha u(z)|\, dz \\ &\le K_0 \sum_{|\alpha|=k} \frac{k}{\alpha!} \epsilon^{-n} \int_{B_\epsilon(x)} |x-z|^k |D^\alpha u(z)|\, dz \int_0^{1-|z-x|/\epsilon} (1-t)^{-n-1}\, dt \\ &< K_0 \sum_{|\alpha|=k} \frac{k}{\alpha!} \epsilon^{-n} \int_{B_\epsilon(x)} |x-z|^k |D^\alpha u(z)|\, dz \\ &\le K_0 \sum_{|\alpha|=k} \frac{k}{\alpha!} \epsilon^{k-n} \int_{B_\epsilon(x)} |D^\alpha u(z)|\, dz. \end{aligned}$$

Estimating the L^p-norm of the last integral above by the previous lemma, we obtain

$$\|u(x) - u_\epsilon(x)\|_p \le K_0 \sum_{|\alpha|=k} \frac{k}{\alpha!} \epsilon^k \|D^\alpha u\|_p \le C\epsilon^k |u|_{k,p}.$$

On the other hand, we have

$$u_\epsilon(x) = \epsilon^{-n} \int_{\mathbb{R}^n} \phi\left(\frac{x-y}{\epsilon}\right) P_{k-1}(u; x, y)\, dy,$$

where

$$P_j(u; x, y) = \sum_{i=0}^{j} T_i(u; x, y),$$

$$T_j(u; x, y) = \sum_{|\alpha|=j} \frac{1}{\alpha!} D^\alpha u(y)(x-y)^\alpha.$$

It is readily verified that

$$\frac{\partial}{\partial x_i} T_j(u; x, y) = \begin{cases} T_{j-1}(D_i u; x, y) & \text{if } j > 0 \\ 0 & \text{if } j = 0 \end{cases}$$

$$\frac{\partial}{\partial x_i} P_j(u; x, y) = \begin{cases} P_{j-1}(D_i u; x, y) & \text{if } j > 0 \\ 0 & \text{if } j = 0 \end{cases}$$

$$\frac{\partial}{\partial y_i} P_j(u; x, y) = T_j(D_i u; x, y) \qquad \text{for} \quad j \ge 0.$$

Since $\frac{\partial}{\partial x_i}\phi\left(\frac{x-y}{\epsilon}\right) = -\frac{\partial}{\partial y_i}\phi\left(\frac{x-y}{\epsilon}\right)$, integration by parts gives

$$D_i u_\epsilon(x) = \epsilon^{-n} \int_{\mathbb{R}^n} \phi\left(\frac{x-y}{\epsilon}\right) P_{k-2}(D_i u; x, y)\, dy + \int_{\mathbb{R}^n} \phi\left(\frac{x-y}{\epsilon}\right) T_{k-1}(D_i u; x, y)\, dy.$$

By induction, if $|\beta| = j \le k$,

$$D^\beta u_\epsilon(x) = \epsilon^{-n} \int_{\mathbb{R}^n} \phi\left(\frac{x-y}{\epsilon}\right) P_{k-1-j}(D^\beta u; x, y)\, dy + j\epsilon^{-n} \int_{\mathbb{R}^n} \phi\left(\frac{x-y}{\epsilon}\right) T_{k-j}(D^\beta u; x, y)\, dy.$$

When $j = k$ the sums P_{k-1-j} are empty, leaving only the second line above, which becomes

$$k\epsilon^{-n} \int_{\mathbb{R}^n} \phi\left(\frac{x-y}{\epsilon}\right) T_0(D^\beta u; x, y)\, dy = k\epsilon^{-n} \int_{\mathbb{R}^n} \phi\left(\frac{x-y}{\epsilon}\right) D^\beta u(y)\, dy.$$

Write any multi-index γ with $|\gamma| > k$ in the form $\beta + \delta$ with $|\beta| = k$ to get that

$$D^\gamma u_\epsilon(x) = k\epsilon^{-n-|\delta|} \int_{\mathbb{R}^n} D^\delta \phi\left(\frac{x-y}{\epsilon}\right) D^\beta u(y)\, dy$$

in these cases. Apply the previous lemma to the various terms above to get that

$$|u_\epsilon|_{j,p} \le \begin{cases} C\, \|u\|_p & \text{if } j \le k-1 \\ C\epsilon^{k-j}\, |u|_{k,p} & \text{if } j \ge k. \end{cases}$$

In deriving this when $j < k$, expand the (nonempty) sums P_{k-1-j} to see that

$$|D^\beta u_\epsilon(x)| \le K_0 \epsilon^{-n} \int_{B_\epsilon(x)} \left[\sum_{i=0}^{k-1-j} |T_i(D^\beta u; x, y)| + j\,|T_{k-j}(D^\beta u; x, y)| \right] dy.$$

This completes the proof. ∎

Boundary Traces

5.34 Of importance in the study of boundary value problems for differential operators defined on a domain Ω is the determination of spaces of functions defined on the boundary of Ω that contain the traces $u\big|_{\text{bdry}\,\Omega}$ of functions u in $W^{m,p}(\Omega)$. For example, if $W^{m,p}(\Omega) \to C^0(\overline{\Omega})$, then clearly $u\big|_{\text{bdry}\,\Omega}$ belongs to $C(\text{bdry}\,\Omega)$. We outline below an L^q-imbedding result for such traces which can be obtained for domains with suitably smooth boundaries as a corollary of Theorem 4.12 via the use of an extension operator.

The more interesting problem of characterizing the image of $W^{m,p}(\Omega)$ under the mapping $u \to u\big|_{\text{bdry}\,\Omega}$ will be dealt with in Chapter 7. See, in particular, Theorem 7.39. The characterization is in terms of Besov spaces which are generalized Sobolev spaces of fractional order.

5.35 Let Ω be a domain in $\mathbb{R}^n$ satisfying the uniform C^m-regularity condition of Paragraph 4.10. Thus there exists a locally finite open cover $\{U_j\}$ of bdry Ω, and corresponding m-smooth transformations Ψ_j mapping $B = \{y \in \mathbb{R}^n : |y| < 1\}$ onto U_j such that $U_j \cap \text{bdry}\,\Omega = \Psi_j(B_0)$, where $B_0 = \{y \in B : y_n = 0\}$. If f is a function having support in U_j, we may define the integral of f over bdry Ω via

$$\int_{\text{bdry}\,\Omega} f(x)\, d\sigma = \int_{U_j \cap \text{bdry}\,\Omega} f(x)\, d\sigma = \int_{B_0} f \circ \Psi_j(y', 0)\, J_j(y')\, dy',$$

where $d\sigma$ is the $(n-1)$-volume element on bdry Ω, $y' = (y_1, \ldots, y_{n-1})$, and, if $x = \Psi_j(y)$, then

$$J_j(y') = \left[\sum_{k=1}^{n} \left(\frac{\partial(x_1, \ldots, \hat{x}_k, \ldots, x_n)}{\partial(y_1, \ldots, y_{n-1})} \right)^2 \right]^{1/2} \Bigg|_{y_n=0} .$$

If f is an arbitrary function defined on $\mathbb{R}^n$, we may set

$$\int_{\text{bdry}\,\Omega} f(x)\, d\sigma = \sum_j \int_{\text{bdry}\,\Omega} f(x) v_j(x)\, d\sigma,$$

where $\{v_j\}$ is a partition of unity for bdry Ω subordinate to $\{U_j\}$.

5.36 THEOREM (A Boundary Trace Imbedding Theorem) Let Ω be a domain in $\mathbb{R}^n$ satisfying the uniform C^m-regularity condition, and suppose there exists a simple (m, p)-extension operator E for Ω. Also suppose that $mp < n$ and $p \le q \le p^* = (n-1)p/(n-mp)$. Then

$$W^{m,p}(\Omega) \to L^q\,(\text{bdry}\,\Omega)\,. \tag{28}$$

If $mp = n$, then imbedding (28) holds for $p \le q < \infty$.

Proof. Imbedding (28) should be interpreted in the following sense. If $u \in W^{m,p}(\Omega)$, then Eu has a trace on bdry Ω in the sense described in Paragraph 4.2, and $\|Eu\|_{0,q,\text{bdry}\,\Omega} \le K\,\|u\|_{m,p,\Omega}$ with K independent of u. Note that since $C_0(R^n)$ is dense in $W^{m,p}(\Omega)$, $\|Eu\|_{0,q,\text{bdry}\,\Omega}$ is independent of the particular extension operator E used.

We prove the special case $mp < n$, $q = p^* = (n-1)p/(n-mp)$ of the theorem; the other cases are similar. We use the notations of the previous Paragraph.

There is a constant K_1 such that for every $u \in W^{m,p}(\Omega)$,

$$\|Eu\|_{m,p,\mathbb{R}^n} \le K_1\,\|u\|_{m,p,\Omega}\,.$$

By the uniform C^m-regularity condition (see Paragraph 4.10) there exists a constant K_2 such that for each j and every $y \in B$ we have $x = \Psi_j(y) \in U_j$,

$$|J_j(y')| \le K_2, \quad \text{and} \quad \left| \frac{\partial(y_1, \ldots, y_n)}{\partial(x_1, \ldots, x_n)} \right| \le K_2.$$

Noting that $0 \le v_j(x) \le 1$ on $\mathbb{R}^n$, and using imbedding (4) of Theorem 4.12 applied over B, we have, for $u \in W^{m,p}(\Omega)$,

$$\int_{\text{bdry}\,\Omega} |Eu(x)|^q\, d\sigma \le \sum_j \int_{U_j \cap \text{bdry}\,\Omega} |Eu(x)|^q\, d\sigma$$

$$
\begin{aligned}
&\le K_2 \sum_j \left\| Eu \circ \Psi_j \right\|_{0,q,B_0}^q \\
&\le K_3 \sum_j \left(\left\| Eu \circ \Psi_j \right\|_{m,p,B}^p \right)^{q/p} \\
&\le K_4 \left(\sum_j \|Eu\|_{m,p,U_j}^p \right)^{q/p} \\
&\le K_4 R \, \|Eu\|_{m,p,\mathbb{R}^n}^q \\
&\le K_5 \, \|u\|_{m,p,\Omega}^q .
\end{aligned}
$$

The second last inequality above makes use of the finite intersection property possessed by the cover $\{U_j\}$. The constant K_4 is independent of j because $|D^\alpha \Psi_{j,i}(y)| \le$ const for all i, j, where $\Psi_j = (\Psi_{j,1}, \ldots, \Psi_{j,n})$. This completes the proof. ∎

Finally, we show that functions in $W^{m,p}(\Omega)$ belong to $W_0^{m,p}(\Omega)$ if and only if they have suitably trivial boundary traces.

5.37 THEOREM (Trivial Traces) Under the same hypotheses as Theorem 5.36, a function u in $W^{m,p}(\Omega)$ belongs to $W_0^{m,p}(\Omega)$ if and only the boundary traces of its derivatives of order less than m all coincide with the 0-function.

Proof. Every function in $C_0^\infty(\Omega)$ has trivial boundary trace, and so do all derivatives of such functions. Since the trace mapping is a continuous linear operator from $W^{m,p}(\Omega)$ to $W^{m-1,p}(\text{bdry}\,\Omega)$, all functions in $W_0^{m,p}(\Omega)$ have trivial boundary traces, and so do their derivatives of order less than m.

To prove the converse, we suppose that $u \in W^{m,p}(\Omega)$ and that u and its derivatives or order less than m have trivial boundary traces. Localization and a suitable change of variables reduces matters to the case where Ω is the half-space $\{x \in R^n : x_n > 0\}$. We then show that the zero-extension $\tilde{u}$ must belong to $W^{m,p}(\mathbb{R}^n)$, forcing u to belong to $W_0^{m,p}(\Omega)$ by Theorem 5.29.

In fact, we claim that if $u \in W^{m,p}(\Omega)$ has trivial boundary traces for u and its derivatives of order less than m, then the distributional derivatives $D^\alpha \tilde{u}$ of order at most m coincide with the zero-extensions $\widetilde{D^\alpha u}$. To verify this, approximate the integrals

$$
\int_{\mathbb{R}^n} \tilde{u}(x) D^\alpha \phi(x)\, dx \qquad \text{and} \qquad (-1)^{|\alpha|} \int_{\mathbb{R}^n} \widetilde{D^\alpha u}(x) \phi(x)\, dx \tag{29}
$$

by approximating u with functions v_j in $C^\infty(\overline{\Omega})$, without requiring that these approximations have trivial traces.

Let e_n be the unit vector $(0, \ldots, 0, 1)$. Since $v_j \in C^\infty(\overline{\Omega})$, integrating by parts with respect to the other variables and then with respect to x_n shows that the difference between the integrals

$$\int_{\mathbb{R}^n} \tilde{v}_j(x) D^\alpha \phi(x)\, dx \qquad \text{and} \qquad (-1)^{|\alpha|} \int_{\mathbb{R}^n} \widetilde{D^\alpha v_j}(x) \phi(x)\, dx$$

is a finite alternating sum of integrals of the form

$$\int_{\mathbb{R}^{n-1}} D^{\alpha - k e_n} v_j(x_1, \ldots, x_{n-1}, 0) D_n^{k-1} \phi(x_1, \ldots, x_{n-1}, 0)\, dx_1 \cdots dx_{n-1} \qquad (30)$$

with $k > 0$. Choose the sequence $\{v_j\}$ to converge to u in $W^{m,p}(\Omega)$. For each multi-index β with $\beta < \alpha$, the trace of $D^\beta v_j$ will converge in $L^p(\mathbb{R}^{n-1})$ to the trace of $D^\beta u$, that is to 0 in that space. Since the restriction of $D_n^{k-1}\phi$ to R^{n-1} belongs to $L^{p'}(\mathbb{R}^{n-1})$, each of the integrals in (30) tends to 0 as $j \to \infty$.

It follows that the two integrals in (29) are equal, and that $\tilde{u} \in W^{m,p}(\mathbb{R}^n)$. This completes the proof. ∎

6

COMPACT IMBEDDINGS OF SOBOLEV SPACES

The Rellich-Kondrachov Theorem

6.1 (Restricted Imbeddings) Let Ω be a domain in $\mathbb{R}^n$ and let Ω_0 be a subdomain of Ω. Let $X(\Omega)$ denote any of the possible target spaces for imbeddings of $W^{m,p}(\Omega)$, that is, $X(\Omega)$ is a space of the form $C_B^j(\Omega)$, $C^j(\overline{\Omega})$, $C^{j,\lambda}(\overline{\Omega})$, $L^q(\Omega_k)$, or $W^{j,q}(\Omega_k)$, where Ω_k, $1 \le k \le n$, is the intersection of Ω with a k-dimensional plane in $\mathbb{R}^n$. Since the linear restriction operator $i_{\Omega_0} : u \to u\big|_{\Omega_0}$ is bounded from $X(\Omega)$ into $X(\Omega_0)$ (in fact $\|i_{\Omega_0}u; X(\Omega_0)\| \le \|u; X(\Omega)\|$) any imbedding of the form

$$W^{m,p}(\Omega) \to X(\Omega) \tag{1}$$

can be composed with this restriction to yield the imbedding

$$W^{m,p}(\Omega) \to X(\Omega_0) \tag{2}$$

and (2) has imbedding constant no larger than (1).

6.2 (Compact Imbeddings) Recall that a set A in a normed space is precompact if every sequence of points in A has a subsequence converging in norm to an element of the space. An operator between normed spaces is called compact if it maps bounded sets into precompact sets, and is called completely continuous if it is continuous and compact. (See Paragraph 1.24; for linear operators compactness and complete continuity are equivalent.) In this chapter we are concerned with the compactness of imbedding operators which are continuous whenever they exist, and so are completely continuous whenever they are compact.

If Ω satisfies the hypotheses of the Sobolev imbedding Theorem 4.12 and if Ω_0 is a bounded subset of Ω, then, with the exception of certain extreme cases, all the restricted imbeddings (1) corresponding to imbeddings asserted in Theorem 4.12 are compact. The most important of these compact imbedding results originated in a lemma of Rellich [Re] and was proved specifically for Sobolev spaces by Kondrachov [K]. Such compact imbeddings have many important applications in analysis, especially to showing that linear elliptic partial differential equations defined over bounded domains have discrete spectra. See, for example, [EE] and [ET] for such applications and further refinements.

We summarize the various compact imbeddings of $W^{m,p}(\Omega)$ in the following theorem

6.3 THEOREM (The Rellich-Kondrachov Theorem) Let Ω be a domain in $\mathbb{R}^n$, let Ω_0 be a bounded subdomain of Ω, and let Ω_0^k be the intersection of Ω_0 with a k-dimensional plane in $\mathbb{R}^n$. Let $j \ge 0$ and $m \ge 1$ be integers, and let $1 \le p < \infty$.

PART I If Ω satisfies the cone condition and $mp \le n$, then the following imbeddings are compact:

$$W^{j+m,p}(\Omega) \to W^{j,q}\left(\Omega_0^k\right) \quad \text{if } 0 < n - mp < k \le n \text{ and } 1 \le q < kp/(n-mp), \tag{3}$$

$$W^{j+m,p}(\Omega) \to W^{j,q}\left(\Omega_0^k\right) \quad \text{if } n = mp,\ 1 \le k \le n \text{ and } 1 \le q < \infty. \tag{4}$$

PART II If Ω satisfies the cone condition and $mp > n$, then the following imbeddings are compact:

$$W^{j+m,p}(\Omega) \to C_B^j(\Omega_0) \tag{5}$$

$$W^{j+m,p}(\Omega) \to W^{j,q}\left(\Omega_0^k\right) \quad \text{if } 1 \le q < \infty. \tag{6}$$

PART III If Ω satisfies the strong local Lipschitz condition, then the following imbeddings are compact:

$$W^{j+m,p}(\Omega) \to C^j(\overline{\Omega_0}) \quad \text{if } mp > m, \tag{7}$$

$$W^{j+m,p}(\Omega) \to C^{j,\lambda}(\overline{\Omega_0}) \quad \text{if } mp > n \ge (m-1)p \text{ and } 0 < \lambda < m - (n/p). \tag{8}$$

PART IV If Ω is an arbitrary domain in $\mathbb{R}^n$, the imbeddings (3)–(8) are compact provided $W^{j+m,p}(\Omega)$ is replaced by $W_0^{j+m,p}(\Omega)$.

6.4 REMARKS

1. Note that if Ω is bounded, we may have $\Omega_0 = \Omega$ in the statement of the theorem.
2. If X, Y, and Z are spaces for which we have the imbeddings $X \to Y$ and $Y \to Z$, and if one of these imbeddings is compact, then the composite imbedding $X \to Z$ is compact. Thus, for example, if $Y \to Z$ is compact, then any sequence $\{u_j\}$ bounded in X will be bounded in Y and will therefore have a subsequence $\{u'_j\}$ convergent in Z.
3. Since the extension operator $u \to \tilde{u}$, where $\tilde{u}(x) = u(x)$ if $x \in \Omega$ and $\tilde{u}(x) = 0$ if $x \notin \Omega$, defines an imbedding $W_0^{j+m,p}(\Omega) \to W^{j+m,p}(\mathbb{R}^n)$ by Lemma 3.27, Part IV of Theorem 6.3 follows from application of Parts I–III to $\mathbb{R}^n$.
4. In proving the compactness of any of the imbeddings (3)–(8) it is sufficient to consider only the case $j = 0$. Suppose, for example, that (3) has been proven compact if $j = 0$. For $j \geq 1$ and $\{u_i\}$ a bounded sequence in $W^{j+m,p}(\Omega)$ it is clear that $\{D^\alpha u_i\}$ is bounded in $W^{m,p}(\Omega)$ for each α such that $|\alpha| \leq j$. Hence $\{D^\alpha u_i|_{\Omega_0^k}\}$ is precompact in $L^q(\Omega_0^k)$ with q specified as in (3). It is possible, therefore, to select (by finite induction) a subsequence $\{u'_i\}$ of $\{u_i\}$ for which $\{D^\alpha u'_i|_{\Omega_0^k}\}$ converges in $L^q(\Omega_0^k)$ for each α such that $|\alpha| \leq j$. Thus $\{u'_i|_{\Omega_0^k}\}$ converges in $W_0^{j,q}(\Omega_0^k)$ and (3) is compact.
5. Since Ω_0 is bounded, $C_B^0(\Omega_0^k) \to L^q(\Omega_0^k)$ for $1 \leq q \leq \infty$; in fact $\|u\|_{0,q,\Omega_0^k} \leq \|u; C_B^0(\Omega_0^k)\|[\mathrm{vol}(\Omega_0^k)]^{1/q}$. Thus the compactness of (6) (for $j = 0$) follows from that of (5).
6. For the purpose of proving Theorem 6.3 the bounded subdomain Ω_0 of Ω may be assumed to satisfy the cone condition in Ω does. If C is a finite cone determining the cone condition for Ω, let $\tilde{\Omega}$ be the union of all finite cones congruent to C, contained in Ω and having nonempty intersection with Ω_0. Then $\Omega_0 \subset \tilde{\Omega} \subset \Omega$ and $\tilde{\Omega}$ is bounded and satisfies the cone condition. If $W^{m,p}(\Omega) \to X(\tilde{\Omega})$ is compact, then so is $W^{m,p}(\Omega) \to X(\Omega_0)$ by restriction.

6.5 (Proof of Theorem 6.3, Part III) If $mp > n \geq (m-1)p$ and if $0 < \lambda < m - (n/p)$, then there exists μ such that $\lambda < \mu < m - (n/p)$. Since Ω_0 is bounded, the imbedding $C^{0,\mu}(\overline{\Omega_0}) \to C^{0,\lambda}(\overline{\Omega_0})$ is compact by Theorem 1.34. Since $W^{m,p}(\Omega) \to C^{0,\mu}(\overline{\Omega}) \to C^{0,\mu}(\overline{\Omega_0})$ by Theorem 4.12 and restriction, imbedding (8) is compact for $j = 0$ by Remark 6.4(2).

If $mp > n$, let j^* be the nonnegative integer satisfying the inequalities $(m - j^*)p > n \geq (m - j^* - 1)p$. Then we have the imbedding chain

$$W^{m,p}(\Omega) \to W^{m-j^*,p}(\Omega) \to C^{0,\mu}(\overline{\Omega_0}) \to C(\overline{\Omega_0}) \qquad (9)$$

where $0 < \mu < m - j^* - (n/p)$. The last imbedding in (9) is compact by Theorem 1.34. Thus (7) is compact for $j = 0$. ∎

6.6 (Proof of Theorem 6.3, Part II) As noted in Remark 6.4(6), Ω_0 may be assumed to satisfy the cone condition. Since Ω_0 is bounded it can, by Lemma 4.22 be written as a finite union, $\Omega_0 = \bigcup_{k=1}^{M} \Omega_k$, where each Ω_k satisfies the strong local Lipschitz condition. If $mp > n$, then

$$W^{m,p}(\Omega) \to W^{m,p}(\Omega_k) \to C(\overline{\Omega_k}),$$

the latter imbedding being compact as proved above. If $\{u_i\}$ is a sequence bounded in $W^{m,p}(\Omega)$, we may select by finite induction on k a subsequence $\{u_i'\}$ whose restriction to Ω_k converges in $C(\overline{\Omega_k})$ for each k, $1 \le k \le M$. But this subsequence then converges in $C_B^0(\Omega_0)$, so proving that (5) is compact for $j = 0$. Therefore (6) is also compact by Remark 6.4(5). ∎

6.7 LEMMA Let Ω be a domain in $\mathbb{R}^n$, Ω_0 a subdomain of Ω, and Ω_0^k the intersection of Ω_0 with a k-dimensional plane in $\mathbb{R}^n$ $(1 \le k \le n)$. Let $1 \le q_1 < q_0$ and suppose that

$$W^{m,p}(\Omega) \to L^{q_0}\left(\Omega_0^k\right)$$

and

$$W^{m,p}(\Omega) \to L^{q_1}\left(\Omega_0^k\right) \qquad \text{compactly.}$$

If $q_1 \le q < q_0$, then

$$W^{m,p}(\Omega) \to L^{q}\left(\Omega_0^k\right) \qquad \text{compactly.}$$

Proof. Let $\lambda = q_1(q_0 - q)/q(q_0 - q_1)$ and $\mu = q_0(q - q_1)/q(q_0 - q_1)$. Then $\lambda > 0$ and $\mu \ge 0$. By Hölder's inequality there exists a constant K such that for all $u \in W^{m,p}(\Omega)$,

$$\|u\|_{0,q,\Omega_0^k} \le \|u\|_{0,q_1,\Omega_0^k}^{\lambda} \, \|u\|_{0,q_0,\Omega_0^k}^{\mu} \le K \, \|u\|_{0,q_1,\Omega_0^k}^{\lambda} \, \|u\|_{m,p,\Omega}^{\mu}.$$

A sequence bounded in $W^{m,p}(\Omega)$ has a subsequence which converges in $L^{q_1}(\Omega_0^k)$ and is therefore a Cauchy sequence in that space. Applying the inequality above to differences between terms of this sequence shows that it is also a Cauchy sequence in $L^q(\Omega_0^k)$, so the imbedding of $W^{m,p}(\Omega)$ into $L^q(\Omega_0^k)$ is compact. ∎

6.8 (Proof of Theorem 6.3, Part I) First we deal with (the case $j = 0$ of) imbedding (3). Assume for the moment that $k = n$ and let $q_0 = np/(n - mp)$. In order to prove that the imbedding

$$W^{m,p}(\Omega) \to L^q(\Omega_0), \qquad 1 \le q < q_0, \qquad (10)$$

is compact, it sufficed, by Lemma 6.7, to do so only for $q = 1$. For $j = 1, 2, 3, \ldots$ let

$$\Omega_j = \{x \in \Omega : \operatorname{dist}(x, \operatorname{bdry}\Omega) > 2/j\}.$$

Let S be a set of functions bounded in $W^{m,p}(\Omega)$. We show that S (when restricted to Ω_0) is precompact in $L^1(\Omega_0)$ by showing that S satisfies the conditions of Theorem 2.32. Accordingly, let $\epsilon > 0$ be given and for each $u \in W^{m,p}(\Omega)$ set

$$\tilde{u}(x) = \begin{cases} u(x) & \text{if } x \in \Omega_0 \\ 0 & \text{otherwise.} \end{cases}$$

By Hölder's inequality and since $W^{m,p}(\Omega) \to L^{q_0}(\Omega_0)$, we have

$$\begin{aligned} \int_{\Omega_0-\Omega_j} |u(x)|\,dx &\le \left(\int_{\Omega_0-\Omega_j} |u(x)|^{q_0}\,dx\right)^{1/q_0} \left(\int_{\Omega_0-\Omega_j} 1\,dx\right)^{1-1/q_0} \\ &\le K_1 \|u\|_{m,p,\Omega} \left[\operatorname{vol}(\Omega_0-\Omega_j)\right]^{1-1/q_0}, \end{aligned}$$

with K_1 independent of u. Since $q_0 > 1$ and Ω_0 has finite volume, j may be selected large enough to ensure that for every $u \in S$,

$$\int_{\Omega_0-\Omega_j} |u(x)|\,dx < \epsilon$$

and also, for every $h \in \mathbb{R}^n$,

$$\int_{\Omega_0-\Omega_j} |\tilde{u}(x+h) - \tilde{u}(x)|\,dx < \frac{\epsilon}{2}.$$

Now if $|h| < 1/j$, then $x + th \in \Omega_{2j}$ provided $x \in \Omega_j$ and $0 \le t \le 1$. If $u \in C^\infty(\Omega)$, it follows that

$$\begin{aligned} \int_{\Omega_j} |u(x+h) - u(x)|\,dx &\le \int_{\Omega_j} dx \int_0^1 \left|\frac{d}{dt}u(x+th)\right| dt \\ &\le |h| \int_0^1 dt \int_{\Omega_{2j}} |\operatorname{grad} u(y)|\,dy \\ &\le |h|\ \|u\|_{1,1,\Omega_0} \le K_2|h|\ \|u\|_{m,p,\Omega}, \end{aligned}$$

where K_2 is independent of u. Since $C^\infty(\Omega)$ is dense in $W^{m,p}(\Omega)$, this estimate holds for any $u \in W^{m,p}(\Omega)$. Hence if $|h|$ is sufficiently small, we have

$$\int_{\Omega_0} |\tilde{u}(x+h) - \tilde{u}(x)|\,dx < \epsilon.$$

Hence S is precompact in $L^1(\Omega_0)$ by Theorem 2.32 and imbedding (10) is compact.

Next suppose that $k < n$ and $p > 1$. The Sobolev Imbedding Theorem 4.12 assures us that $W^{m,p}(\Omega) \to L^{kp/(n-mp)}(\Omega_0^k)$. For any $q < kp/(n-mp)$ we can choose r such that $1 \le r < p, n-mr < k$, and $q \le kr/(n-mr) < kp/(n-mp)$. Since Ω_0 is bounded, the imbeddings

$$W^{m,p}(\Omega) \to W^{m,p}(\Omega_0) \to W^{m,r}(\Omega_0)$$

exist. By Theorem 5.10 we have

$$\begin{aligned}\|u\|_{q,\Omega_0^k} &\le K_1 \|u\|_{kr/(n-mr),\Omega_0^k} \\ &\le K_2 \|u\|^{1-\theta}_{nr/(n-mr),\Omega_0} \|u\|^{\theta}_{m,r,\Omega_0} \\ &\le K_3 \|u\|^{1-\theta}_{nr/(n-mr),\Omega_0} \|u\|^{\theta}_{m,p,\Omega},\end{aligned}$$

where K_j and θ are constants (independent of $u \in W^{m,p}(\Omega)$) and θ satisfies $0 < \theta < 1$. Since $nr/(n-mr) < np/(n-mp)$, a sequence bounded in $W^{m,p}(\Omega)$ must have a subsequence convergent in $L^{nr/(n-mr)}(\Omega_0)$ by the earlier part of this proof. That sequence is therefore a Cauchy sequence in $L^{nr/(n-mr)}(\Omega_0)$, and by the above inequality it is therefore a Cauchy sequence in $L^q(\Omega_0^k)$, so the imbedding $W^{m,p}(\Omega) \to L^q(\Omega_0^k)$ is compact and so is $W^{m,p}(\Omega) \to L^1(\Omega_0^k)$.

If $p = 1$ and $0 \le n - m < k < n$, then necessarily $m \ge 2$. Composing the continuous imbedding $W^{m,1}(\Omega) \to W^{m-1,r}(\Omega)$, where $r = n/(n-1) > 1$, with the compact imbedding $W^{m-1,r}(\Omega) \to L^1(\Omega_0^k)$, (which is compact because $k \ge n-(m-1) > n-(m-1)r$), completes the proof of the compactness of (3).

To prove that imbedding (4) is compact we proceed as follows. If $n = mp$, $p > 1$, and $1 \le q < \infty$, then we may select r so that $1 \le r < p$, $k > n - mr > 0$, and $kr/(n-mr) > q$. Assuming again that Ω_0 satisfies the cone condition, we have

$$W^{m,p}(\Omega) \to W^{m,r}(\Omega_0) \to L^q(\Omega_0^k).$$

The latter imbedding is compact by (3). If $p = 1$ and $n = m \ge 2$, then, setting $r = n/(n-1) > 1$ so that $n = (n-1)r$, we have for $1 \le q < \infty$,

$$W^{n,1}(\Omega) \to W^{n-1,r}(\Omega) \to L^q(\Omega_0^k),$$

the latter imbedding being compact as shown immediately above. Finally, if $n = m = p = 1$, then $k = 1$ also. Letting $q_0 > 1$ be arbitrarily chosen we prove the compactness of $W^{1,1}(\Omega) \to L^1(\Omega_0)$ exactly as in the case $k = n$ considered at the beginning of this proof. Since $W^{1,1}(\Omega) \to L^q(\Omega_0)$ for $1 \le q < \infty$, all these imbeddings are compact by Lemma 6.7. ∎

Two Counterexamples

6.9 (Quasibounded Domains) We say that an unbounded domain $\Omega \subset \mathbb{R}^n$ is *quasibounded* if

$$\lim_{\substack{x\in\Omega \\ |x|\to\infty}} \operatorname{dist}(x, \operatorname{bdry}\Omega) = 0.$$

An unbounded domain is not quasibounded if and only if it contains infinitely many pairwise disjoint congruent balls.

6.10 Two obvious questions arise from consideration of the statement of the Rellich-Kondrachov Theorem 6.3. First, can the theorem be extended to cover unbounded Ω_0? Second, can the *extreme cases*

$$W^{j+m,p}(\Omega) \to W^{j,q}\left(\Omega_0^k\right), \qquad 0 < n-, p < k \le n, \quad q = kp/(n-mp)$$

and

$$W^{j+m,p}(\Omega) \to C^{j,\lambda}(\overline{\Omega_0}), \qquad mp > n > (m-1)p, \quad \lambda = m - (n/p)$$

ever be compact? The first of these questions will be investigated later in this chapter. For the moment though we show that the answer is negative if $k = n$ and Ω_0 is not quasibounded. However, the situation changes (see [Lp]) for subspaces of symmetric functions.

6.11 EXAMPLE Let Ω be an unbounded domain in $\mathbb{R}^n$ that is not quasibounded. Then there exists a sequence $\{B_i\}$ of mutually disjoint open balls contained in Ω and all having the same positive radius. Let $\phi_1 \in C_0^\infty(B_1)$ satisfy $\|\phi_1\|_{j,p,B_1} = A_{j,p} > 0$ for each $j = 0, 1, 2, \ldots$ and each $p \ge 1$. Let ϕ_i be a translate of ϕ_1 having support in B_i. Then $\{\phi_i\}$ is a bounded sequence in $W_0^{m,p}(\Omega)$ for any fixed m and p. But for any q,

$$\|\phi_i - \phi_k\|_{j,q,\Omega} = \left(\|\phi_i\|_{j,q,B_i}^q + \|\phi_k\|_{j,q,B_i}^q\right)^{1/q} = 2^{1/q} A_{j,q} > 0$$

so that $\{\phi_i\}$ cannot have a sequence converging in $W^{j,q}(()\,\Omega)$ for any $j \ge 0$. Thus no compact imbedding of the form $W_0^{j+m,p}(\Omega) \to W^{j,q}(\Omega)$ is possible. The non-compactness of the other imbeddings of Theorem 6.3 is proved similarly. ∎

Now we provide an example showing that the answer to the second question raised in Paragraph 6.10 is always negative.

6.12 EXAMPLE Let integers j, m, n be given with $j \ge 0$ and $m, n \ge 1$. Let $p \ge 1$. If $mp < n$, let k be an integer such that $n - mp < k \le n$ and let $q = kp/(n - mp)$. If $(m-1)p < n < mp$, let $\lambda = m - (n/p)$. Let Ω

be a domain in $\mathbb{R}^n$ and let Ω_0 be a nonempty bounded subdomain of Ω having nonempty intersection Ω_0^k with a k-dimensional plane H in $\mathbb{R}^n$ which, without loss of generality, we can take to be the plane $\mathbb{R}^k$ spanned by the $x_1, x_2, \dots, x_k$ coordinate axes. We show that the imbeddings

$$W^{j+m,p}(\Omega) \to W^{j,q}(\Omega_0^k) \qquad \text{if} \quad mp < n \tag{11}$$

$$W^{j+m,p}(\Omega) \to C^{j,\lambda}(\overline{\Omega_0}) \qquad \text{if} \quad (m-1)p \le n < mp \tag{12}$$

cannot be compact.

Let $B_r(x)$ be the open ball of radius r in $\mathbb{R}^n$ centred at x and let ϕ be a nontrivial function in $C_0^\infty(B_1(0))$. Let $\{a_i\}$ be a sequence of distinct points in Ω_0^k, and let $B_i = B_{r_i}(a_i)$ where the positive radii r_i satisfy $r_i \le 1$ and are chosen so that the balls B_i are pairwise disjoint and contained in Ω_0. We define a scaled, translated dilation ϕ_i of ϕ with support in B_i by

$$\phi_i(x) = r_i^{j+m-(n/p)}\phi(y), \qquad \text{where} \quad x = a_i + r_i y.$$

The functions ϕ_i have disjoint supports in Ω_0 and, since $D^\alpha \phi_i(x) = r^{-|\alpha|} D^\alpha \phi(y)$ and $dx = r_i^n\, dy$, we have, for $|\alpha| \le j+m$,

$$\int_\Omega |D^\alpha \phi_i(x)|^p\, dx = r_i^{(j+m-|\alpha|)p} \int_\Omega |\mathscr{D}^\alpha \phi(y)|^p\, dy.$$

Therefore, $\{\phi\}$ is bounded in $W^{j+m,p}(\Omega)$.

On the other hand, $dx_1 \cdots dx_k = r_i^k\, dy_1 \cdots dy_k$, so that if $|\alpha| = j$, then

$$\int_{\Omega_0^k} |D^\alpha \phi_i(x)|^q\, dx_1 \cdots dx_k = r_i^{k+q[m-(n/p)]} \int_{\mathbb{R}^k} |D^\alpha \phi(y)|^q\, dy_1 \cdots dy_k.$$

Since $k + q\big[m - (n/p)\big] = 0$, this shows that

$$\|\phi_i\|_{j,q,\Omega_0^k} \ge |\phi_i|_{j,q,\Omega_0^k} = C_1\, |\phi|_{j,q,\mathbb{R}^k} > 0$$

for all i, and $\{\phi_i\}$ is bounded away from zero in $W^{j,q}(\Omega_0^k)$. The disjointness of the supports of the functions ϕ_i now implies that $\{\phi\}$ can have no subsequence converging in $W^{j,q}(\Omega_0^k)$, so the imbedding (11) cannot be compact.

Now suppose that $(m-1)p \le n < mp$. Let a be a point in $B_1(0)$ and β be a particular multiindex satisfying $|\beta| = j$ such that $|D^\beta \phi(a)| = C_2 > 0$. Let $b_i = a_i + r_i a$ and let c_i be the point on the boundary of B_i closest to b_i. We have

$$\big|D^\beta \phi_i(b_i)\big| = r_i^{m-(n/p)} C_2 = r_i^\lambda C_2,$$

and, since $D^\beta \phi_i(c_i) = 0$,

$$\|\phi_i\,;C^{j,\lambda}(\overline{\Omega_0})\| \ge \frac{|D^\beta \phi_i(b_i) - D^\beta \phi_i(c_i)|}{|b_i - c_i|^\lambda} = C_2 > 0.$$

Again, this precludes the existence of a subsequence of of $\{\phi_i\}$ convergent in $C^{j,\lambda}(\overline{\Omega_0})$, so the imbedding (12) cannot be compact. ∎

6.13 REMARK Observe that the above examples in fact showed that no imbeddings of $W_0^{j+m,p}(\Omega)$, not just of the larger space $W^{j+m,p}(\Omega)$, into the appropriate target space can be compact. We now examine the possibility of obtaining compact imbeddings of $W_0^{m,p}(\Omega)$ for certain unbounded domains.

Unbounded Domains — Compact Imbeddings of $W_0^{m,p}(\Omega)$

6.14 Let Ω be an unbounded domain in $\mathbb{R}^n$. We shall be concerned below with determining whether the imbedding

$$W_0^{m,p}(\Omega) \to L^p(\Omega) \tag{13}$$

is compact. If it is, then it will follow by Remark 6.4(4), Lemma 6.7, and the second part of the proof in Paragraph 6.8 that the imbeddings

$$W_0^{j+m,p}(\Omega) \to W^{j,q}(\Omega_k), \qquad 0 < n - mp < k \le n, \quad p \le q < kp/(n - mp),$$

$$W_0^{j+m,p}(\Omega) \to W^{j,q}(\Omega_k), \qquad n = mp, \quad 1 \le k \le n, p \le q < \infty$$

are also compact. See Theorem 6.28 for the corresponding compactness of imbeddings into continuous function spaces.

As was shown in Example 6.11, imbedding (13) cannot be compact unless Ω is quasibounded. In Theorem 6.16 we give a geometric condition on Ω that is sufficient to guarantee the compactness of (13), and in Theorem 6.19 we give an analytic condition that is necessary and sufficient for the compactness of (13). Both theorems are from [A2].

6.15 Let Ω_r denote the set $\{x \in \Omega : |x| \ge r\}$. In the following discussion any cube H referred to will have its edges parallel to the coordinate axes. For a domain Ω, a cube H, and an integer ν satisfying $1 \le \nu \le n$, we define the quantity $\mu_{n-\nu}(H, \Omega)$ to be the maximum of the $(n-\nu)$-measure of $P(H - \Omega)$ taken over all projections P onto $(n-\nu)$-dimensional faces of H.

6.16 THEOREM Let ν be an integer such that $1 \le \nu \le n$ and $mp > \nu$ (or $m = p = \nu = 1$). Suppose that for every $\epsilon > 0$ there exist numbers h and r with $0 < h \le 1$ and $r \ge 0$ such that for every cube $H \subset \mathbb{R}^n$ having side h and nonempty intersection with Ω_r we have

$$\frac{\mu_{n-\nu}(H, \Omega)}{h^{n-\nu}} \ge \frac{h^p}{\epsilon}.$$

Then imbedding (13) is compact.

6.17 REMARKS

1. We will deduce this theorem from Theorem 6.19 later in this section.
2. The above theorem shows that for quasibounded Ω the compactness of (13) may depend in an essential way on the dimension of bdry Ω.
3. For $\nu = n$, the condition of Theorem 6.16 places only the minimum restriction of quasiboundedness on Ω; if $mp > n$ then (13) is compact for any quasibounded Ω. It can also be shown that if $p > 1$ and Ω is quasibounded with boundary having no finite accumulation points, then (13) cannot be compact unless $mp > n$.
4. If $\nu = 1$, the condition of Theorem 6.16 places no restrictions on m and p but requires that bdry Ω be "essentially $(n-1)$-dimensional." Any quasibounded domain whose boundary consists of reasonably regular $(n-1)$-dimensional surfaces will satisfy that condition. An example of such a domain is the "spiny urchin" of Figure 4, a domain in $\mathbb{R}^2$ obtained by deleting from the plane the union of the sets S_k, $(k = 1, 2, \ldots)$, specified in polar coordinates by

$$S_k = \left\{(r, \theta) : r \geq k,\ \theta = n\pi/2^k,\ n = 1, 2, \ldots, 2^{k+1}\right\}.$$

Note that this domain, though quasibounded, is simply connected and has empty exterior.

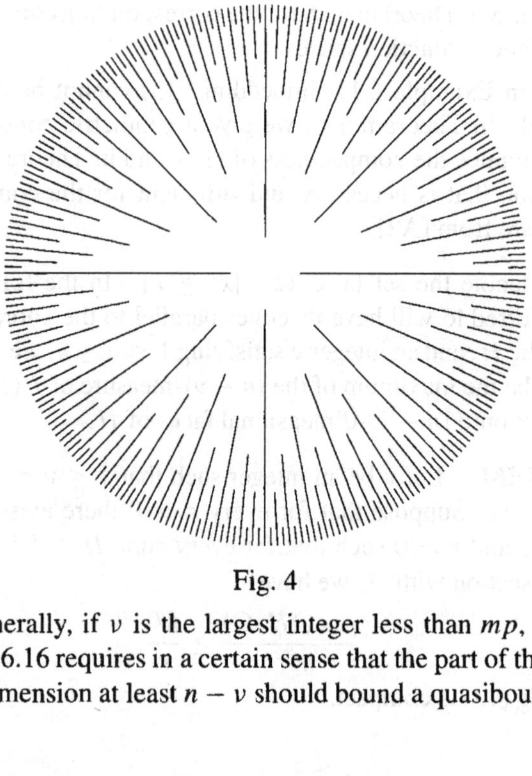

Fig. 4

5. More generally, if ν is the largest integer less than mp, the condition of Theorem 6.16 requires in a certain sense that the part of the boundary of Ω having dimension at least $n - \nu$ should bound a quasibounded domain.

6.18 (A Definition of Capacity) Let H be a cube of edge length h in $\mathbb{R}^n$ and let E be a closed subset of H. Given m and p we define a functional $I_H^{m,p}$ on $C^\infty(H)$ by

$$I_H^{m,p}(u) = \sum_{1\le j\le m} h^{jp}\,|u|_{j,p,H}^p = \sum_{1\le|\alpha|\le m} h^{|\alpha|p}\int_H |D^\alpha u(x)|^p\,dx.$$

Let $C^\infty(H,E)$ denote the set of all nontrivial functions $u \in C^\infty(H)$ that vanish identically in a neighbourhood of E. We define the (m,p)-capacity $Q^{m,p}(H,E)$ of E in H by

$$Q^{m,p}(H,E) = \inf\left\{\frac{I_H^{m,p}(u)}{\|u\|_{0,p,H}^p} : u \in C^\infty(H,E)\right\}.$$

Clearly $Q^{m,p}(H,E) \le Q^{m+1,p}(H,E)$ and, whenever $E \subset F \subset H$, we have $Q^{m,p}(H,E) \le Q^{m,p}(H,F)$.

The following theorem characterizes those domains for which imbedding (13) is compact in terms of this capacity.

6.19 THEOREM Imbedding (13) is compact if and only if Ω satisfies the following condition: For every $\epsilon > 0$ there exists $h \le 1$ and $r \ge 0$ such that the inequality

$$Q^{m,p}(H, H-\Omega) \ge h^p/\epsilon$$

holds for every n-cube H of edge length h having nonempty intersection with Ω_r. (This condition clearly implies that Ω is quasibounded.)

Prior to proving this theorem we prepare the following lemma.

6.20 LEMMA There exists a constant $K(m,p)$ such that for any n-cube H of edge length h, any measurable subset A of H with positive volume, and any $u \in C^1(H)$, we have

$$\|u\|_{0,p,H}^p \le \frac{2^{p-1}h^n}{\operatorname{vol}(A)}\|u\|_{0,p,A}^p + K\frac{h^{n+p}}{\operatorname{vol}(A)}\|\operatorname{grad} u\|_{0,p,H}^p.$$

Proof. Let $y \in A$ and $x = (\rho,\phi) \in H$, where (ρ,ϕ) denote spherical coordinates centred at y, in terms of which the volume element is given by $dx = \omega(\phi)\,\rho^{n-1}\,d\rho\,d\phi$. Let bdry H be specified by $\rho = f(\phi)$, $\phi \in \Sigma$. Clearly $f(\phi) \le \sqrt{n}h$. Since

$$u(x) = u(y) + \int_0^\rho \frac{d}{dr}u(r,\phi)\,dr,$$

we have by Lemma 2.2 and Hölder's inequality

$$\begin{aligned}
&\int_H |u(x)|^p\,dx \\
&\le 2^{p-1}h^n|u(y)|^p + 2^{p-1}\int_H \left|\int_0^\rho \frac{d}{dr}u(r,\phi)\,dr\right|^p dx \\
&\le 2^{p-1}h^n|u(y)|^p + 2^{p-1}\int_\Sigma \omega(\phi)\,d\phi\int_0^{f(\phi)}\rho^{n+p-2}\,d\rho\int_0^\rho |\operatorname{grad} u(r,\phi)|^p\,dr \\
&\le 2^{p-1}h^n|u(y)|^p + \frac{2^{p-1}}{n+p-1}\left(\sqrt{n}h\right)^{n+p-1}\int_H \frac{|\operatorname{grad} u(z)|^p}{|z-y|^{n-1}}\,dz.
\end{aligned}$$

Integrating y over A and using Lemma 4.64 we obtain

$$(\operatorname{vol}(A))\,\|u\|_{0,p,H}^p \le 2^{p-1}h^n\,\|u\|_{0,p,A}^p + Kh^{n+p}\,\|\operatorname{grad} u\|_{0,p,H}^p,$$

as required. ∎

6.21 (Proof of Theorem 6.19 — Necessity) Suppose that Ω does not satisfy the condition stated in the theorem. Then there exists a finite constant $K_1 = 1/\epsilon$ such that for every h with $0 < h \le 1$ there exists a sequence $\{H_j\}$ of mutually disjoint cubes of edge length h which intersect Ω and for which

$$Q^{m,p}(H_j, H_j - \Omega) < K_1 h^p.$$

By the definition of capacity, for each such cube H_j there exists a function $u_j \in C^\infty(H_j, H_j - \Omega)$ such that $\|u_j\|_{0,p,H_j}^p = h^n$, $\|\operatorname{grad} u_j\|_{0,p,H_j}^p \le K_1 h^n$, and $\|u_j\|_{m,p,H_j}^p \le K_2(h)$. Let $A_j = \{x \in H_j : |u_j(x)| < \frac{1}{2}\}$. By the previous Lemma we have

$$h^n \le \frac{2^{p-1}h^n}{\operatorname{vol}(A_j)}\cdot\frac{\operatorname{vol}(A_j)}{2^p} + \frac{KK_1}{\operatorname{vol}(A_j)}h^{2n+p}$$

from which it follows that $\operatorname{vol}(A_j) \le K_3 h^{n+p}$. Let us choose h so small that $K_3 h^p \le \frac{1}{3}$, whence $\operatorname{vol}(A_j) \le \frac{1}{3}\operatorname{vol}(H_j)$. Choose functions $w_j \in C_0^\infty(H_j)$ such that $w_j(x) = 1$ on a subset S_j of H_j having volume no less than $\frac{2}{3}\operatorname{vol}(H_j)$, and such that

$$\sup_j \max_{|\alpha|\le m} \sup_{x\in H_j} |D^\alpha w_j(x)| = K_4(h) < \infty.$$

Then $v_j = u_j w_j \in C_0^\infty(H_j \cap \Omega) \subset C_0^\infty(\Omega)$ and $|v_j(x)| \ge \frac{1}{2}$ on $S_j \cap (H_j - A_j)$, a set of volume not less than $h^n/3$. Hence $\|v_j\|_{0,p,H_j}^p \ge h^n/3\cdot 2^p$. On the other hand

$$\int_{H_j} |D^\alpha u_j(x)|^p \cdot |D^\beta w_j(x)|^p\,dx \le K_4(h)\,K_2(h)$$

provided $|\alpha|, |\beta| \le m$. Hence $\{v_j\}$ is a bounded sequence in $W_0^{m,p}(\Omega)$. Since the supports of the functions v_j are disjoint, $\|v_i - v_j\|_{0,p,\Omega}^p \ge 2h^n/3 \cdot 2^p$ so the imbedding (13) cannot be compact. ∎

6.22 (Proof of Theorem 6.19 — Sufficiency) Suppose Ω satisfies the condition stated in the theorem. Let $\epsilon > 0$ be given and choose $r \ge 0$ and $h \le 1$ such that for every cube H of edge h intersecting Ω_r we have $Q^{m,p}(H, H - \Omega) \ge h^p/\epsilon^p$. Then for every $u \in C_0^\infty(\Omega)$ we obtain

$$\|u\|_{0,p,H}^p \le \frac{\epsilon^p}{h^p} I_H^{m,p}(u) \le \epsilon^p \|u\|_{m,p,H}^p .$$

Since a neighbourhood of Ω_r can be tessellated by such cubes H we have by summation

$$\|u\|_{0,p,\Omega_r} \le \epsilon \|u\|_{m,p,\Omega} .$$

That any bounded set S in $W_0^{m,p}(\Omega)$ is precompact in $L^p(\Omega)$ now follows from Theorems 2.33 and 6.3. ∎

6.23 LEMMA There is a constant K independent of h such that for any cube H in $\mathbb{R}^n$ having edge length h, for every q satisfying $p \le q \le np/(n - mp)$ (or $p \le q < \infty$ if $mp = n$, or $p \le q \le \infty$ if $mp > n$), and for every $u \in C^\infty(H)$ we have

$$\|u\|_{0,q,H} \le K \left(\sum_{|\alpha| \le m} h^{|\alpha|p - n + np/q} \|D^\alpha u\|_{0,p,H}^p \right)^{1/p} .$$

Proof. We may suppose H to be centred at the origin and let $\tilde{H}$ be the cube of unit edge concentric with H and having edges parallel to those of H. The stated inequality holds for $\tilde{u} \in C^\infty(\tilde{H})$ by the Sobolev imbedding theorem. It then follows for H via the dilation $u(x) = \tilde{u}(x/h)$. ∎

6.24 LEMMA If $mp > n$ (or if $m = p = n = 1$), there exists a constant $K = K(m, p, n)$ such that for every cube H of edge length h in $\mathbb{R}^n$ and every $u \in C^\infty(H)$ that vanishes in a neighbourhood of some point $y \in H$, we have

$$\|u\|_{0,p,H}^p \le K\, I_H^{m,p}(u).$$

Proof. Let (ρ, ϕ) be spherical coordinates centred at y. Then

$$u(\rho, \phi) = \int_0^\rho \frac{d}{dt} u(t, \phi)\, dt.$$

If $n > (m - 1)p$, then let $q = np/(n - mp + p)$, so that $q > n$. Otherwise let $q > \max\{n, p\}$ be an arbitrary and finite. If $(\rho, \phi) \in H$, then by Hölder's inequality

$$|u(\rho,\phi)|^q\rho^{n-1} \le (\sqrt{n}h)^{n-1}\int_0^\rho \left|\frac{d}{dt}u(t,\phi)\right|^q t^{n-1}\,dt\left(\int_0^{\sqrt{n}h} t^{-(n-1)/(q-1)}\,dt\right)^{q-1}$$
$$\le K_1h^{q-1}\int_0^\rho \left|\frac{d}{dt}u(t,\phi)\right|^q t^{n-1}\,dt.$$

It follows, using the previous lemma with $m-1$ in place of m, that

$$\begin{aligned}\|u\|_{0,q,H}^q &\le K_2h^q\int_H |\operatorname{grad} u(x)|^q\,dx \qquad (14)\\ &\le K_2h^q\sum_{|\alpha|=1}\|D^\alpha u\|_{0,q,H}^q\\ &\le K_3h^q\sum_{|\alpha|=1}\left(\sum_{|\beta|\le m-1} h^{|\beta|p-n+n/q}\,\|D^{\alpha+\beta}u\|_{0,p,H}^p\right)^{q/p}.\end{aligned}$$

If $p>n$ (or $p=n=1$) the desired result follows directly from (14) with $q=p$:

$$\|u\|_{0,p,H}^p \le K\,I_H^{1,p}(u)\le K\,I_H^{m,p}(u).$$

Otherwise, a further application of Hölder's inequality yields

$$\begin{aligned}\|u\|_{0,p,H}^p &\le \|u\|_{0,q,H}^p\,(\mathrm{vol}(H))^{(q-p)/q}\\ &\le K_2^{p/q}\sum_{1\le|\gamma|\le m} h^{|\gamma|p}\,\|D^\gamma u\|_{0,p,H}^p = K\,I_H^{m,p}(u). \quad \blacksquare\end{aligned}$$

6.25 (Proof of Theorem 6.16) Let $mp>\nu$ (or $m=p=\nu=1$) and let H be a cube in $\mathbb{R}^n$ for which $\mu_{n-\nu}(H,\Omega)\ge h^p/\epsilon$. Let P be the maximal projection of $H-\Omega$ onto an $(n-\nu)$-dimensional face of H and let $E=P(H-\Omega)$. Without loss of generality we may assume that the face F of H containing E is parallel to the $x_{\nu+1},\dots,x_n$ coordinate plane. For each point $x=(x',x'')$ in E, where $x'=(x_1,\dots,x_\nu)$ and $x''=(x_{\nu+1},\dots,x_n)$ let $H_{x''}$ be the ν-dimensional cube of edge length h in which H intersects the ν-plane through x normal to F. By the definition of P there exists $y\in H_{x''}-\Omega$. If $u\in C^\infty(H,H-\Omega)$, then $u(\cdot,x'')\in C^\infty(H_{x''},y)$. Applying the previous lemma to $u(\cdot,x'')$ we obtain

$$\int_{H_{x''}}|u(x',x'')|^p\,dx'\le K_1\sum_{1\le|\alpha|\le m}h^{|\alpha|p}\int_{H_{x''}}|D^\alpha u(x',x'')|^p\,dx',$$

where K_1 is independent of H, x'', and u. Integrating this inequality over E and denoting $H' = \{x' : x = (x', x'') \in H \text{ for some } x''\}$, we obtain

$$\|u\|_{0,p,H'\times E}^p \le K_1 I_{H'\times E}^{m,p}(u) \le K_1 I_H^{m,p}(u).$$

Now we apply Lemma 6.20 with $A = H' \times E$ so that $\mathrm{vol}(A) = h^\nu \mu_{n-\nu}(H, \Omega)$. This yields

$$\|u\|_{0,p,H}^p \le K_2 \frac{h^{n-\nu}}{\mu_{n-\nu}(H,\Omega)} I_H^{m,p}(u),$$

where K_2 is independent of H. It follows that

$$Q^{m,p}(H, H-\Omega) \ge \frac{\mu_{n-\nu}(H,\Omega)}{K_2 h^{n-\nu}} \ge \frac{h^p}{\epsilon K_2}.$$

Hence Ω satisfies the hypothesis of Theorem 6.19 if it satisfies that of Theorem 6.16. ∎

The following two interpolation lemmas enable us to extend Theorem 6.16 to cover imbeddings into spaces of continuous functions.

6.26 LEMMA Let $1 \le p < \infty$ and $0 < \mu \le 1$. There exists a constant $K = K(n, p, \mu)$ such that for every $u \in C_0^\infty(\mathbb{R}^n)$ we have

$$\sup_{x\in\mathbb{R}^n} |u(x)| \le K \|u\|_{0,p,\mathbb{R}^n}^\lambda \left(\sup_{\substack{x,y\in\mathbb{R}^n \\ x\ne y}} \frac{|u(x)-u(y)|}{|x-y|^\mu} \right)^{1-\lambda}, \tag{15}$$

where $\lambda = p\mu/(n+p\mu)$.

Proof. We may assume

$$\sup_{x\in\mathbb{R}^n} |u(x)| = N > 0 \qquad \text{and} \qquad \sup_{\substack{x,y\in\mathbb{R}^n \\ x\ne y}} \frac{|u(x)-u(y)|}{|x-y|^\mu} = M < \infty.$$

Let ϵ satisfy $0 < \epsilon \le N/2$. Then there exists a point x_0 in $\mathbb{R}^n$ such that we have $|u(x_0)| \ge N - \epsilon \ge N/2$. Now $|u(x_0 - u(x)|/|x_0 - x|^\mu \le M$ for all x, so

$$|u(x)| \ge |u(x_0)| - M|x_0 - x|^\mu \ge \frac{1}{2}|u(x_0)|$$

provided $|x - x_0| \le (N/4M)^{1/\mu} = r$. Hence

$$\int_{\mathbb{R}^n} |u(x)|^p\, dx \ge \int_{B_r(x_0)} \left(\frac{|u(x_0)|}{2}\right)^p dx \ge K_1 \left(\frac{N-\epsilon}{2}\right)^p \left(\frac{N}{4M}\right)^{n/\mu}.$$

Since this holds for arbitrarily small ϵ we have

$$\|u\|_{0,p,\mathbb{R}^n} \geq \left(\frac{K_1^{1/p}}{2 \cdot 4^{n/\mu p}}\right) N^{1+(n/\mu p)} M^{-n/\mu p}$$

from which (15) follows at once. ∎

6.27 LEMMA Let Ω be an arbitrary domain in $\mathbb{R}^n$, and let $0 < \lambda < \mu \leq 1$. For every function $u \in C^{0,\mu}(\overline{\Omega})$ we have

$$\left\|u; C^{0,\lambda}(\overline{\Omega})\right\| \leq 3^{1-\lambda/\mu} \left\|u; C(\overline{\Omega})\right\|^{1-\lambda/\mu} \left\|u; C^{0,\mu}(\overline{\Omega})\right\|^{\lambda/\mu}. \tag{16}$$

Proof. Let $p = \mu/\lambda$ and $p' = p/(p-1)$. Let

$$A_1 = \left\|u; C(\overline{\Omega})\right\|^{1/p}, \qquad B_1 = \sup_{\substack{x,y\in\Omega \\ x\neq y}} \left(\frac{|u(x) - u(y)|}{|x-y|^{\mu}}\right)^{1/p},$$

$$A_2 = \left\|u; C(\overline{\Omega})\right\|^{1/p'}, \qquad B_2 = \sup_{\substack{x,y\in\Omega \\ x\neq y}} |u(x) - u(y)|^{1/p'}.$$

Clearly $A_1^p + B_1^p = \left\|u; C^{0,\mu}(\overline{\Omega})\right\|$ and $B_2^{p'} \leq 2\left\|u; C(\overline{\Omega})\right\|$. By Hölder's inequality for sums we have

$$\begin{aligned}
\left\|u; C^{0,\lambda}(\overline{\Omega})\right\| &= \left\|u; C(\overline{\Omega})\right\| + \sup_{\substack{x,y\in\Omega \\ x\neq y}} \frac{|u(x) - u(y)|}{|x-y|^{\lambda}} \\
&\leq A_1 A_2 + B_1 B_2 \\
&\leq \left(A_1^p + B_1^p\right)^{1/p} \left(A_2^{p'} + B_2^{p'}\right)^{1/p'} \\
&\leq \left\|u; C^{0,\mu}(\overline{\Omega})\right\|^{\lambda/\mu} \left(3\left\|u; C(\overline{\Omega})\right\|\right)^{1-\lambda/\mu}
\end{aligned}$$

as required. ∎

6.28 THEOREM Let Ω satisfy the hypotheses of Theorem 6.16. Then the following imbeddings are compact:

$$W_0^{j+m,p}(\Omega) \to C^j(\overline{\Omega}) \quad \text{if} \quad mp > n \tag{17}$$

$$W_0^{j+m,p}(\Omega) \to C^{j,\lambda}(\overline{\Omega}) \quad \text{if} \quad mp > n \geq (m-1)p \quad \text{and} \quad 0 < \lambda < m - (n/p). \tag{18}$$

Proof. It is sufficient to deal with the case $j = 0$. If $mp > n$, let j^* be the nonnegative integer satisfying $(m - j^*)p > n \geq (m - j^* - 1)p$. Then we have the chain of imbeddings

$$W_0^{m,p}(\Omega) \to W_0^{m-j^*,p}(\Omega) \to C^{0,\mu}(\overline{\Omega}) \to C(\overline{\Omega}),$$

where $0 < \mu < m - j^* - (n/p)$. If $\{u_i\}$ is a bounded sequence in $W_0^{m,p}(\Omega)$, then it is also bounded in $C^{0,\mu}(\overline{\Omega})$. By Theorem 6.16, $\{u_i\}$ has a subsequence $\{u_i'\}$ converging in $L^p(\Omega)$. By (15), which applies by completion to the functions u_i, this subsequence is a Cauchy sequence in $C(\overline{\Omega})$ and so converges there. Hence (17) is compact for $j = 0$. Furthermore, if $mp > n \geq (m-1)p$ (that is, if $j^* = 0$) and $0 < \lambda < \mu$, then by (16) $\{u_i'\}$ is also a Cauchy sequence in $C^{0,\lambda}(\overline{\Omega})$ whence (18) is also compact. ∎

An Equivalent Norm for $W_0^{m,p}(\Omega)$

6.29 (Domains of Finite Width) Consider the problem of determining for what domains Ω in $\mathbb{R}^n$ is the seminorm

$$|u|_{m,p,\Omega} = \left(\sum_{|\alpha|=m} \|D^\alpha u\|_{0,p,\Omega}^p \right)^{1/p}$$

actually a norm on $W_0^{m,p}(\Omega)$ equivalent to the standard norm

$$\|u\|_{m,p,\Omega} = \left(\sum_{|\alpha|\leq m} \|D^\alpha u\|_{0,p,\Omega}^p \right)^{1/p}.$$

This problem is closely related to the problem of determining for which unbounded domains Ω the imbedding $W_0^{m,p}(\Omega) \to L^p(\Omega)$ is compact because both problems depend on estimates for the L^p norm of a function in terms of L^p estimates for its derivatives.

We can easily show the equivalence of the above seminorm and norm for a domain of *finite width*, that is, a domain in $\mathbb{R}^n$ that lies between two parallel planes of dimension $(n-1)$. In particular, this is true for any bounded domain.

6.30 THEOREM (Poincaré's Inequality) If domain $\Omega \subset \mathbb{R}^n$ has finite width, then there exists a constant $K = K(p)$ such that for all $\phi \in C_0^\infty(\Omega)$

$$\|\phi\|_{0,p,\Omega} \leq K\, |\phi|_{1,p,\Omega}. \tag{19}$$

This inequality is known as *Poincare's Inequality.*

Proof. Without loss of generality we can assume that Ω lies between the hyperplanes $x_n = 0$ and $x_n = c > 0$. Denoting $x = (x', x_n)$, where $x' = (x_1, \ldots, x_{n-1})$, we have for any $\phi \in C_0^\infty(\Omega)$,

$$\phi(x) = \int_0^{x_n} \frac{d}{dt}\phi(x', t)\, dt$$

so that, by Hölder's inequality,

$$\begin{aligned}\|\phi\|_{0,p,\Omega}^p &= \int_{\mathbb{R}^{n-1}} dx' \int_0^c |\phi(x)|^p \, dx_n \\ &\le \int_{\mathbb{R}^{n-1}} dx' \int_0^c x_n^{p-1} \, dx_n \int_0^c |D_n\phi(x',t)|^p \, dt \\ &\le \frac{c^p}{p} |\phi|_{1,p,\Omega}^p .\end{aligned}$$

Inequality (19) follows with $K = c/p^{1/p}$. ∎

6.31 COROLLARY If Ω has finite width, $|\cdot|_{m,p,\Omega}$ is a norm on $W_0^{m,p}(\Omega)$ equivalent to the standard norm $\|\cdot\|_{m,p,\Omega}$.

Proof. If $\phi \in C_0^\infty(\Omega)$ then any derivative of ϕ also belongs to $C_0^\infty(\Omega)$. Now (19) implies

$$|\phi|_{1,p,\Omega}^p \le \|\phi\|_{1,p,\Omega}^p = \|\phi\|_{0,p,\Omega}^p + |\phi|_{1,p,\Omega}^p \le (1 + K^p)\, |\phi|_{1,p,\Omega}^p ,$$

and successive iterations of this inequality to derivatives $D^\alpha\phi$, ($|\alpha| \le m-1$) leads to

$$|\phi|_{m,p,\Omega}^p \le \|\phi\|_{m,p,\Omega}^p \le K_1 \, |\phi|_{m,p,\Omega}^p .$$

By completion, this holds for all u in $W_0^{m,p}(\Omega)$. ∎

6.32 (Quasicylindrical Domains) An unbounded domain Ω in $\mathbb{R}^n$ is called *quasicylindrical* if

$$\limsup_{x\in\Omega,\ |x|\to\infty} \operatorname{dist}(x, \operatorname{bdry}\Omega) < \infty.$$

Every quasibounded domain is quasicylindrical, as is every (unbounded) domain of finite width. The seminorm $|\cdot|_{m,p,\Omega}$ is not equivalent to the norm $\|\cdot\|_{m,p,\Omega}$ on $W_0^{m,p}(\Omega)$ for unbounded Ω if Ω is not quasicylindrical. We leave it to the reader to construct a suitable counterexample.

The following theorem is clearly analogous to Theorem 6.16.

6.33 THEOREM Suppose there exist an integer ν and constants K, R, and h such that $1 \le \nu \le n$, $0 < K \le 1$, $0 \le R < \infty$, and $0 < h < \infty$. Suppose also that either $\nu < p$ or $\nu = p = 1$, and that for every cube H in $\mathbb{R}^n$ having edge length h and nonempty intersection with $\Omega_R = \{x \in \Omega : |x| \ge R\}$ we have

$$\frac{\mu_{n-\nu}(H,\Omega)}{h^{n-\nu}} \ge K,$$

where $\mu_{n-\nu}(H,\Omega)$ is as defined prior to the statement of Theorem 6.16. Then $|\cdot|_{m,p,\Omega}$ is a norm on $W_0^{m,p}(\Omega)$ equivalent to the standard norm $\|\cdot\|_{m,p,\Omega}$.

Proof. As observed in the previous Corollary, it is again sufficient to prove that $\|u\|_{0,p,\Omega} \le K_1 |u|_{1,p,\Omega}$ holds for all $u \in C_0^\infty(\Omega)$. Let H be a cube of edge length h having nonempty intersection with Ω_R. Since $\nu < p$ (or $\nu = p = 1$) the proof of Theorem 6.16 shows that

$$Q^{1,p}(H, H - \Omega) \ge \frac{\mu_{n-\nu}(H, \Omega)}{K_2 h^{n-\nu}} \ge \frac{K}{K_2}$$

for all $u \in C_0^\infty(\Omega)$, K_2 being independent of u. Hence

$$\|u\|_{0,p,H}^p \le (K_2/K) I_H^{1,p} = K_3 |u|_{1,p,H}^p .$$

By summing this inequality over the cubes comprising a tessellation of some neighbourhood of Ω_R, we obtain

$$\|u\|_{0,p,\Omega_R}^p \le K_3 |u|_{1,p,\Omega}^p . \tag{20}$$

It remains to be proven that

$$\|u\|_{0,p,B_R}^p \le K_3 |u|_{1,p,\Omega}^p ,$$

where $B_R = \{x \in \mathbb{R}^n : |x| < R\}$. Let (ρ, ϕ) denote the spherical coordinates of the point $x \in \mathbb{R}^n$ $(\rho \ge 0,\ \phi \in \Sigma)$ so that $dx = \rho^{n-1}\omega(\phi)\, d\rho\, d\phi$. For any $u \in C^\infty(\mathbb{R}^n)$ we have

$$u(\rho, \phi) = u(\rho + R, \phi) - \int_\rho^{R+\rho} \frac{d}{dt} u(t, \phi)\, dt$$

so that (by Lemma 2.2)

$$|u(\rho, \phi)|^p \le 2^{p-1} |u(\rho + R, \phi)|^p + 2^{p-1} R^{p-1} \rho^{1-n} \int_\rho^{R+\rho} |\operatorname{grad} u(t, \phi)|^p t^{n-1}\, dt.$$

Hence

$$\begin{aligned}
\|u\|_{0,p,B_R}^p &= \int_\Sigma \omega(\phi)\, d\phi \int_0^R |u(\rho, \phi)|^p \rho^{n-1}\, d\rho \\
&\le 2^{p-1} \int_\Sigma \omega(\phi)\, d\phi \int_0^R |u(\rho + R, \phi)|^p (\rho + R)^{n-1}\, d\rho \\
&\quad + 2^{p-1} R^p \int_\Sigma \omega(\phi)\, d\phi \int_0^{2R} |\operatorname{grad} u(t, \phi)|^p t^{n-1}\, dt.
\end{aligned}$$

Therefore, we have for $u \in C_0^\infty(\Omega)$

$$\begin{aligned}\|u\|_{0,p,B_R}^p &\le 2^{p-1}\|u\|_{0,p,B_{2R}-B_R}^p + 2^{p-1}R^p|u|_{1,p,B_{2R}}^p \\ &\le 2^{p-1}\|u\|_{0,p,\Omega_R}^p + 2^{p-1}R^p|u|_{1,p,\Omega}^p \le K_4|u|_{1,p,\Omega}^p\end{aligned}$$

by (20). ∎

Unbounded Domains — Decay at Infinity

6.34 The fact that elements of $W_0^{m,p}(\Omega)$ vanish in a generalized sense on the boundary of Ω played a critical role in our showing that the imbedding

$$W_0^{m,p}(\Omega) \to L^p(\Omega) \qquad (21)$$

is compact for certain unbounded domains Ω. Since elements of $W^{m,p}(\Omega)$ do not have this property, there remains a question of whether an imbedding of the form

$$W^{m,p}(\Omega) \to L^p(\Omega) \qquad (22)$$

can ever be compact for unbounded Ω, or even for bounded Ω which are sufficiently irregular that no imbedding of the form

$$W^{m,p}(\Omega) \to L^q(\Omega) \qquad (23)$$

can exist for any $q > p$. Note that if Ω has finite volume, the existence of imbedding (23) for some $q > p$ implies the compactness of imbedding (22) by the method of the first part of the proof in Paragraph 6.8. By Theorem 4.46 imbedding (23) cannot, however, exist if $q > p$ and Ω is unbounded but has finite volume.

6.35 EXAMPLE For $j = 1, 2, \ldots$ let B_j be an open ball in $\mathbb{R}^n$ having radius r_j, and suppose that $\overline{B_j} \cap \overline{B_i}$ is empty whenever $j \neq i$. Let $\Omega = \bigcup_{j=1}^\infty B_j$. Note that Ω may be bounded or unbounded. The sequence $\{u_j\}$ defined by

$$u_j(x) = \begin{cases} (\mathrm{vol}(B_j))^{-1/p} & \text{if } x \in \overline{B_j} \\ 0 & \text{if } x \notin \overline{B_j}\end{cases}$$

is bounded in $W^{m,p}(\Omega)$ for every integer $m \ge 0$, but is not precompact in $L^p(\Omega)$ no matter how fast $r_j \to 0$ as $j \to \infty$. (Of course, imbedding (21) is compact by Theorem 6.16 provided $\lim_{j\to\infty} r_j = 0$.) Even if Ω is bounded, imbedding (23) cannot exist for any $q > p$.

6.36 Let us state at once that there do exist unbounded domains Ω for which the imbedding (22) is compact. See Example 6.53. An example of such a domain was given by the authors in [AF2] and it provided a basis for an investigation of the general problem in [AF3]. The approach of this latter paper is used in the following discussion.

First we concern ourselves with necessary conditions for the compactness of (23) for $q \geq p$. These conditions involve rapid decay at infinity for any unbounded domain (see Theorem 6.45). The techniques involved in the proof also yield a strengthened version of Theorem 4.46, namely Theorem 6.41, and a converse of the assertion [see Remark 4.13(3)] that $W^{m,p}(\Omega) \to L^q(\Omega)$ for $1 \leq q < p$ if Ω has finite volume.

A sufficient condition for the compactness of (22) is given in Theorem 6.52. It applies to many domains, bounded and unbounded, to which neither the Rellich-Kondrachov theorem nor any generalization of that theorem obtained by the same methods can be applied. (e.g. exponential cusps — see Example 6.54).

6.37 (Tessellations and λ-fat Cubes) Let T be a tessellation of $\mathbb{R}^n$ by closed n-cubes of edge length h. If H is one of the cubes in T, let $N(H)$ denote the cube of edge length $3h$ concentric with H and therefore consisting of the 3^n elements of T that intersect H. We call $N(H)$ the *neighbourhood* of H. By the *fringe* of H we shall mean the shell $F(H) = N(H) - H$.

Let Ω be a given domain in $\mathbb{R}^n$ and T a given tessellation as above. Let $\lambda > 0$. A cube $H \in T$ will be called λ-*fat* (with respect to Ω) if

$$\mu(H \cap \Omega) > \lambda\, \mu(F(H) \cap \Omega),$$

where μ denotes the n-dimensional Lebesgue measure in $\mathbb{R}^n$. (We use μ instead of "vol" for notational simplicity in the following discussion where the symbol must be used many times.) If H is not λ-fat then we will say it is λ-*thin.*

6.38 THEOREM Suppose there exists a compact imbedding of the form

$$W^{m,p}(\Omega) \to L^q(\Omega)$$

for some $q \geq p$. Then for every $\lambda > 0$ and every tessellation T of $\mathbb{R}^n$ by cubes of fixed size, T can have only finitely many λ-fat cubes.

Proof. Suppose, to the contrary, that for some $\lambda > 0$ there exists a tessellation T of $\mathbb{R}^n$ by cubes of edge length h containing a sequence $\{H_j\}_{j=1}^{\infty}$ of λ-fat cubes. Passing to a subsequence if necessary we may assume that $N(H_j) \cap N(H_i)$ is empty whenever $j \neq i$. For each j there exists $\phi_j \in C_0^\infty(N(H_j))$ such that

(i) $|\phi_j(x)| \leq 1$ for all $x \in \mathbb{R}^n$,

(ii) $\phi_j(x) = 1$ for $x \in H_j$, and

(iii) $|D^\alpha \phi_j(x)| \leq M$ for all j, all $x \in \mathbb{R}^n$, and all α satisfying $0 \leq |\alpha| \leq m$.

In fact, all the ϕ_j can be taken to be translates of one of them. Let $\psi_j = c_j\phi_j$, where the positive constants c_j are chosen so that

$$\|\psi_j\|_{0,q,\Omega}^q \geq c_j^q \int_{H_j\cap\Omega} |\phi_j(x)|^q\,dx = c_j^q \mu(H_j \cap \Omega) = 1.$$

But then

$$\begin{aligned}
\|\psi_j\|_{m,p,\Omega}^p &= c_j^p \sum_{0\leq|\alpha|\leq m} \int_{N(H_j)\cap\Omega} |D^\alpha \phi_j(x)|^p\,dx \\
&\leq M^p c_j^p \mu\big(N(H_j)\cap\Omega\big) \\
&< M^p c_j^p \mu\big(H_j\cap\Omega\big)\left(1+\frac{1}{\lambda}\right) = M^p\left(1+\frac{1}{\lambda}\right)c_j^{p-q},
\end{aligned}$$

since H_j is λ-fat. Now $\mu(H_j \cap \Omega) \leq \mu(H_j) = h^n$ so $c_j \geq h^{-n/q}$. Since $p - q \leq 0$, $\{\psi_j\}$ is bounded in $W^{m,p}(\Omega)$. But the functions ψ_j have disjoint supports, so $\{\psi_j\}$ cannot be precompact in $L^q(\Omega)$, contradicting the assumption that $W^{m,p}(\Omega) \to L^q(\Omega)$ is compact. Thus every T can possess at most finitely many λ-fat cubes. ∎

6.39 COROLLARY Suppose that $W^{m,p}(\Omega) \to L^q(\Omega)$ for some $q > p$. If T is a tessellation of $\mathbb{R}^n$ by cubes of fixed edge-length, and if $\lambda > 0$ is given, then there exists $\epsilon > 0$ such that $\mu(H \cap \Omega) \geq \epsilon$ for every λ-fat $H \in T$.

Proof. Suppose, to the contrary, that there exists a sequence $\{H_j\}$ of λ-fat cubes with $\lim_{j\to\infty} \mu(H_j \cap \Omega) = 0$. If c_j is defined as in the above proof, we have $\lim_{j\to\infty} c_j = \infty$. But then $\lim_{j\to\infty} \|\psi_j\|_{m,p,\Omega} = 0$ since $p < q$. Since $\{\psi_j\}$ is bounded away from 0 in $L^q(\Omega)$, we have contradicted the continuity of the imbedding $W^{m,p}(\Omega) \to L^q(\Omega)$. ∎

6.40 REMARK It follows from the above corollary that if there exists an imbedding

$$W^{m,p}(\Omega) \to L^q(\Omega) \tag{24}$$

for some $q > p$ then one of the following alternatives must hold:

(a) There exists $\epsilon > 0$ and a tessellation T of $\mathbb{R}^n$ consisting of cubes of fixed size such that $\mu(H \cap \Omega) \geq \epsilon$ for infinitely many cubes $H \in T$.

(b) For every $\lambda > 0$, every tessellation T of $\mathbb{R}^n$ consisting of cubes of fixed size contains only finitely many λ-fat cubes.

We will show in Theorem 6.42 that (b) implies that Ω has finite volume. By Theorem 4.46, (b) is therefore inconsistent with the existence of (24) for $q > p$. On the other hand, (a) implies that $\mu\big(\{x \in \Omega : N \leq |x| \leq N+1\}\big)$ does not approach zero as N tends to infinity. We have therefore proved the following strengthening of Theorem 4.46.

6.41 THEOREM Let Ω be an unbounded domain in $\mathbb{R}^n$ satisfying

$$\limsup_{N\to\infty} \operatorname{vol}(\{x \in \Omega : N \le |x| \le N+1\}) = 0.$$

Then there can be no imbedding of the form (24) for any $q > p$.

6.42 THEOREM Suppose that imbedding (24) is compact for some $q \ge p$. Then Ω has finite volume.

Proof. Let T be a tessellation of $\mathbb{R}^n$ by cubes of unit edge length, and let $\lambda = 1/[2(3^n - 1)]$. Let P be the union of the finitely many λ-fat cubes in T. Clearly $\mu(P \cap \Omega) \le \mu(P) < \infty$.

Let H be a λ-thin cube in T. Let H_1 be one of the $3^n - 1$ cubes in T constituting the fringe of H selected so that $\mu(H_1 \cap \Omega)$ is maximal. Thus

$$\mu(H \cap \Omega) \le \lambda\, \mu\big(F(H) \cap \Omega\big) \le \lambda(3^n - 1)\mu(H_1 \cap \Omega) = \frac{1}{2}\mu(H_1 \cap \Omega).$$

If H_1 is also λ-thin, then we may select a cube $H_2 \in T$ with $H_2 \subset F(H_1)$ such that $\mu(H_1 \cap \Omega) \le \frac{1}{2}\mu(H_2 \cap \Omega)$.

Suppose an infinite chain $\{H_1, H_2, \dots\}$ of λ-thin cubes can be constructed in the above manner. Then

$$\mu(H \cap \Omega) \le \frac{1}{2}\mu(H_1 \cap \Omega) \le \cdots \le \frac{1}{2^j}\mu(H_j \cap \Omega) \le \frac{1}{2^j}$$

for each j since $\mu(H_j \cap \Omega) \le \mu(H_j) = 1$. Hence $\mu(H \cap \Omega) = 0$. Denoting by P_∞ the union of λ-thin cubes $H \in T$ for which such an infinite chain can be constructed, we have $\mu(P_\infty \cap \Omega) = 0$.

Let P_j denote the union of λ-thin cubes $H \in T$ for which some such chain ends on the jth step; that is, H_j is λ-fat. Any particular λ-fat cube H' can occur as the end H_j of a chain beginning at H only if H is contained in the cube of edge $2j+1$ centred on H'. Hence there are at most $(2j+1)^n$ such cubes $H \subset P_j$ having H' as chain endpoint. Thus

$$\begin{aligned}
\mu(P_j \cap \Omega) &= \sum_{H \subset P_j} \mu(H \cap \Omega) \\
&\le \frac{1}{2^j} \sum_{H \subset P_j} \mu(H_j \cap \Omega) \\
&\le \frac{(2j+1)^n}{2^j} \sum_{H' \subset P} \mu(H' \cap \Omega) = \frac{(2j+1)^n}{2^j}\mu(P \cap \Omega),
\end{aligned}$$

so that $\mu(\Omega) = \mu(P \cap \Omega) + \mu(P_\infty \cap \Omega) + \sum_{j=1}^\infty \mu(P_j \cap \Omega) < \infty$. ∎

Suppose $1 \le q < p$. If $\mathrm{vol}(\Omega) < \infty$, then the imbedding

$$W^{m,p}(\Omega) \to L^q(\Omega)$$

exists because $W^{m,p}(\Omega) \to L^p(\Omega)$ trivially and $L^p(\Omega) \to L^q(\Omega)$ by Theorem 2.14.

We are now in a position to prove the converse.

6.43 THEOREM If the imbedding $W^{m,p}(\Omega) \to L^q(\Omega)$ exists for some p and q satisfying $1 \le q < p$, then Ω has finite volume.

Proof. Let T, λ, and again let P denote the union of the λ-fat cubes in T. If we can show that $\mu(P \cap \Omega)$ is finite, it will follow by the same argument used in that theorem that $\mu(\Omega)$ is finite.

Accordingly, suppose that $\mu(P \cap \Omega)$ is not finite. Then there exists a sequence $\{H_j\}$ of λ-fat cubes in T such that $\sum_{j=1}^\infty \mu(H_j \cap \Omega) = \infty$. If L is the lattice of centres of cubes in T, we may break up L into 3^n mutually disjoint sublattices $\{L_i\}_{i=1}^{3^n}$ each having period 3 in each coordinate direction. For each i let T_i be the set of all cubes in T that have centres in L_i. For some i we must have $\sum_{\lambda-\mathrm{fat}\, H \in T_i} \mu(H \cap \Omega) = \infty$. Thus we may assume that the cubes of the sequence $\{H_j\}$ all belong to T_i for some i, so that $N(H_j) \cap N(H_k)$ is empty if $j \ne k$.

Choose integer j_1 so that

$$2 \le \sum_{j=1}^{j_1} \mu(H_j \cap \Omega) < 4.$$

Let ϕ_j be as in the proof of Theorem 6.38, and let

$$\psi_1(x) = 2^{-1/p} \sum_{j=1}^{j_1} \phi_j(x).$$

Since the supports of the functions ϕ_j are mutually disjoint and since the cubes H_j are λ-fat, for $|\alpha| \le m$ we have

$$\begin{aligned} \|D^\alpha \psi_1\|_{0,p,\Omega}^p &= \frac{1}{2} \sum_{j=1}^{j_1} \int_\Omega |D^\alpha \phi_j(x)|^p \, dx \\ &\le \frac{1}{2} M^p \sum_{j=1}^{j_1} \mu(N(H_j) \cap \Omega) \\ &< \frac{1}{2} M^p \left(1 + \frac{1}{\lambda}\right) \sum_{j=1}^{j_1} \mu(H_j \cap \Omega) < 2M^p \left(1 + \frac{1}{\lambda}\right). \end{aligned}$$

On the other hand,

$$\|\psi_1\|_{0,q,\Omega}^q \geq 2^{-q/p} \sum_{j=1}^{j_1} \mu(H_j \cap \Omega) \geq 2^{1-(q/p)}.$$

Having so defined j_1 and ψ_1, we can now define $j_2, j_3, \ldots$ and $\psi_2, \psi_3, \ldots$ inductively so that

$$2^k \leq \sum_{j=j_{k-1}+1}^{j_k} \mu(H_j \cap \Omega) < 2^{k+1}$$

and

$$\psi_k(x) = 2^{-k/p} k^{-2/p} \sum_{j=j_{k-1}+1}^{j_k} \phi_j(x).$$

As above, we have for $|\alpha| \leq m$,

$$\|D^\alpha \psi_k\|_{0,p,\Omega}^p \leq \frac{2}{k^2} M^p \left(1 + \frac{1}{\lambda}\right)$$

and

$$\|\psi_k\|_{0,q,\Omega}^q \geq 2^{k(1-q/p)} M^p \left(\frac{1}{k}\right)^{2q/p}.$$

Thus $\psi = \sum_{k=1}^{\infty} \psi_k$ belongs to $W^{m,p}(\Omega)$ but not to $L^q(\Omega)$ contradicting the imbedding $W^{m,p}(\Omega) \to L^q(\Omega)$. Hence $\mu(P \cap \Omega) < \infty$ as required. ∎

6.44 If there exists a compact imbedding of the form $W^{m,p}(\Omega) \to L^q(\Omega)$ for some $q \geq p$, then, as we have shown, Ω has finite volume. In fact, considerably more is true; $\mu(\{x \in \Omega : |x| \geq R\})$ must approach zero very rapidly as $R \to \infty$, as we show in Theorem 6.45 below.

If Q is a union of cubes H in some tessellation T of $\mathbb{R}^n$ by cubes of fixed edge length, we extend the notions of neighbourhood and fringe to Q in an obvious manner:

$$N(Q) = \bigcup_{\substack{H \in T \\ H \subset Q}} N(H), \qquad F(Q) = N(Q) - Q.$$

Given $\delta > 0$, let $\lambda = \delta/[3^n(1+\delta)]$. If all the cubes $H \in T$ satisfying $H \subset Q$ are λ-thin, then Q is itself δ-thin in the sense that

$$\mu(Q \cap \Omega) \leq \delta \mu(F(Q) \cap \Omega).$$

To see this note that as H runs through the cubes comprising Q, $F(H)$ covers $N(Q)$ at most 3^n times. Hence

$$\begin{aligned}\mu(Q \cap \Omega) &= \sum_{H \subset Q} \mu(H \cap \Omega) \leq \lambda \sum_{H \subset Q} \mu(F(H) \cap \Omega) \\ &\leq 3^n \lambda \mu(N(Q) \cap \Omega) = 3^n \lambda[\mu(Q \cap \Omega) + \mu(F(Q) \cap \Omega)]\end{aligned}$$

and the fact that Q is δ-thin follows by transposition (permissible since $\mu(\Omega) < \infty$) and since $3^n\lambda/(1 - 3^n\lambda) = \delta$.

For any measurable set $S \subset \mathbb{R}^n$ let Q be the union of all cubes in T whose interiors intersect S, and define $F(S) = f(Q)$. If S is at a positive distance from the finitely many λ-fat cubes in T, then Q consists of λ-thin cubes and we obtain

$$\mu(S \cap \Omega) \le \mu(Q \cap \Omega) \le \delta\mu\big(F(S) \cap \Omega\big). \tag{25}$$

6.45 THEOREM (Rapid Decay) Suppose there exists a compact imbedding of the form

$$W^{m,p}(\Omega) \to L^q(\Omega) \tag{26}$$

for some $q \ge p$. For each $r \ge 0$ let $\Omega_r = \{x \in \Omega : |x| > r\}$, let $S_r = \{x \in \Omega : |x| = r\}$, and let A_r denote the surface area (Lebesgue $(n-1)$-measure) of S_r. Then

(a) For given $\epsilon, \delta > 0$ there exists R such that if $r \ge R$, then

$$\mu(\Omega_r) \le \delta\,\mu\big(\{x \in \Omega : r - \epsilon \le |x| \le r\}\big).$$

(b) If A_r is positive and ultimately nonincreasing as $r \to \infty$, then for each $\epsilon > 0$

$$\lim_{r\to\infty} \frac{A_{r+\epsilon}}{A_r} = 0.$$

Proof. Given $\epsilon > 0$ and $\delta > 0$, let $\lambda = \delta/[3^n(1+\delta)]$ and let T be a tessellation of $\mathbb{R}^n$ by cubes of edge length $\epsilon/(2\sqrt{n})$. Let R be large enough that the finitely many λ-fat cubes in T lie in the ball of radius $R - \epsilon/2$ about the origin. If $r \ge R$ and $S = \Omega_r$, then any $H \in T$ whose interior intersects S is λ-thin. Moreover, any cube in the fringe of S can only intersect Ω at points x satisfying $r - \epsilon/2 \le |x| \le r$. By (25),

$$\mu(\Omega_r) = \mu(S \cap \Omega) \le \delta\,\mu\big(F(S) \cap \Omega\big) = \delta\,\mu\big(\{x \in \Omega : r - \epsilon \le |x| \le r\}\big),$$

which proves (a).

For (b) choose R_0 so that A_r is nonincreasing for $r \ge R_0$. Fix $\epsilon', \delta > 0$ and let $\epsilon = \epsilon'/2$. Let R be as in (a). If $r \ge \max\{R, R_0 + \epsilon'\}$, then

$$\begin{aligned} A_{r+\epsilon'} &\le \frac{1}{\epsilon}\int_{r+\epsilon}^{r+2\epsilon} A_s\,ds \le \frac{1}{\epsilon}\mu(\Omega_{r+\epsilon}) \\ &\le \frac{\delta}{\epsilon}\mu\big(\{x \in \Omega : r \le |x| \le r+\epsilon\}\big) = \frac{\delta}{\epsilon}\int_r^{r+\epsilon} A_s\,ds \le \delta A_r. \end{aligned}$$

Since ϵ' and δ are arbitrary, (b) follows. ∎

6.46 COROLLARY If there exists a compact imbedding of the form (26) for some $q \geq p$, then for every $k > 0$ we have

$$\lim_{r\to\infty} e^{kr}\mu(\Omega_r) = 0. \tag{27}$$

Proof. Fix k and let $\delta = e^{-(k+1)}$. From conclusion (a) of Theorem 6.45 there exists R such that $r \geq R$ implies $\mu(\Omega_{r+1}) \leq \delta\mu(\Omega_r)$. Thus if j is a positive integer and $0 \leq t < 1$, we have

$$\begin{aligned} e^{k(R+j+t)}\mu(\Omega_{R+j+t}) &\leq e^{k(R+j+1)}\mu(\Omega_{R+j}) \\ &\leq e^{k(R+1)}\, e^{kj}\, \delta^j\, \mu(\Omega_R) = e^{k(R+1)}\, \mu(\Omega_R)\, e^{-j}. \end{aligned}$$

The last term approaches zero as j tends to infinity. ∎

6.47 REMARKS

1. We work with Sobolev spaces defined intrinsically in domains. If instead, we had defined $W^{m,p}(\Omega)$ to consist of all restrictions to Ω of functions in $W^{m,p}(R^n)$, then the outcome for Corollary 6.46 would have been different. With that definition, it is shown in [BSc] that $W^{m,p}(\Omega)$ imbeds compactly in $L^p(\Omega)$ if and only if the volume of the intersection of Ω with cubes of fixed edge-length tends to 0 as the centres of those cubes tend to ∞. There are many domains Ω satisfying the latter condition but not satisfying (27). None of these domains can have any Sobolev extension property.
2. The argument used in the proof of Theorem 6.45(a) works for any norm ρ on $\mathbb{R}^n$ in place of the usual Euclidean norm $\rho(x) = |x|$. The same holds for Theorem 6.45(b) provided A_r is well defined (with respect to the norm ρ) and provided

$$\mu(\{x \in \Omega : r \leq \rho(x) \leq r + \epsilon\}) = \int_r^{r+\epsilon} A_s\, ds.$$

This is true, for example, if $\rho(x) = \max_{1\leq i\leq n} |x_i|$.
3. For the proof of Theorem 6.45(b) it is sufficient that A_r have an equivalent nonincreasing majorant, that is, there should exist a positive, nonincreasing function $f(r)$ and a constant $M > 0$ such that for sufficiently large r

$$A_r \leq f(r) \leq M\, A_r.$$

4. Theorem 6.38 is sharper than Theorem 6.45, because the conclusions of the latter theorem are global whereas the compactness of (26) depends on local properties of Ω. We illustrate this by means of two examples.

6.48 EXAMPLE Let $f \in C^1([0,\infty))$ be positive and nonincreasing with bounded derivative f'. We consider the planar domain (Figure 5)

$$\Omega = \{(x,y) \in \mathbb{R}^2 : x > 0,\ 0 < y < f(x)\}.$$

With respect to the supremum norm on $\mathbb{R}^2$, that is $\rho(x,y) = \max\{|x|,|y|\}$, we have $A_s = f(s)$ for sufficiently large s. Hence Ω satisfies conclusion (b) of Theorem 6.45 (and, since f is monotonic, conclusion (a) as well) if and only if

$$\lim_{s\to\infty} \frac{f(s+\epsilon)}{f(s)} = 0 \tag{28}$$

holds for every $\epsilon > 0$. For example, $f(x) = \exp(-x^2)$ satisfies this condition but $f(x) = \exp(-x)$ does not. We shall show in Example 6.53 that the imbedding

$$W^{m,p}(\Omega) \to L^p(\Omega) \tag{29}$$

is compact if (28) holds. Thus (28) is necessary and sufficient for compactness of the above imbedding for domains of this type. ∎

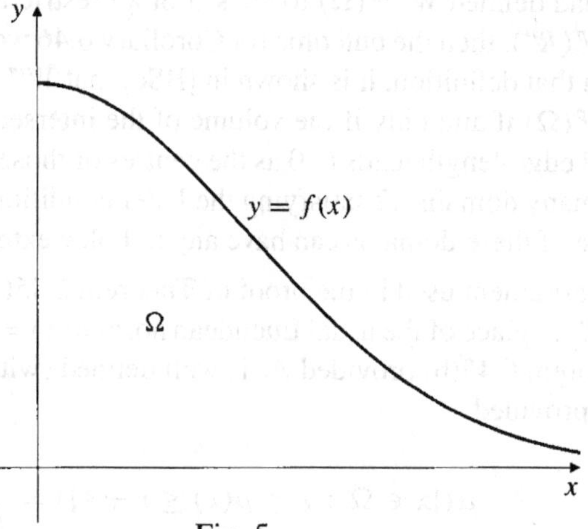

Fig. 5

6.49 EXAMPLE Let f be as in the previous example, and assume also that $f'(0) = 0$. Let g be a positive, nonincreasing function in $C^1([0,\infty))$ satisfying

(i) $g(0) = \frac{1}{2}f(0)$, and $g'(0) = 0$,

(ii) $g(x) < f(x)$ for all $x \geq 0$,

(iii) $g(x)$ is constant on infinitely many disjoint intervals of unit length.

Let $h(x) = f(x) - g(x)$ and consider the domain (Figure 6)

$$\tilde{\Omega} = \{(x,y) \in \mathbb{R}^2 : 0 < y < g(x) \text{ if } x \geq 0,\ 0 < y < h(-x) \text{ if } x < 0\}.$$

Again we have $A_s = f(s)$ for sufficiently large s, so $\tilde{\Omega}$ satisfies the conclusions of Theorem 6.45 if (28) holds.

If, however, T is a tessellation of $\mathbb{R}^2$ by squares of edge length $\frac{1}{4}$ having edges parallel to the coordinate axes, and if one of the squares in T has centre at the origin, then T has infinitely many $\frac{1}{3}$-fat squares with centres on the positive x-axis. By Theorem 6.38 the imbedding (29) cannot be compact for the domain $\tilde{\Omega}$. ∎

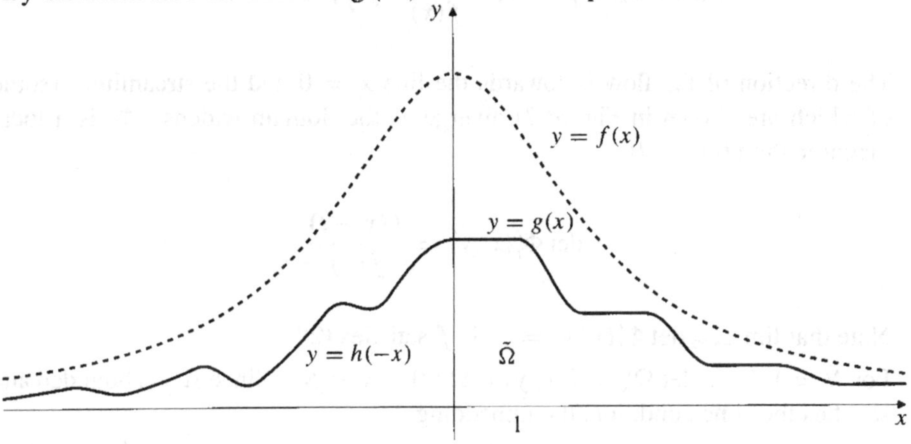

Fig. 6

Unbounded Domains — Compact Imbeddings of $W^{m,p}(\Omega)$

6.50 (Flows) The above examples suggest that any sufficient condition for the compactness of the imbedding

$$W^{m,p}(\Omega) \to L^p(\Omega)$$

for unbounded domains Ω must involve the rapid decay of volume locally in each branch of Ω_r as r tends to infinity. A convenient way to express such local decay is in terms of flows on Ω.

By a *flow* on Ω we mean a continuously differentiable map $\Phi : U \to \Omega$ where U is an open set in $\Omega \times \mathbb{R}$ containing $\Omega \times \{0\}$, and where $\Phi(x, 0) = x$ for every $x \in \Omega$.

For fixed $x \in \Omega$ the curve $t \to \Phi(x, t)$ is called a *streamline* of the flow. For fixed t the map $\Phi_t : x \to \Phi(x, t)$ sends a subset of Ω into Ω. We shall be concerned with the Jacobian of this map:

$$\det \Phi_t'(x) = \left. \frac{\partial(\Phi_1, \ldots, \Phi_n)}{\partial(x_1, \ldots, x_n)} \right|_{(x,t)} .$$

It is sometimes required of a flow Φ that $\Phi_{s+t} = \Phi_s \circ \Phi_t$ but we do not need this property and so do not assume it.

6.51 EXAMPLE Let Ω be the domain of Example 6.48. Define the flow

$$\Phi(x, y, t) = \left(x - t, \frac{f(x-t)}{f(x)}\, y\right), \qquad 0 < t < x.$$

The direction of the flow is towards the line $x = 0$ and the streamlines (some of which are shown in Figure 7) diverge as the domain widens. Φ_t is a local magnification for $t > 0$:

$$\det \Phi_t'(x, y) = \frac{f(x-t)}{f(x)}.$$

Note that $\lim_{x\to\infty} \det \Phi_t'(x, y) = \infty$ if f satisfies (28).
For $N = 1, 2, \ldots$ let $\Omega_N^* = \{(x, y) \in \Omega : 0 < x < N\}$. Since Ω_N^* is bounded and satisfies the cone condition, the imbedding

$$W^{1,p}\left(\Omega_N^*\right) \to L^p(\Omega_N^*)$$

is compact. This compactness, together with properties of the flow Φ are sufficient to force the compactness of $W^{m,p}(\Omega) \to L^p(\Omega)$ as we now show.

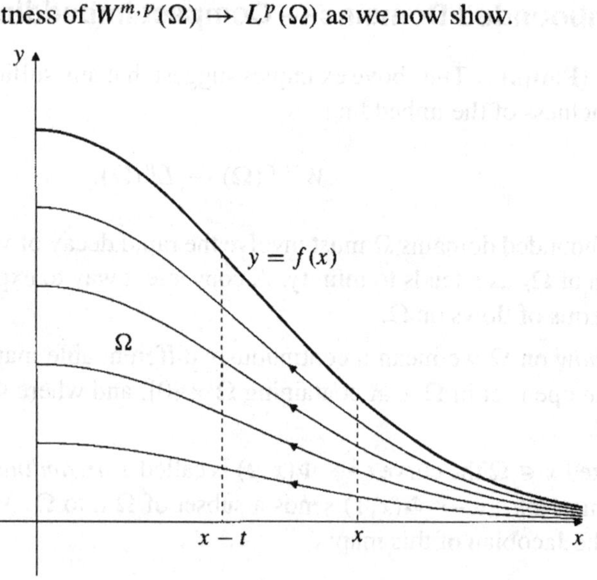

Fig. 7

6.52 THEOREM Let Ω be an open set in $\mathbb{R}^n$ having the following properties:

(a) There exists an infinite sequence $\{\Omega_N^*\}_{N=1}^\infty$ of open subsets of Ω such that $\Omega_N^* \subset \Omega_{N+1}^*$ and such that for each N the imbedding

$$W^{1,p}\left(\Omega_N^*\right) \to L^p(\Omega_N^*)$$

is compact.

(b) There exists a flow $\Phi : U \to \Omega$ such that if $\Omega_N = \Omega - \Omega_N^*$, then

  (i) $\Omega_N \times [0, 1] \subset U$ for each N,

  (ii) Φ_t is one-to-one for all t,

  (iii) $|(\partial/\partial t)\Phi(x, t)| \le M$ (constant) for all $(x, t) \in U$.

(c) The functions $d_N(t) = \sup_{x\in\Omega_N} |\det \Phi_t'(x)|^{-1}$ satisfy

  (i) $\lim_{N\to\infty} d_N(1) = 0$,

  (ii) $\lim_{N\to\infty} \int_0^1 d_N(t)\, dt = 0$.

Then the imbedding $W^{m,p}(\Omega) \to L^p(\Omega)$ is compact.

Proof. Since we have $W^{m,p}(\Omega) \to W^{1,p}(\Omega) \to L^p(\Omega)$ it is sufficient to prove that the latter imbedding is compact. Let $u \in C^1(\Omega)$. For each $x \in \Omega_N$ we have

$$u(x) = u\big(\Phi_1(x)\big) - \int_0^1 \frac{\partial}{\partial t} u\big(\Phi_t(x)\big)\, dt.$$

Now

$$\begin{aligned}\int_{\Omega_N} \left|u\big(\Phi_1(x)\big)\right| dx &\le d_N(1) \int_{\Omega_N} \left|u\big(\Phi_1(x)\big)\right| \left|\det \Phi_1'(x)\right| dx \\ &= d_N(1) \int_{\Phi_1(\Omega_N)} |u(y)|\, dy \\ &\le d_N(1) \int_\Omega |u(y)|\, dy.\end{aligned}$$

Also

$$\begin{aligned}\int_{\Omega_N} \left| \int_0^1 \frac{\partial}{\partial t} u\big(\Phi_t(x)\big)\, dt \right| dx &\le \int_{\Omega_N} dx \int_0^1 \left|\operatorname{grad} u\big(\Phi_t(x)\big)\right| \left|\frac{\partial}{\partial t}\Phi_t(x)\right| dt \\ &\le M \int_0^1 d_N(t)\, dt \int_{\Omega_N} \left|\operatorname{grad} u\big(\Phi_t(x)\big)\right| \left|\det \Phi_t'(x)\right| dx \\ &\le M \left(\int_0^1 d_N(t)\, dt\right) \left(\int_\Omega |\operatorname{grad} u(y)|\, dy\right).\end{aligned}$$

Putting $\delta_N = \max\left\{d_N(1), M\int_0^1 d_N(t)\, dt\right\}$, we have

$$\int_{\Omega_N} |u(x)|\, dx \le \delta_N \int_\Omega \big(|u(y)| + |\operatorname{grad} u(y)|\big)\, dy \le \delta_N \, \|u\|_{1,1,\Omega} \tag{30}$$

and $\lim_{N\to\infty} \delta_N = 0$.

Now suppose u is real-valued and belongs to $C^1(\Omega) \cap W^{1,p}(\Omega)$. By Hölder's inequality, the distributional derivatives of $|u|^p$

$$D_j|u|^p = p \cdot |u|^{p-1} \cdot \operatorname{sgn} u \cdot D_j u,$$

satisfy

$$\int_\Omega \left|D_j|u(x)|^p\right| dx \le p \left\|D_j u\right\|_{0,p,\Omega} \|u\|_{0,p,\Omega}^{p-1} \le p \|u\|_{1,p,\Omega}^p .$$

Thus $|u|^p \in W^{1,1}(\Omega)$ and by Theorem 3.17 there is a sequence $\{\phi_j\}$ of functions in $C^1(\Omega) \cap W^{1,1}(\Omega)$ such that $\lim_{j\to\infty} \left\|\phi_j - |u|^p\right\|_{1,1,\Omega} = 0$. Thus, by (30)

$$\begin{aligned}\int_{\Omega_N} |u(x)|^p \, dx &= \lim_{j\to\infty} \int_{\Omega_N} \phi_j(x)\, dx \le \limsup_{j\to\infty} \delta_N \left\|\phi_j\right\|_{1,1,\Omega} \\ &\le \delta_N \left\||u|^p\right\|_{1,1,\Omega} \le K\delta_N \|u\|_{1,p,\Omega}^p ,\end{aligned}$$

where $K = K(n, p)$. This inequality holds for arbitrary complex-valued function $u \in C^1(\Omega) \cap W^{1,p}(\Omega)$ by virtue of its separate applications to the real and imaginary parts of u.

If S is a bounded set in $W^{1,p}(\Omega)$ and $\epsilon > 0$, we may, by the above inequality, select N so that for all $u \in S$

$$\int_{\Omega_N} |u(x)|^p \, dx < \epsilon.$$

Since $W^{1,p}(\Omega - \Omega_N) \to L^p(\Omega - \Omega_N)$ is compact, the precompactness of S in $L^p(\Omega)$ follows by Theorem 2.33. Hence $W^{1,p}(\Omega) \to L^p(\Omega)$ is compact. ▮

6.53 EXAMPLE Consider again the domain of Examples 6.48 and 6.51 and the flow Φ given in the latter example. We have

$$d_N(t) = \sup_{x\ge N} \frac{f(x)}{f(x-t)} \le 1 \qquad \text{if} \quad 0 \le t \le 1$$

and by (28)

$$\lim_{N\to\infty} d_N(t) = 0 \qquad \text{if} \quad t > 0.$$

Thus by dominated convergence

$$\lim_{N\to\infty} \int_0^1 d_N(t)\, dt = 0.$$

The assumption that f' is bounded guarantees that the speed $|(\partial/\partial t)\Phi(x, y, t)|$ is bounded on U. Thus Ω satisfies the hypotheses of Theorem 6.52 and the imbedding $W^{m,p}(\Omega) \to L^p(\Omega)$ is compact for this domain. ∎

6.54 EXAMPLE Theorem 6.52 can also be used to show the compactness of $W^{m,p}(\Omega) \to L^p(\Omega)$ for some bounded domains to which neither the Rellich-Kondrachov theorem nor the techniques used in its proof can be applied. For example, we consider

$$\Omega = \big\{(x, y) \in \mathbb{R}^2 \ : 0 < x < 2,\ 0 < y < f(x)\big\},$$

where $f \in C^1([0, 2])$ is positive, nondecreasing, has bounded derivative f', and satisfies $\lim_{x\to 0+} f(x) = 0$. Let

$$U = \big\{(x, y, t) \in \mathbb{R}^3 : (x, y) \in \Omega,\ -x < t < 2 - x\big\}$$

and define the flow $\Phi : U \to \Omega$ by

$$\Phi(x, y, t) = \left(x + t,\ \frac{f(x+t)}{f(x)}\, y\right).$$

Then $\det \Phi_t'(x, y) = f(x + t)/f(x)$. If $\Omega_N^* = \{(x, y) \in \Omega : x > 1/N\}$, then

$$d_N(t) = \sup_{0<x\le 1/N} \left|\frac{f(x)}{f(x+t)}\right|$$

satisfies $d_N(t) \le 1$ for $0 \le t \le 1$, and $\lim_{N\to\infty} d_N(t) = 0$ if $t > 0$. Hence also $\lim_{N\to\infty} \int_0^1 d_N(t)\, dt = 0$ by dominated convergence. Since Ω_N^* is bounded and satisfies the cone condition, and since the boundedness of $\partial\Phi/\partial t$ is assured by that of f', we have, by Theorem 6.52 the compactness of

$$W^{m,p}(\Omega) \to L^p(\Omega). \tag{31}$$

However, suppose that $\lim_{x\to 0+} f(x)/x^k = 0$ for every k. (For example, this is true if $f(x) = e^{-1/x}$.) Then Ω has an exponential cusp at the origin and by Theorem 4.48 there exists no imbedding of the form $W^{m,p}(\Omega) \to L^q(\Omega)$ for any $q > p$ so the method of proof of the Rellich-Kondrachov theorem cannot be used to show the compactness of (31).

6.55 REMARKS

1. It is easy to imagine domains more general than those in the above examples to which Theorem 6.52 applies, although it may be difficult to specify an appropriate flow. A domain with many (perhaps infinitely many) unbounded branches can, if connected, admit a suitable flow provided volume decays sufficiently rapidly in each branch, a condition not fulfilled by the domain $\tilde{\Omega}$ in Example 6.49. For unbounded domains in which volume

decays monotonically in each branch Theorem 6.45 is essentially a converse of Theorem 6.52 in that the proof of Theorem 6.45 can be applied separately to show that the volume decays in each branch in the required way.

2. Since the only unbounded domains for which $W^{m,p}(\Omega)$ imbeds compactly into $L^p(\Omega)$ have finite volume there can be no extensions of Theorem 6.52 to give compact imbeddings into $L^q(\Omega)$ (where $q > p$), or $C_B(\Omega)$ etc.; there do not exist such imbeddings.

Hilbert-Schmidt Imbeddings

6.56 (Complete Orthonormal Systems) A *complete orthonormal system* in a separable Hilbert space X is a sequence $\{e_i\}_{i=1}^{\infty}$ of elements of X satisfying

$$(e_i, e_j)_X = \begin{cases} 1 & \text{if } i = j \\ 0 & \text{if } i \neq j, \end{cases}$$

(where $(\cdot, \cdot)_X$ is the inner product on X), and such that for each $x \in X$ we have

$$\lim_{k \to \infty} \left\| x - \sum_{i=1}^{k} (x, e_i)_X e_i \, ; X \right\| = 0. \tag{32}$$

Thus $x = \sum_{i=1}^{\infty} (x, e_i) e_i$, the series converging with respect to the norm in X. It is well known that every separable Hilbert space possesses such a complete orthonormal system. There follows from (32) the Parseval identity

$$\|x \, ; X\|^2 = \sum_{i=1}^{\infty} |(x, e_i)_X|^2 .$$

6.57 (Hilbert-Schmidt Operators) Let X and Y be two separable Hilbert spaces and let $\{e_i\}_{i=1}^{\infty}$ and $\{f_i\}_{i=1}^{\infty}$ be given complete orthornomal systems in X and Y respectively. Let A be a bounded linear operator with domain X taking values in Y, and let A^* be the adjoint of A taking Y into X and defined by

$$(x, A^* y)_X = (Ax, y)_Y, \qquad x \in X, \quad y \in Y.$$

Define

$$|||A|||^2 = \sum_{i=1}^{\infty} \|Ae_i \, ; Y\|^2, \qquad |||A^*|||^2 = \sum_{i=1}^{\infty} \|A^* f_i \, ; X\|^2 .$$

If $|||A|||$ is finite, A is called a *Hilbert-Schmidt operator* and we call $|||A|||$ its *Hilbert-Schmidt norm.* Recall that the operator norm of A is given by

$$\|A\| = \sup\{\|Ax\,;Y\| \ :\ \|x\,;X\| \le 1\}.$$

We must justify the definition of the Hilbert-Schmidt norm.

6.58 LEMMA The norms $|||A|||$ and $|||A^*|||$ are independent of the particular orthonormal systems $\{e_i\}$ and $\{f_i\}$ used to define them. Moreover

$$|||A||| = |||A^*||| \ge \|A\|.$$

Proof. By Parseval's identity

$$|||A|||^2 = \sum_{i=1}^{\infty} \|Ae_i\,;Y\|^2 = \sum_{i=1}^{\infty}\sum_{j=1}^{\infty} |(Ae_i, f_j)_Y|^2$$
$$= \sum_{j=1}^{\infty}\sum_{i=1}^{\infty} |(e_i, A^*f_j)_X|^2 = \sum_{j=1}^{\infty} \left\|A^*f_j\,;X\right\|^2 = |||A^*|||^2.$$

Hence each expression is independent of $\{e_i\}$ and $\{f_j\}$. For any $x \in X$ we have

$$\|Ax\,;Y\|^2 = \left\|\sum_{i=1}^{\infty}(x, e_i)_X Ae_i\,;Y\right\|^2 \le \left(\sum_{i=1}^{\infty} |(x, e_i)_X|\,\|Ae_i\,;Y\|\right)^2$$
$$\le \left(\sum_{i=1}^{\infty} |(x, e_i)_X|^2\right)\left(\sum_{j=1}^{\infty} \left\|Ae_j\,;Y\right\|^2\right) = \|x\,;X\|^2\,|||A|||^2.$$

Hence $\|A\| \le |||A|||$ as required. ∎

6.59 REMARK Consider the scalars (Ae_i, f_j) for $1 \le i, j < \infty$; they are the entries in an infinite matrix representing the operator A. The lemma above shows that the Hilbert-Schmidt norm of A is the sum of the squares of the absolute values of the elements of this matrix. Similarly, the numbers (A^*f_j, e_i) are the entries in a matrix representing A^*. Since these matrices are adjoints of each other, the equality of the corresponding Hilbert-Schmidt norms of the operators is assured.

6.60 We leave to the reader the task of verifying the following assertions.

(a) If X, Y, and Z are separable Hilbert spaces and A and B are bounded linear operators from X into Y and Y into Z, respectively, then $B \circ A$, which maps X into Z, is a Hilbert-Schmidt operator if either A or B is. If A is Hilbert-Schmidt, then $|||B \circ A||| \le \|B\|\,|||A|||$.

(b) Every Hilbert-Schmidt operator is compact.

The following Theorem, due to Maurin [Mr] has far-reaching implications for eigenfunction expansions corresponding to differential operators.

6.61 THEOREM (Maurin's Theorem) Let Ω be a bounded domain in $\mathbb{R}^n$ satisfying the cone condition. Let m and k be nonnegative integers with $k > n/2$. Then the imbedding map

$$W^{m+k,2}(\Omega) \to W^{m,2}(\Omega) \tag{33}$$

is a Hilbert-Schmidt operator. Similarly the imbedding map

$$W_0^{m+k,2}(\Omega) \to W_0^{m,2}(\Omega) \tag{34}$$

is a Hilbert-Schmidt operator for any bounded domain Ω.

Proof. Given $y \in \Omega$ and α with $|\alpha| \le m$ we define a linear functional T_y^α on $W^{m+k,2}(\Omega)$ by

$$T_y^\alpha(u) = D^\alpha u(y).$$

Since $2k > m$, the Sobolev Imbedding Theorem 4.12 implies that T_y^α is bounded on $W^{m+k,2}(\Omega)$ and has norm bounded by a constant K independent of y and α:

$$|T_y^\alpha(u)| \le \max_{0\le|\alpha|\le m} \sup_{x\in\Omega} |D^\alpha u(x)| \le K\, \|u\|_{m+k,2,\Omega}\,.$$

By the Riesz representation theorem for Hilbert spaces there exists $v_y^\alpha \in W^{m+k,2}(\Omega)$ such that

$$D^\alpha u(y) = T_y^\alpha(u) = \left(u, v_y^\alpha\right)_{m+k},$$

where $(\cdot,\cdot)_{m+k}$ is the inner product on $W^{m+k,2}(\Omega)$. Moreover

$$\left\|v_y^\alpha\right\|_{m+k,2,\Omega}^2 = \left\|T_y^\alpha ; \left[W^{m+k,2}(\Omega)\right]'\right\| \le K.$$

If $\{e_i\}_{i=1}^\infty$ is a complete orthonormal system in $W^{m+k,2}(\Omega)$, then

$$\left\|v_y^\alpha\right\|_{m+k,2,\Omega}^2 = \sum_{i=1}^\infty \left|\left(e_i, v_y^\alpha\right)_{m+k}\right|^2 = \sum_{i=1}^\infty |D^\alpha e_i(y)|^2.$$

Consequently,

$$\sum_{i=1}^\infty \|e_i\|_{m,2,\Omega}^2 \le \sum_{|\alpha|\le m} \int_\Omega \left\|v_y^\alpha\right\|_{m+k,2,\Omega}^2 dy \le \sum_{|\alpha|\le m} K \operatorname{vol}(\Omega) < \infty.$$

Hence imbedding (33) is Hilbert-Schmidt. The corresponding imbedding (34) is also Hilbert-Schmidt without the cone-condition requirement as it is not needed for the application of Theorem 4.12 in this case. ∎

The following generalization of Maurin's theorem is due to Clark [Ck].

6.62 THEOREM Let μ be a nonnegative, measurable function defined on the domain Ω in $\mathbb{R}^n$. Let $W_0^{m,2;\mu}(\Omega)$ be the Hilbert space obtained by completing $C_0^\infty(\Omega)$ with respect to the weighted norm

$$\|u\|_{m,2;\mu} = \left(\sum_{|\alpha|\le m} \int_\Omega |D^\alpha u(x)|^2 \mu(x)\, dx \right)^{1/2}.$$

For $y \in \Omega$ let $\tau(y) = \operatorname{dist}(y, \operatorname{bdry}\Omega)$. Suppose that

$$\int_\Omega \left(\tau(y)\right)^{2\nu} \mu(y)\, dy < \infty \tag{35}$$

for some nonnegative integer ν. If $k > \nu + n/2$, then the imbedding

$$W_0^{m+k,2}(\Omega) \to W_0^{m,2;\mu}(\Omega) \tag{36}$$

(exists and) is Hilbert-Schmidt.

Proof. The argument is parallel to that given in the proof of Maurin's theorem above. Let $\{e_i\}$, T_y^α, and v_y^α be defined as there. If $y \in \Omega$, let y_0 be chosen in $\operatorname{bdry}\Omega$ such that $\tau(y) = |y - y_0|$. If ν is a positive integer and $u \in C_0^\infty(\Omega)$, we have by Taylor's formula with remainder

$$D^\alpha u(y) = \sum_{|\beta|=\nu} \frac{1}{\beta!} D^{\alpha+\beta} u(y_\beta)(y - y_\beta)^\beta$$

for some points y_β satisfying $|y - y_\beta| \le \tau(y)$. If $|\alpha| \le m$ and $k > \nu + n/2$, we obtain from Theorem 4.12

$$|D^\alpha u(y)| \le K\, \|u\|_{m+k,2,\Omega} \left(\tau(y)\right)^\nu.$$

By completion this inequality holds for any $u \in W_0^{m+k,2}(\Omega)$. As in the proof of Maurin's theorem, it follows that

$$\left\|v_y^\alpha\right\|_{m+k,2,\Omega} = \sup_{\|u\|_{m+k,2,\Omega}=1} |D^\alpha u(y)| \le K\left(\tau(y)\right)^\nu,$$

and hence also that

$$\sum_{i=1}^{\infty} \|e_i\|_{m,2;\mu}^2 \le \sum_{|\alpha| \le m} \int_\Omega \left\|v_y^\alpha\right\|_{m+k,2,\Omega}^2 \mu(y)\,dy$$

$$\le K^2 \sum_{|\alpha| \le m} \int_\Omega \left(\tau(y)\right)^{2\nu} \mu(y)\,dy < \infty$$

by (35). Hence imbedding (36) is Hilbert-Schmidt. ∎

6.63 REMARK Various choices of μ and ν lead to generalizations of Maurin's theorem for imbeddings of the sort (34). If $\mu(x) = 1$ and $\nu = 0$ we obtain the obvious generalization to unbounded domains of finite volume. If $\mu(x) = 1$ and $\nu > 0$, Ω may be unbounded and even have infinite volume, but it must be quasibounded by (35). Of course quasiboundedness may not be sufficient to guarantee (35). If μ is the characteristic function of a bounded subdomain Ω_0 of Ω, and $\nu = 0$, we obtain the Hilbert-Schmidt imbedding

$$W_0^{m+k,2}(\Omega) \to W^{m,2}(\Omega_0), \qquad k > n/2.$$

7

FRACTIONAL ORDER SPACES

Introduction

7.1 This chapter is concerned with extending the notion of the standard Sobolev space $W^{m,p}(\Omega)$ to include spaces where m need not be an integer. There are various ways to define such *fractional order* spaces; many of them depend on using interpolation to construct scales of spaces suitably intermediate between two extreme spaces, say $L^p(\Omega)$ and $W^{m,p}(\Omega)$.

Interpolation methods themselves come in two flavours: real methods and complex methods. We have already seen an example of the real method in the Marcinkiewicz theorem of Paragraph 2.58. Although the details of the real method can be found in several sources, for example, [BB], [BL], and [BSh], we shall provide a treatment here in sufficient detail to make clear its application to the development of the Besov spaces, one of the scales of fractional order Sobolev spaces that particularly lends itself to characterizing the spaces of traces of functions in $W^{m,p}(\Omega)$ on the boundaries of smoothly bounded domains Ω; such characterizations are useful in the study of boundary-value problems. Several older interpolation methods are known [BL, pp. 70–75] to be equivalent to the now-standard real interpolation method that we use here. In the the corresponding chapter of the previous edition [A] of this book, the older method of traces was used rather than the method presented in this edition. Later in this Chapter, we prove a trace theorem (Theorem 7.39) giving an instance of that equivalence.

After that we shall describe more briefly other scales of fractional order Sobolev spaces, some obtained by complex methods and some by Fourier decompositions.

The Bochner Integral

7.2 In developing the real interpolation method below we will use the concept of the integral of a Banach-space-valued function defined on an interval on the real line $\mathbb{R}$. (For the complex method we will use the concept of analytic Banach-space-valued functions of a complex variable.) We present here a brief description of the Bochner integral, referring the reader to [Y] or [BB] for more details.

Let X be a Banach space with norm $\|\cdot\|_X$ and let f be a function defined on an interval (a, b) in $\mathbb{R}$ (which may be infinite) and having values in X. In addition, let μ be a measure on (a, b) given by $d\mu(t) = w(t)\,dt$ where w is continuous and positive on (a, b). Of special concern to us later will be the case where $a = 0$, $b = \infty$, and $w(t) = 1/t$. In this case μ is the Haar measure on $(0, \infty)$, which is invariant under scaling in the multiplicative group $(0, \infty)$: if $(c, d) \subset (0, \infty)$ and $\lambda > 0$, then $\mu(\lambda c, \lambda d) = \mu(c, d)$.

We want to define the integral of f over (a, b).

7.3 (Definition of the Bochner Integral) If $\{A_1, \dots, A_k\}$ is a finite collection of mutually disjoint subsets of (a, b) each having finite μ-measure, and if $\{x_1, \dots, x_k\}$ is a corresponding set of elements of X, we call the function f defined by

$$f(t) = \sum_{i=1}^{k} \chi_{A_i}(t) x_i, \qquad a < t < b,$$

a *simple function* on (a, b) into X. For such simple functions we define, obviously,

$$\int_a^b f(t)\,d\mu(t) = \sum_{i=1}^{k} \mu(A_i) x_i = \sum_{i=1}^{k} \left(\int_{A_i} w(t)\,dt \right) x_i.$$

Of course, a different representation of the simple function f using a different collection of subsets of (a, b) will yield the same value for the integral; the subsets in the collections need not be mutually disjoint, and given two such collections we can always form an equivalent mutually disjoint collection consisting of pairwise intersections of the elements of the two collections.

Now let f an arbitrary function defined on (a, b) into X. We say that f is *(strongly) measurable* on (a, b) if there exists a sequence $\{f_j\}$ of simple functions with supports in (a, b) such that

$$\lim_{j\to\infty} \left\| f_j(t) - f(t) \right\|_X \qquad \text{a.e. in } (a, b). \tag{1}$$

It can be shown that f is measurable if its range is separable and if, for each x' in the dual of X, the scalar-valued function $x'\big(f(\cdot)\big)$ is measurable on (a, b).

Suppose that a sequence of simple functions $\{f_j\}$ satisfying (1) can be chosen in such a way that

$$\lim_{j\to\infty}\int_a^b \big\|f_j(t) - f(t)\big\|_X \, d\mu(t) = 0.$$

Then we say that f is *Bochner integrable* on (a, b) and we define

$$\int_a^b f(t)\, d\mu(t) = \lim_{j\to\infty}\int_a^b f_j(t)\, d\mu(t).$$

Again we observe that the limit does not depend on the choice of the approximating simple functions.

A measurable function f is integrable on (a, b) if and only if the scalar-valued function $\|f(\cdot)\|_X$ is integrable on (a, b). In fact, there holds the "triangle inequality"

$$\left\|\int_a^b f(t)\, d\mu(t)\right\|_X \le \int_a^b \|f(t)\|_X \, d\mu(t).$$

7.4 (The Spaces $L^q(a, b; d\mu, X)$) If $1 \le q \le \infty$, we say that $f \in L^q(a, b; d\mu, X)$ provided $\|f\,;\, L^q(a, b; d\mu, X)\| < \infty$, where

$$\big\|f\,;\, L^q(a, b; d\mu, X)\big\| = \begin{cases} \left(\displaystyle\int_a^b \|f(t)\|_X^q \, d\mu(t)\right)^{1/q} & \text{if } 1 \le q < \infty \\ \operatorname{ess\,sup}_{a<t<b}\{\|f(t)\|_X\} & \text{if } q = \infty. \end{cases}$$

In particular, if $X = \mathbb{R}$ or $X = \mathbb{C}$, we will denote $L^q(a, b; d\mu, X)$ simply by $L^q(a, b; d\mu)$.

7.5 (The spaces L^q_*) Of much importance below is the special case where $X = \mathbb{R}$ or $\mathbb{C}$, $(a, b) = (0, \infty)$, and $d\mu = dt/t$; we will further abbreviate the notation for this , denoting $L^q(a, b; d\mu, X)$ simply L^q_*. Note that L^q_* is equivalent to $L^q(\mathbb{R})$ with Lebesgue measure via a change of variable: if $t = e^s$ and $f(t) = f(e^s) = F(s)$, then $\big\|f\,;\, L^q_*\big\| = \|F\|_{q,\mathbb{R}}$. Most of the properties of $L^q(\mathbb{R})$ transfer to properties of L^q_*. In particular Hölder's and Young's inequalities hold; we will need both of them below. It should be noted that the convolution of two functions f and g defined on $(0, \infty)$ and integrated with respect to the Haar measure dt/t is given by

$$f * g(t) = \int_0^\infty f\left(\frac{t}{s}\right) g(s)\, \frac{ds}{s},$$

and Young's inequality proclaims $\|f * g\,;L^r_*\| \le \|f\,;L^p_*\|\,\|g\,;L^q_*\|$ provided $p, q, r \ge 1$ and $1 + (1/r) = (1/p) + (1/q)$.

Intermediate Spaces and Interpolation — The Real Method

7.6 In this Section we will be discussing the construction of Banach spaces X that are suitably intermediate between two Banach spaces X_0 and X_1, each of which is (continuously) imbedded in a Hausdorff topological vector space $\mathcal{X}$, and whose intersection is nontrivial. (Such a pair of spaces $\{X_0, X_1\}$ is called an *interpolation pair* and X is called an intermediate space of the pair. In some of our later applications, we will have $X_1 \to X_0$ (for example, $X_0 = L^p(\Omega)$ and $X_1 = W^{m,p}(\Omega)$), in which case we can clearly take $\mathcal{X} = X_0$. We shall, in fact, be constructing families of such intermediate spaces $X_{\theta,q}$ between X_0 and X_1, such that if $Y_{\theta,q}$ is the corresponding intermediate space for another such interpolation pair $\{Y_0, Y_1\}$ with Y_0 and Y_1 imbedded in $\mathcal{Y}$, and if T is a linear operator from $\mathcal{X}$ into $\mathcal{Y}$ (for example an imbedding operator) such that T is bounded from X_i into Y_i, $i = 0, 1$, then T will also be bounded from $X_{\theta,q}$ into $Y_{\theta,q}$.

There are many different ways of constructing such intermediate spaces, mostly leading to the same spaces with equivalent norms. We examine here two such methods, the J-method and the K-method, (together called the real method) due to Lions and Peetre. The theory is developed in several texts, in particular [BB] and [BL]. Our approach follows [BB] and we will omit some aspects of the theory for which we have no future need.

7.7 (Intermediate Spaces) Let $\|\cdot\|_{X_i}$ denote the norm in X_i, $i = 0, 1$. The intersection $X_0 \cap X_1$ and the algebraic sum $X_0 + X_1 = \{u = u_0 + u_1 : u_0 \in X_0, u_1 \in X_1\}$ are themselves Banach spaces with respect to the norms

$$\|u\|_{X_0\cap X_1} = \max\left\{\|u\|_{X_0}, \|u\|_{X_1}\right\}$$
$$\|u\|_{X_0+X_1} = \inf\left\{\|u_0\|_{X_0} + \|u_1\|_{X_1} : u = u_0 + u_1, u_0 \in X_0, u_1 \in X_1\right\}.$$

and $X_0 \cap X_1 \to X_i \to X_0 + X_1$ for $i = 0, 1$.

In general, we say that a Banach space X is *intermediate* between X_0 and X_1 if there exist the imbeddings

$$X_0 \cap X_1 \to X \to X_0 + X_1.$$

7.8 (The J and K norms) For each fixed $t > 0$ the following functionals define norms on $X_0 \cap X_1$ and $X_0 + X_1$ respectively, equivalent to the norms defined above:

$$J(t; u) = \max\left\{\|u\|_{X_0}, t\,\|u\|_{X_1}\right\}$$
$$K(t; u) = \inf\left\{\|u_0\|_{X_0} + t\,\|u_1\|_{X_1} : u = u_0 + u_1, u_0 \in X_0, u_1 \in X_1\right\}.$$

Evidently $J(1; u) = \|u\|_{X_0 \cap X_1}$, $K(1; u) = \|u\|_{X_0 + X_1}$, and $J(t; u)$ and $K(t; u)$ are continuous and monotonically increasing functions of t on $(0, \infty)$. Moreover

$$\min\{1, t\} \|u\|_{X_0 \cap X_1} \le J(t; u) \le \max\{1, t\} \|u\|_{X_0 \cap X_1} \tag{2}$$
$$\min\{1, t\} \|u\|_{X_0 + X_1} \le K(t; u) \le \max\{1, t\} \|u\|_{X_0 + X_1} . \tag{3}$$

$J(t; u)$ is a convex function of t because, if $0 < a < b$ and $0 < \theta < 1$,

$$\begin{aligned} J\big((1-\theta)a + \theta b; u\big) &= \max\big\{\|u\|_{X_0}, (1-\theta)a \|u\|_{X_1} + \theta b \|u\|_{X_1}\big\} \\ &\le (1-\theta) \max\big\{\|u\|_{X_0}, a \|u\|_{X_1}\big\} + \theta \max\big\{\|u\|_{X_0}, b \|u\|_{X_1}\big\} \\ &= (1-\theta) J(a; u) + \theta J(b; u). \end{aligned}$$

Also for such a, b, θ and any $u_0 \in X_0$ and $u_1 \in X_1$ for which $u = u_0 + u_1$ we have

$$\begin{aligned} &\|u_0\|_{X_0} + \big((1-\theta)a + \theta b\big) \|u_1\|_{X_1} \\ &\qquad = (1-\theta)\big(\|u_0\|_{X_0} + a \|u_1\|_{X_1}\big) + \theta\big(\|u_0\|_{X_0} + b \|u_1\|_{X_1}\big) \\ &\qquad \ge (1-\theta) K(a; u) + \theta K(b; u), \end{aligned}$$

so that $K\big((1-\theta)a + \theta b); u\big) \ge (1-\theta)K(a; u) + \theta K(b; u)$ and $K(t; u)$ is a concave function of t.

Finally we observe that if $u \in X_0 \cap X_1$, then for any positive t and s we have $K(t; u) \le \|u\|_{X_0} \le J(s; u)$ and $K(t; u) \le t \|u\|_{X_1} = (t/s)s \|u\|_{X_1} \le (t/s) J(s; u)$. Accordingly,

$$K(t; u) \le \min\left\{1, \frac{t}{s}\right\} J(s; u). \tag{4}$$

7.9 (The K-method) If $0 \le \theta \le 1$ and $1 \le q \le \infty$ we denote by $(X_0, X_1)_{\theta,q;K}$ the space of all $u \in X_0 + X_1$ such that the function $t \to t^{-\theta} K(t; u)$ belongs to $L^q_* = L^q(0, \infty; dt/t)$.

Of course, the zero element $u = 0$ of $X_0 + X_1$ always belongs to $(X_0, X_1)_{\theta,q;K}$. The following theorem shows that if $1 \le q < \infty$ and either $\theta = 0$ or $\theta = 1$, then $(X_0, X_1)_{\theta,q;K}$ contains only this trivial element. Otherwise $(X_0, X_1)_{\theta,q;K}$ is an intermediate space between X_0 and X_1.

7.10 THEOREM If and only if either $1 \le q < \infty$ and $0 < \theta < 1$ or $q = \infty$ and $0 \le \theta \le 1$, then the space $(X_0, X_1)_{\theta,q;K}$ is a nontrivial Banach space with norm

$$\|u\|_{\theta,q;K} = \begin{cases} \left(\displaystyle\int_0^\infty \big(t^{-\theta} K(t; u)\big)^q \frac{dt}{t}\right)^{1/q} & \text{if } 1 \le q < \infty \\ \operatorname{ess\,sup}_{0<t<\infty} \{t^{-\theta} K(t; u)\} & \text{if } q = \infty. \end{cases}$$

Furthermore,

$$\|u\|_{X_0+X_1} \le \frac{\|u\|_{\theta,q;K}}{\|t^{-\theta}\min\{1,t\}\,;\,L^q_*\|} \le \|u\|_{X_0\cap X_1} \tag{5}$$

so there hold the imbeddings

$$X_0 \cap X_1 \to (X_0, X_1)_{\theta,q;K} \to X_0 + X_1$$

and $(X_0, X_1)_{\theta,q;K}$ is an intermediate space between X_0 and X_1.
Otherwise $(X_0, X_1)_{\theta,q;K} = \{0\}$.

Proof. It is easily checked that the function $t \to t^{-\theta}\min\{1,t\}$ belongs to L^q_* if and only if θ and q satisfy the conditions of the theorem. Since (3) shows that

$$\|t^{-\theta}\min\{1,t\}\,;\,L^q_*\|\,\|u\|_{X_0+X_1} \le \|t^{-\theta}K(t;u)\,;\,L^q_*\| = \|u\|_{\theta,q;K},$$

there can be no nonzero elements of $(X_0, X_1)_{\theta,q;K}$ unless those conditions are satisfied. If so, then the left inequality in (5) holds and $(X_0, X_1)_{\theta,q;K} \to X_0 + X_1$. Also, by (4) we have $K(t;u) \le \min\{1,t\}J(1;u) = \min\{1,t\}\,\|u\|_{X_0\cap X_1}$ so the right inequality in (5) holds and $X_0 \cap X_1 \to (X_0, X_1)_{\theta,q;K}$.
Verification that $\|u\|_{\theta,q;K}$ is a norm and that $(X_0, X_1)_{\theta,q;K}$ is complete under it are left as exercises for the reader. ∎

Note that $u \in X_0$ and $\theta = 0$ implies that $t^{-\theta}K(t;u) \le \|u\|_{X_0}$. Also, $u \in X_1$ and $\theta = 1$ implies that $t^{-\theta}K(t;u) \le \|u\|_{X_1}$. Thus we also have

$$X_0 \to (X_0, X_1)_{0,\infty;K} \quad \text{and} \quad X_1 \to (X_0, X_1)_{1,\infty;K}. \tag{6}$$

7.11 THEOREM (A Discrete Version of the K-method) For each integer i let $K_i(u) = K(2^i;u)$. Then $u \in (X_0, X_1)_{\theta,q;K}$ if and only if the sequence $\{2^{-i\theta}K_i(u)\}_{i=-\infty}^{\infty}$ belongs to the space ℓ^q (defined in Paragraph 2.27). Moreover, the ℓ^q-norm of that sequence is equivalent to $\|u\|_{\theta,q;K}$.

Proof. First write (for $1 \le q < \infty$)

$$\int_0^\infty \left(t^{-\theta}K(t;u)\right)^q \frac{dt}{t} = \sum_{i=-\infty}^{\infty} \int_{2^i}^{2^{i+1}} \left(t^{-\theta}K(t;u)\right)^q \frac{dt}{t}.$$

Since $K(t;u)$ increases and $t^{-\theta}$ decreases as t increases, we have for $2^i \le t \le 2^{i+1}$,

$$2^{-(i+1)\theta}K_i(u) \le t^{-\theta}K(t;u) \le 2^{-i\theta}K_{i+1}(u),$$

so that

$$2^{-\theta q} \ln 2\big[2^{-i\theta} K_i(u)\big]^q \le \int_{2^i}^{2^{i+1}} \big(t^{-\theta} K(t; u)\big)^q \frac{dt}{t} \le 2^{\theta q} \ln 2\big[2^{-(i+1)\theta} K_{i+1}(u)\big]^q.$$

Summing on i and taking qth roots then gives

$$2^{-\theta} (\ln 2)^{1/q} \left\| \{2^{-i\theta} K_i(u)\}\, ;\, \ell^q \right\| \le \|u\|_{\theta,q;K} \le 2^{\theta} (\ln 2)^{1/q} \left\| \{2^{-i\theta} K_i(u)\}\, ;\, \ell^q \right\|.$$

The proof for $q = \infty$ is easier and left for the reader. ∎

7.12 (The J-method) If $0 \le \theta \le 1$ and $1 \le q \le \infty$ we denote by $(X_0, X_1)_{\theta,q;J}$ the space of all $u \in X_0 + X_1$ such that

$$u = \int_0^\infty f(t) \frac{dt}{t}$$

(Bochner integral) for some $f \in L^1(0, \infty; dt/t, X_0+X_1)$ having values in $X_0 \cap X_1$ and such that the real-valued function $t \to t^{-\theta} J(t; f)$ belongs to L^q_*.

7.13 THEOREM If either $1 < q \le \infty$ and $0 < \theta < 1$ or $q = 1$ and $0 \le \theta \le 1$, then $(X_0, X_1)_{\theta,q;J}$ is a nontrivial Banach space with norm

$$\begin{aligned} \|u\|_{\theta,q;J} &= \inf_{f \in S(u)} \left\| t^{-\theta} J\big(t; f(t)\big)\, ;\, L^q_* \right\| \\ &= \inf_{f \in S(u)} \left(\int_0^\infty \big[t^{-\theta} J\big(t; f(t)\big)\big]^q \frac{dt}{t} \right)^{1/q}, \qquad (\text{if } q < \infty), \end{aligned}$$

where

$$S(u) = \left\{ f \in L^1(0, \infty; dt/t, X_0 + X_1) \,:\, u = \int_0^\infty f(t) \frac{dt}{t} \right\}.$$

Furthermore,

$$\|u\|_{X_0+X_1} \le \left(\left\| t^{-\theta} \min\{1, t\}\, ;\, L^{q'}_* \right\| \right) \|u\|_{\theta,q;J} \le \|u\|_{X_0 \cap X_1} \tag{7}$$

so that

$$X_0 \cap X_1 \to (X_0, X_1)_{\theta,q;J} \to X_0 + X_1$$

and $(X_0, X_1)_{\theta,q;J}$ is an intermediate space between X_0 and X_1.

Proof. Again we leave verification of the norm and completeness properties to the reader and we concentrate on the imbeddings.

Let $f \in S(u)$. By (3) and (4) with $t = 1$ and $s = \tau$ we have

$$\|f(\tau)\|_{X_0+X_1} \le K\big(1, f(\tau)\big) \le \min\left\{1, \frac{1}{\tau}\right\} J\big(\tau, f(\tau)\big).$$

Accordingly, If $(1/q) + (1/q') = 1$, then by Hölder's inequality

$$\begin{aligned}\|u\|_{X_0+X_1} &\le \int_0^\infty \|f(\tau)\|_{X_0+X_1} \frac{d\tau}{\tau} \le \int_0^\infty \min\left\{1, \frac{1}{\tau}\right\} J\big(\tau, f(\tau)\big) \frac{d\tau}{\tau} \\ &\le \left\|\tau^\theta \min\left\{1, \frac{1}{\tau}\right\} ; L_*^{q'}\right\| \left\|t^{-\theta} J\big(t; f(t)\big) ; L_*^q\right\|.\end{aligned}$$

The first factor in this product of norms is finite if θ and q satisfy the conditions of the theorem, and if we replace τ with $1/t$ in it, we can see that it is equal to $\left\|t^{-\theta} \min\{1, t\} ; L_*^{q'}\right\|$. Since the above inequality holds for all $f \in S(u)$, the left inequality in (7) is established and $(X_0, X_1)_{\theta,q;J} \to X_0 + X_1$.

To verify the right inequality in (7), let $u \in X_0 \cap X_1$. Let $\phi(t) \ge 0$ satisfy $\|t^{-\theta}\phi(t) ; L_*^q\| = 1$. Hölder's inequality shows that

$$\int_0^\infty \phi(\tau) \min\{1, 1/\tau\} \frac{d\tau}{\tau} < \infty.$$

If

$$f(t) = \frac{\phi(t) \min\{1, 1/t\}}{\displaystyle\int_0^\infty \phi(\tau) \min\{1, 1/\tau\} \frac{d\tau}{\tau}} u,$$

then $f \in S(u)$ and

$$\begin{aligned}J\big(t; f(t)\big) &= \frac{\phi(t) \min\{1, 1/t\}}{\displaystyle\int_0^\infty \phi(\tau) \min\{1, 1/\tau\} \frac{d\tau}{\tau}} J(t; u) \\ &\le \frac{\phi(t)}{\displaystyle\int_0^\infty \phi(\tau) \min\{1, 1/\tau\} \frac{d\tau}{\tau}} \|u\|_{X_0\cap X_1},\end{aligned}$$

the latter inequality following from (2) since $\max\{1, t\} = \big(\min\{1, 1/t\}\big)^{-1}$. Therefore,

$$\begin{aligned}&\left(\int_0^\infty \phi(\tau) \min\{1, 1/\tau\} \frac{d\tau}{\tau}\right) \|u\|_{\theta,q;J} \\ &\le \left(\int_0^\infty \phi(\tau) \min\{1, 1/\tau\} \frac{d\tau}{\tau}\right) \left(\int_0^\infty \big(t^{-\theta} J\big(t; f(t)\big)\big)^q \frac{dt}{t}\right)^{1/q} \\ &\le \left(\int_0^\infty \big(t^{-\theta}\phi(t) \|u\|_{X_0\cap X_1}\big)^q \frac{dt}{t}\right)^{1/q} = \|u\|_{X_0\cap X_1}.\end{aligned}$$

By the converse to Hölder's inequality,

$$\sup\left\{\int_0^\infty \phi(\tau)\min\{1,1/\tau\}\,\frac{d\tau}{\tau} \;:\; \left\|\tau^{-\theta}\phi(\tau)\,;\,L_*^q\right\| = 1\right\}$$
$$= \left\|\tau^{\theta}\min\{1,1/\tau\}\,;\,L_*^{q'}\right\| = \left\|t^{-\theta}\min\{1,t\}\,;\,L_*^{q'}\right\|.$$

Thus the right inequality in (7) is established and $X_0 \cap X_1 \to (X_0, X_1)_{\theta,q;J}$. ∎

7.14 Observe that if $u = \int_0^\infty f(t)\,dt/t$ where $f(t) \in X_0 \cap X_1$, then

$$\|u\|_{X_0} \le \int_0^\infty \|f(t)\|_{X_0}\,\frac{dt}{t} \le \int_0^\infty J\big(t, f(t)\big)\,\frac{dt}{t}$$
$$\|u\|_{X_1} \le \int_0^\infty \|f(t)\|_{X_1}\,\frac{dt}{t} \le \int_0^\infty t^{-1}J\big(t, f(t)\big)\,\frac{dt}{t}.$$

Each of these estimates holds for all such representations of u, so $\|u\|_{X_0} \le \|u\|_{0,1;J}$ and $\|u\|_{X_1} \le \|u\|_{1,1;J}$. Combining these with (6) we obtain

$$\begin{aligned}(X_0, X_1)_{0,1;J} &\to X_0 \to (X_0, X_1)_{0,\infty;K}\\ (X_0, X_1)_{1,1;J} &\to X_1 \to (X_0, X_1)_{1,\infty;K}.\end{aligned} \tag{8}$$

There is also a discrete version of the J-method leading to an equivalent norm for $(X_0, X_1)_{\theta,q;J}$.

7.15 THEOREM (A Discrete Version of the J-method) An element u of $X_0 + X_1$ belongs to $(X_0, X_1)_{\theta,q;J}$ if and only if $u = \sum_{i=-\infty}^{\infty} u_i$ where the series converges in $X_0 + X_1$ and the sequence $\{2^{-\theta i}J(2^i, u_i)\}$ belongs to ℓ^q. In this case

$$\inf\left\{\left\|\{2^{-\theta i}J(2^i; u_i)\}\,;\,\ell^q\right\| \;:\; u = \sum_{i=-\infty}^{\infty} u_i\right\}$$

is a norm on $(X_0, X_1)_{\theta,q;J}$ equivalent to $\|u\|_{\theta,q;J}$.

Proof. Again we will show this for $1 \le q < \infty$ and leave the easier case $q = \infty$ to the reader.

First suppose that $u \in (X_0, X_1)_{\theta,q;J}$ and let $\epsilon > 0$. Then there exists a function $f \in L^1(0,\infty; dt/t, X_0 + X_1)$ such that

$$u = \int_0^\infty f(t)\,\frac{dt}{t}$$

and

$$\int_0^\infty \left[t^{-\theta}J\big(t; f(t)\big)\right]^q \frac{dt}{t} \le (1+\epsilon)\,\|u\|_{\theta,q;J}.$$

Let the sequence $\{u_i\}_{i=-\infty}^{\infty}$ be defined by

$$u_i = \int_{2^i}^{2^{i+1}} f(t)\,\frac{dt}{t}.$$

then $\sum_{i=-\infty}^{\infty} u_i$ converges to u in $X_0 + X_1$ because the integral representation converges to u there. Moreover,

$$\begin{aligned}
2^{-i\theta} J(2^i; u_i) &\le \int_{2^i}^{2^{i+1}} 2^{-i\theta} J\big(t; f(t)\big)\,\frac{dt}{t} \\
&= 2^{\theta} \int_{2^i}^{2^{i+1}} 2^{-(i+1)\theta} J\big(t; f(t)\big)\,\frac{dt}{t} \\
&\le 2^{\theta} \int_{2^i}^{2^{i+1}} t^{-\theta} J\big(t; f(t)\big)\,\frac{dt}{t} \\
&\le 2^{\theta} (\ln 2)^{1/q'} \left(\int_{2^i}^{2^{i+1}} \big[t^{-\theta} J\big(t; f(t)\big)\big]^q \,\frac{dt}{t} \right)^{1/q},
\end{aligned}$$

where $q' = q/(q-1)$ and Hölder's inequality was used in the last line. Thus

$$\sum_{i=-\infty}^{\infty} \big[2^{-i\theta} J(2^i; u_i)\big]^q \le 2^{\theta q} (\ln 2)^{q/q'} \int_0^{\infty} \big[t^{-\theta} J\big(t; f(t)\big)\big]^q \,\frac{dt}{t}$$

and, since ϵ is arbitrary,

$$\big\| \{2^{-i\theta} J(2^i; u_i)\} ; \ell^q \big\| \le 2^{\theta} (\ln 2)^{1/q'} \|u\|_{\theta,q;J}.$$

Conversely, if $u = \sum_{i=-\infty}^{\infty} u_i$ where the series converges in $X_0 + X_1$, we can define a function $f \in L^1(0, \infty; dt/t, X_0 + X_1)$ by

$$f(t) = \frac{1}{\ln 2} u_i, \qquad \text{for} \quad 2^i \le t < 2^{i+1}, \qquad -\infty < i < \infty,$$

and we will have

$$\int_{2^i}^{2^{i+1}} f(t)\,\frac{dt}{t} = u_i \qquad \text{and} \qquad u = \int_0^{\infty} f(t)\,\frac{dt}{t}.$$

Moreover,

$$\begin{aligned}
\int_{2^i}^{2^{i+1}} \big[t^{-\theta} J\big(t; f(t)\big)\big]^q \,\frac{dt}{t} &\le \int_{2^i}^{2^{i+1}} \big[2^{-i\theta} J\big(2^{i+1}; f(t)\big)\big]^q \,\frac{dt}{t} \\
&\le \left(\frac{2}{\ln 2}\right)^q \int_{2^i}^{2^{i+1}} \big[2^{-i\theta} J(2^i; u_i)\big]^q \,\frac{dt}{t} \\
&= \frac{2^q}{(\ln 2)^{q-1}} \big[2^{-i\theta} J(2^i; u_i)\big]^q.
\end{aligned}$$

Summing on i then gives

$$\|u\|_{\theta,q;J} \le \left(\frac{2}{(\ln 2)^{1/q'}}\right) \left\| \{2^{-i\theta} J(2^i; u_i)\} ; \ell^q \right\| . \blacksquare$$

Next we prove that for $0 < \theta < 1$ the J- and K-methods generate the same intermediate spaces with equivalent norms.

7.16 THEOREM (Equivalence Theorem) If $0 < \theta < 1$ and $1 \le q \le \infty$, then

(a) $(X_0, X_1)_{\theta,q;J} \to (X_0, X_1)_{\theta,q;K}$, and

(b) $(X_0, X_1)_{\theta,q;K} \to (X_0, X_1)_{\theta,q;J}$. Therefore

(c) $(X_0, X_1)_{\theta,q;J} = (X_0, X_1)_{\theta,q;K}$, the two spaces having equivalent norms.

Proof. Conclusion (a) is a consequence of the somewhat stronger result

$$(X_0, X_1)_{\theta,p;J} \to (X_0, X_1)_{\theta,q;K}, \qquad \text{if } 1 \le p \le q \tag{9}$$

which we now prove. Let $u = \int_0^\infty f(s)\, ds/s \in (X_0, X_1)_{\theta,p;J}$. Since $K(t; \cdot)$ is a norm on $X_0 + X_1$, we have by the triangle inequality and (4)

$$\begin{aligned} t^{-\theta} K(t; u) &\le t^{-\theta} \int_0^\infty K\big(t; f(s)\big) \frac{ds}{s} \\ &\le \int_0^\infty \left(\frac{t}{s}\right)^{-\theta} \min\left\{1, \frac{t}{s}\right\} s^{-\theta} J\big(s; f(s)\big) \frac{ds}{s} \\ &= \left[t^{-\theta} \min\{1, t\}\right] * \left[t^{-\theta} J\big(t; f(t)\big)\right]. \end{aligned}$$

By Young's inequality with $1 + (1/q) = (1/r) + (1/p)$ (so $r \ge 1$)

$$\begin{aligned} \|u\|_{\theta,q;K} &= \left\| t^{-\theta} K(t; u) ; L_*^q \right\| \\ &\le \left\| t^{-\theta} \min\{1, t\} ; L_*^r \right\| \left\| t^{-\theta} J\big(t; f(t)\big) ; L_*^p \right\| \\ &\le C_{\theta,p,q} \|u\|_{\theta,p;K}, \end{aligned}$$

which confirms (9) and hence (a).

Now we prove (b) by using the discrete versions of the J and K methods. Let $u \in (X_0, X_1)_{\theta,p;K}$. By the definition of $K(t; u)$, for each integer i there exist $v_i \in X_0$ and $w_i \in X_1$ such that

$$u = v_i + w_i \qquad \text{and} \qquad \|v_i\|_{X_0} + 2^i \|w_i\|_{X_1} \le 2K(2^i; u).$$

Then the sequences $\{2^{-i\theta} \|v_i\|_{X_0}\}$ and $\{2^{i(1-\theta)} \|w_i\|_{X_1}\}$ both belong to ℓ^q and each has ℓ^q-norm bounded by a constant times $\|u\|_{\theta,q;K}$. For each index i let $u_i = v_{i+1} - v_i$. Since

$$0 = u - u = \big(v_{i+1} + w_{i+1}\big) - \big(v_i + w_i\big) = \big(v_{i+1} - v_i\big) + \big(w_{i+1} - w_i\big),$$

we have, in fact,

$$u_i = v_{i+1} - v_i = w_i - w_{i+1}.$$

The first of these representations of u_i shows that $\{2^{-i\theta}\,\|u_i\|_{X_0}\}$ belongs to ℓ^q; the second representations shows that $\{2^{i(1-\theta)}\,\|u_i\|_{X_1}\}$ also belongs to ℓ^q. Therefore $\{2^{-i\theta}J(2^i; u_i)\} \in \ell^q$ and has ℓ^q-norm bounded by a constant times $\|u\|_{\theta,q;K}$. Since $\ell^q \subset \ell^\infty$, the sequence $\left\{2^{j(1-\theta)}\left\|w_j\right\|_{X_1}\right\}$ is bounded even though $2^{j(1-\theta)} \to \infty$ as $j \to \infty$. Thus $\left\|w_j\right\|_{X_1} \to 0$ as $j \to \infty$. Since $\sum_{i=0}^{j} u_i = w_0 - w_{j+1}$, the half series $\sum_{i=0}^{\infty}$ converges to w_0 in X_1 and hence in $X_0 + X_1$. Similarly, the half-series $\sum_{i=-\infty}^{-1} u_i$ converges to v_0 in X_0, and thus in $X_0 + X_1$. Thus the full series $\sum_{i=-\infty}^{\infty} u_i$ converges to $v_0 + w_0 = u$ in $X_0 + X_1$ and we have

$$\|u\|_{\theta,q;J} \le \text{const.}\ \|u\|_{\theta,q;K}\,.$$

This completes the proof of (b) and hence (c). ∎

7.17 COROLLARY If $0 < \theta < 1$ and $1 \le p \le q \le \infty$, then

$$(X_0, X_1)_{\theta,p;K} \to (X_0, X_1)_{\theta,q;K}. \tag{10}$$

Proof. $(X_0, X_1)_{\theta,p;K} \to (X_0, X_1)_{\theta,p;J} \to (X_0, X_1)_{\theta,q;K}$ by part (b) and imbedding (9). ∎

7.18 (Classes of Intermediate Spaces) We define three classes of intermediate spaces X between X_0 and X_1 as follows:

(a) X belongs to class $\mathscr{K}(\theta; X_0, X_1)$ if for all $u \in X$

$$K(t; u) \le C_1 t^\theta\,\|u\|_X\,,$$

where C_1 is a constant.

(b) X belongs to class $\mathscr{J}(\theta; X_0, X_1)$ if for all $u \in X_0 \cap X_1$

$$\|u\|_X \le C_2 t^{-\theta} J(t; u),$$

where C_2 is a constant.

(c) X belongs to class $\mathscr{H}(\theta; X_0, X_1)$ if X belongs to both $\mathscr{K}(\theta; X_0, X_1)$ and $\mathscr{J}(\theta; X_0, X_1)$.

The following lemma gives necessary and sufficient conditions for membership in these classes.

7.19 LEMMA Let $0 \le \theta \le 1$ and let X be an intermediate space between X_0 and X_1.

(a) $X \in \mathscr{K}(\theta; X_0, X_1)$ if and only if $X \to (X_0, X_1)_{\theta,\infty;K}$.

(b) $X \in \mathscr{J}(\theta; X_0, X_1)$ if and only if $(X_0, X_1)_{\theta,1;J} \to X$.

(c) $X \in \mathscr{H}(\theta; X_0, X_1)$ if and only if $(X_0, X_1)_{\theta,1;J} \to X \to (X_0, X_1)_{\theta,\infty;K}$.

Proof. Conclusion (a) is immediate since $\|u\|_{\theta,\infty;K} = \sup_{0<t<\infty}\big(t^{-\theta}k(t;u)\big)$. Since (c) follows from (a) and (b), only (b) requires proof.

First suppose $X \in \mathscr{J}(\theta; X_0, X_1)$. Let $u \in (X_0, X_1)_{\theta,1;J}$. If $f(t)$ is any function on $(0,\infty)$ with values in $X_0 \cap X_1$ such that $u = \int_0^\infty f(t)\,dt/t$, then

$$\|u\|_X \le \int_0^\infty \|f(t)\|_X \frac{dt}{t} \le C_2 \int_0^\infty t^{-\theta} J\big(t; f(t)\big)\frac{dt}{t}.$$

Since this holds for all such representations of u we have

$$\|u\|_X \le C_2 (X_0, X_1)_{\theta,1;J}, \tag{11}$$

and so $(X_0, X_1)_{\theta,1;J} \to X$.

Conversely, suppose that $(X_0, X_1)_{\theta,1;J} \to X$; therefore (11) holds with some constant C_2. Let $u \in X_0 \cap X_1$, let $\lambda > 0$ and $t > 0$, and let

$$f_\lambda(s) = \begin{cases} (1/\lambda)u & \text{if } te^{-\lambda} \le s \le t \\ 0 & \text{otherwise.} \end{cases}$$

Then

$$\int_0^\infty f_\lambda(s)\frac{ds}{s} = \left(\int_{te^{-\lambda}}^t \frac{ds}{s}\right)\left(\frac{1}{\lambda}\right)u = u.$$

Since $J\big(s;(1/\lambda)u\big) = (1/\lambda)J(s;u)$ we have

$$\|u\|_{\theta,1;J} \le \int_0^\infty s^{-\theta} J\big(s; f_\lambda(s)\big)\frac{ds}{s} = \frac{1}{\lambda}\int_{te^{-\lambda}}^t s^{-\theta} J(s;u)\frac{ds}{s}.$$

Since $s^{-\theta}J(s;u)$ is continuous in s and $\int_{te^{-\lambda}}^t ds/s = \lambda$, we can let $\lambda \to 0+$ in the above inequality and obtain $\|u\|_{\theta,1;J} \le t^{-\theta} J(t;u)$. Hence

$$\|u\|_X \le C_2 (X_0, X_1)_{\theta,1;J} \le C_2 t^{-\theta} J(t;u)$$

and the proof of (b) is complete. ∎

The following corollary follows immediately, using the equivalence theorem, (10), and (8).

7.20 COROLLARY If $0 < \theta < 1$ and $1 \le q \le \infty$, then

$$(X_0, X_1)_{\theta,q;J} = (X_0, X_1)_{\theta,q;K} \in \mathscr{H}(\theta; X_0, X_1).$$

Moreover, $X_0 \in \mathscr{H}(0; X_0, X_1)$ and $X_1 \in \mathscr{H}(1; X_0, X_1)$. ∎

Next we examine the result of constructing intermediate spaces between two intermediate spaces.

7.21 THEOREM (The Reiteration Theorem) Let $0 \le \theta_0 < \theta_1 \le 1$ and let X_{θ_0} and X_{θ_1} be intermediate spaces between X_0 and X_1. For $0 \le \lambda \le 1$, let $\theta = (1-\lambda)\theta_0 + \lambda\theta_1$.

(a) If $X_{\theta_i} \in \mathscr{K}(\theta_i; X_0, X_1)$ for $i = 0, 1$, and if either $0 < \lambda < 1$ and $1 \le q < \infty$ or $0 \le \lambda \le 1$ and $q = \infty$, then

$$\left(X_{\theta_0}, X_{\theta_1}\right)_{\lambda,q;K} \to (X_0, X_1)_{\theta,q;K}.$$

(b) If $X_{\theta_i} \in \mathscr{J}(\theta_i; X_0, X_1)$ for $i = 0, 1$, and if either $0 < \lambda < 1$ and $1 < q \le \infty$ or $0 \le \lambda \le 1$ and $q = 1$, then

$$(X_0, X_1)_{\theta,q;J} \to \left(X_{\theta_0}, X_{\theta_1}\right)_{\lambda,q;J}.$$

(c) If $X_{\theta_i} \in \mathscr{H}(\theta_i; X_0, X_1)$ for $i = 0, 1$, and if $0 < \lambda < 1$ and $1 \le q \le \infty$, then

$$\left(X_{\theta_0}, X_{\theta_1}\right)_{\lambda,q;J} = \left(X_{\theta_0}, X_{\theta_1}\right)_{\lambda,q;K} = (X_0, X_1)_{\theta,q;K} = (X_0, X_1)_{\theta,q;J}.$$

(d) Moreover,

$$(X_0, X_1)_{\theta_0,1;J} \to \left(X_{\theta_0}, X_{\theta_1}\right)_{0,1;J} \to X_{\theta_0} \to \left(X_{\theta_0}, X_{\theta_1}\right)_{0,\infty;K} \to (X_0, X_1)_{\theta_0,\infty;K}$$
$$(X_0, X_1)_{\theta_1,1;J} \to \left(X_{\theta_0}, X_{\theta_1}\right)_{1,1;J} \to X_{\theta_1} \to \left(X_{\theta_0}, X_{\theta_1}\right)_{1,\infty;K} \to (X_0, X_1)_{\theta_1,\infty;K}.$$

Proof. The important conclusions here are (c) and (d) and these follow from (a) and (b) which we must prove. In both proofs we need to distinguish the function norms $K(t; u)$ and $J(t; u)$ used in the construction of the intermediate spaces between X_0 and X_1 from those used for the intermediate spaces between X_{θ_0} and X_{θ_1}. We will use K^* and J^* for the latter.

Proof of (a) If $u \in \left(X_{\theta_0}, X_{\theta_1}\right)_{\lambda,q;K}$, then $u = u_0 + u_1$ where $u_i \in X_{\theta_i}$. Since $X_{\theta_i} \in \mathscr{K}(\theta_i; X_0, X_1)$, we have

$$\begin{aligned}K(t; u) &\le K(t; u_0) + K(t; u_1)\\ &\le C_0 t^{\theta_0} \left\|u_0 ; X_{\theta_0}\right\| + C_1 t^{\theta_1} \left\|u_1 ; X_{\theta_1}\right\|\\ &\le C_0 t^{\theta_0} \left(\left\|u_0 ; X_{\theta_0}\right\| + \frac{C_1}{C_0} t^{\theta_1-\theta_0} \left\|u_1 ; X_{\theta_1}\right\|\right).\end{aligned}$$

Since this estimate holds for all such representations of u, we have

$$K(t;u) \le C_0 t^{\theta_0} K^*\left(\frac{C_1}{C_0} t^{\theta_1-\theta_0}; u\right).$$

If $\theta = (1-\lambda)\theta_0 + \lambda\theta_1$, then $\lambda = (\theta-\theta_0)/(\theta_1-\theta_0)$, and (assuming $q < \infty$)

$$\begin{aligned}\left\|t^{-\theta}K(t;u)\,;\,L^q_*\right\| &\le C_0\left[\int_0^\infty \left(t^{-(\theta-\theta_0)}K^*\left(\frac{C_1}{C_0}t^{\theta_1-\theta_0}; u\right)\right)^q \frac{dt}{t}\right]^{1/q}\\ &= \frac{C_0^{1-\lambda}C_1^{\lambda}}{(\theta_1-\theta_0)^{1/q}}\left[\int_0^\infty \left(s^{-\lambda}K^*(s;u)\right)^q \frac{ds}{s}\right]^{1/q}\end{aligned}$$

via the transformation $s = (C_1/C_0)t^{\theta_1-\theta_0}$. Hence

$$\|u\|_{\theta,q;K} \le \frac{C_0^{1-\lambda}C_1^{\lambda}}{(\theta_1-\theta_0)^{1/q}}\|u\|_{\lambda,q;K}$$

and so $\left(X_{\theta_0}, X_{\theta_1}\right)_{\lambda,q;K} \to (X_0, X_1)_{\theta,q;K}$.

Proof of (b) Let $u \in (X_0, X_1)_{\theta,q;J}$. Then $u = \int_0^\infty f(s)\,ds/s$ for some f taking values in $X_0 \cap X_1$ satisfying $s^{-\theta}J\big((s; f(s)\big) \in L^q_*$. Clearly $f(s) \in X_{\theta_0} \cap X_{\theta_1}$. Since $X_{\theta_i} \in \mathscr{J}(\theta_i; X_0, X_1)$ we have

$$\begin{aligned}J^*\big(s; f(s)\big) &= \max\left\{\left\|f(s)\,;\,X_{\theta_0}\right\|, s\left\|f(s)\,;\,X_{\theta_1}\right\|\right\}\\ &\le \max\left\{C_0t^{-\theta_0}J\big(t; f(s)\big), C_1t^{-\theta_1}sJ\big(t; f(s)\big)\right\}\\ &= C_0t^{-\theta_0}\max\left\{1, \frac{C_1}{C_0}t^{-(\theta_1-\theta_0)}s\right\}J\big(t; f(s)\big).\end{aligned}$$

This estimate holds for all $t > 0$ so we can choose t so that $t^{-(\theta_1-\theta_0)}s = C_0/C_1$ and obtain

$$J^*\big(s; f(s)\big) \le C_0\left(\frac{C_1}{C_0}s\right)^{-\theta_0/(\theta_1-\theta_0)} J\left(\left(\frac{C_1}{C_0}s\right)^{1/(\theta_1-\theta_0)}; f(s)\right).$$

If $\theta = (1-\lambda)\theta_0 + \lambda\theta_1$, then

$$\begin{aligned}&\left\|s^{-\lambda}J^*\big(s; f(s)\big)\,;\,L^q_*\right\|\\ &\le C_0^{1-\lambda}C_1^{\lambda}\left(\int_0^\infty \left[\left(\frac{C_1}{C_0}s\right)^{-\theta/(\theta_1-\theta_0)} J\left(\left(\frac{C_1}{C_0}s\right)^{1/(\theta_1-\theta_0)}; f(s)\right)\right]^q \frac{ds}{s}\right)^{1/q}\\ &\le C_0^{1-\lambda}C_1^{\lambda}(\theta_1-\theta_0)^{1/q}\left(\int_0^\infty \left[t^{-\theta}J\big(t; g(t)\big)\right]^q \frac{dt}{t}\right)^{1/q}\\ &= C_0^{1-\lambda}C_1^{\lambda}(\theta_1-\theta_0)^{1/q}\left\|t^{-\theta}J\big(t; g(t)\big)\,;\,L^q_*\right\|,\end{aligned}$$

where $g(t) = f\big((C_0/C_1)t^{\theta_1-\theta_0}\big) = f(s) \in X_0 \cap X_1$. Since

$$\int_0^\infty g(t)\,\frac{dt}{t} = \frac{1}{\theta_1-\theta_0}\int_0^\infty f(s)\,\frac{ds}{s} = \frac{1}{\theta_1-\theta_0}u,$$

we have

$$\|u\|_{\lambda,q;J} \le \frac{C_0^{1-\lambda}C_1^{\lambda}}{(\theta_1-\theta_0)^{(q-1)/q}}\,\|u\|_{\theta,q;J}$$

and so $(X_0, X_1)_{\theta,q;J} \to \big(X_{\theta_0}, X_{\theta_1}\big)_{\lambda,q;J}$. ∎

7.22 (Interpolation Spaces) Let $P = \{X_0, X_1\}$ and $Q = \{Y_0, Y_1\}$ be two interpolation pairs of Banach spaces, and let T be a bounded linear operator from $X_0 + X_1$ into $Y_0 + Y_1$ having the property that T is bounded from X_i into Y_i, with norm at most M_i, $i = 0, 1$; that is,

$$\|Tu_i\|_{Y_i} \le M_i\,\|u_i\|_{X_i}, \quad \text{for all } u_i \in X_i), \qquad (i = 1, 2).$$

If X and Y are intermediate spaces for $\{X_0, X_1\}$ and $\{Y_0, Y_1\}$, respectively, we call X and Y *interpolation spaces of type* θ for P and Q, where $0 \le \theta \le 1$, if every such linear operator T maps X into Y with norm M satisfying

$$M \le C M_0^{1-\theta} M_1^{\theta}, \tag{12}$$

where constant $C \ge 1$ is independent of T. We say that the interpolation spaces X and Y are *exact* if inequality (12) holds with $C = 1$. If $X_0 = Y_0$, $X_1 = Y_1$, $X = Y$ and $T = I$, the identity operator on $X_0 + X_1$, then $C = 1$ for all $0 \le \theta \le 1$, so no smaller C is possible in (12).

7.23 THEOREM (An Exact Interpolation Theorem) Let $P = \{X_0, X_1\}$ and $Q = \{Y_0, Y_1\}$ be two interpolation pairs.

(a) If either $0 < \theta < 1$ and $1 \le q \le \infty$ or $0 \le \theta \le 1$ and $q = \infty$, then the intermediate spaces $(X_0, X_1)_{\theta,q;K}$ and $(Y_0, Y_1)_{\theta,q;K}$ are exact interpolation spaces of type θ for P and Q

(b) If either $0 < \theta < 1$ and $1 < q \le \infty$ or $0 \le \theta \le 1$ and $q = 1$, then the intermediate spaces $(X_0, X_1)_{\theta,q;J}$ and $(Y_0, Y_1)_{\theta,q;J}$ are exact interpolation spaces of type θ for P and Q.

Proof. Let $T : X_0 + X_1 \to Y_0 + Y_1$ satisfy $\|Tu_i\|_{Y_i} \le M_i\,\|u_i\|_{X_i}$, $i = 0, 1$. If $u \in X_0 + X_1$, then

$$\begin{aligned} K(t; Tu) &= \inf\Big\{\|Tu_0\|_{Y_0} + t\,\|Tu_1\|_{Y_1} \,:\, u = u_0 + u_1,\, u_i \in X_i\Big\} \\ &\le M_0 \inf_{\substack{u=u_0+u_1\\ u_i\in X_i}} \left(\|u_0\|_{X_0} + \frac{M_1}{M_0}t\,\|u_1\|_{X_1}\right) = M_0 K\big((M_1/M_0)t; u\big). \end{aligned}$$

If $u \in (X_0, X_1)_{\theta,q;K}$, then

$$\|Tu\|_{\theta,q;K} = \left\|t^{-\theta}K(t;Tu)\,;L^q_*\right\| \le M_0\left\|t^{-\theta}K\big((M_1/M_0)t;u\big)\,;L^q_*\right\|$$
$$= M_0\left(\frac{M_0}{M_1}\right)^{-\theta}\left\|s^{-\theta}K(s;u)\,;L^q_*\right\| = M_0^{1-\theta}M_1^{\theta}\,\|u\|_{\theta,q;K},$$

which proves (a).

If $u \in X_0 \cap X_1$, then

$$J(t;Tu) = \max\left\{\|Tu\|_{Y_0}, t\,\|Tu\|_{Y_1}\right\}$$
$$\le M_0 \max\left\{\|u\|_{X_0}, (M_1/M_0)t\,\|u\|_{X_1}\right\} = M_0 J\big((M_1/M_0)t;u\big).$$

If $u = \int_0^\infty f(t)\,dt/t$, where $f(t) \in X_0 \cap X_1$ and $t^{-\theta}J\big(t;f(t)\big) \in L^q_*$, then

$$\|Tu\|_{\theta,q;J} = \left\|t^{-\theta}J\big(t;Tf(t)\big)\,;L^q_*\right\|$$
$$\le M_0\left\|t^{-\theta}J\big((M_1/M_0)t;f(t)\,;L^q_*\right\| = M_0\left(\frac{M_0}{M_1}\right)^{-\theta}\left\|s^{-\theta}J\big(s;g(s)\big)\,;L^q_*\right\|,$$

where $g(s) = f\big((M_0/M_1)s\big) = f(t)$. Since this estimate holds for all representations of $u = \int_0^\infty g(s)\,ds/s$, we have

$$\|Tu\|_{\theta,q;J} \le M_0^{1-\theta}M_1^{\theta}\,\|u\|_{\theta,q;J}$$

and the proof of (b) is complete. ∎

The Lorentz Spaces

7.24 (Equimeasurable Decreasing Rearrangement) Recall that, as defined in Paragraph 2.53, the distribution function δ_u corresponding to a measurable function u finite a.e. in a domain $\Omega \subset \mathbb{R}^n$ is given by

$$\delta_u(t) = \mu\{x \in \Omega : |u(x)| > t\}$$

and is nonincreasing on $[0, \infty)$. (It is also right continuous there, but that is of no relevance for integrals involving the distribution function since a nonincreasing function can have at most countably many points of discontinuity.) Moreover, if $u \in L^p(\Omega)$, then

$$\|u\|_p = \begin{cases} \left(p\displaystyle\int_0^\infty t^p\delta_u(y)\,\frac{dt}{t}\right)^{1/p} & \text{if } 1 \le p < \infty, \\ \inf\{t : \delta_u(t) = 0\} & \text{if } p = \infty. \end{cases}$$

The *equimeasurable decreasing rearrangement of u* is the function u^* defined by

$$u^*(s) = \inf\{t : \delta_u(t) \le s\}.$$

This definition and the fact that δ_u is nonincreasing imply that $u*$ is nonincreasing too. Moreover, $u^*(s) > t$ if and only if $\delta_u(t) > s$, and this latter condition is trivially equivalent to $s < \delta_u(t)$. Therefore,

$$\delta_{u^*}(t) = \mu\left\{s : u^*(s) > t\right\} = \mu\{s : 0 \le s < \delta_u(t)\} = \mu\left\{[0, \delta_u(t))\right\} = \delta_u(t).$$

This justifies our calling u^* and u equimeasurable; the size of both functions exceeds any number s on sets having the same measure. Also,

$$\delta_{u^*}(t) = \mu\{s : u^*(s) > t\} = \inf\{s : u^*(s) \le t\}$$

so that

$$\delta_u(t) = \inf\{s : u^*(s) \le t\}.$$

This further illustrates the symmetry between δ_u and u^*.

Note also that

$$u^*\big(\delta_u(t)\big) = \inf\big\{s : \delta_u(s) \le \delta_u(t)\big\} \le t.$$

If $u^*\big(\delta_u(t)\big) = s < t$, then δ_u is constant on the interval (s, t) in which case u^* has a jump discontinuity of magnitude at least $t - s$ at $\delta_u(t)$.

Similarly, $\delta_u\big(u^*(s)\big) \le s$, with equality if δ_u is continuous at $t = u^*(s)$. The relationship between δ_u and u^* is illustrated in Figure 8. Except at points where either function is discontinuous (and the other is constant on an interval), each is the inverse of the other.

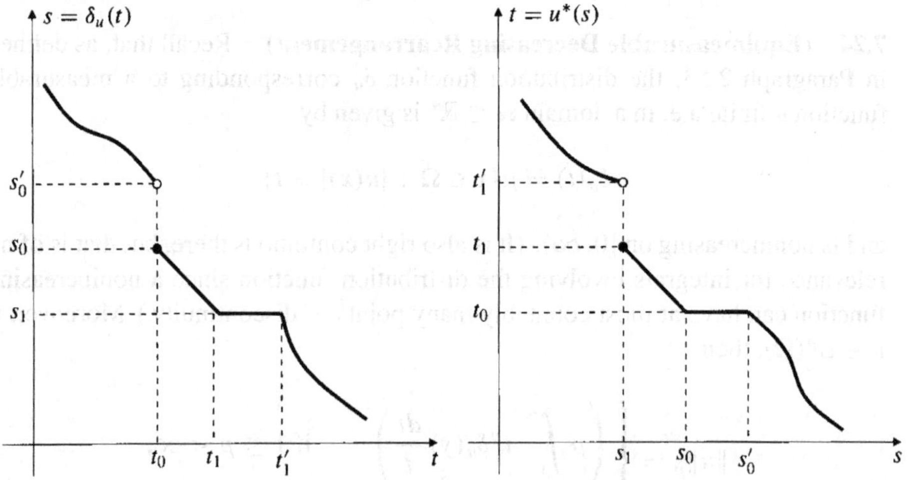

Fig. 8

If $S_t = \{x \in \Omega : |u(x)| > t\}$, then

$$\int_{S_t} |u(x)|\, dx = \int_0^{\delta_u(t)} u^*(s)\, ds, \tag{13}$$

and if $u \in L^p(\Omega)$, then

$$\|u\|_p = \begin{cases} \left(\int_0^\infty \left(u^*(s)\right)^p dx\right)^{1/p} & \text{if } 1 \le p < \infty \\ \sup\limits_{0<s<\infty} u^*(s) & \text{if } p = \infty. \end{cases}$$

7.25 (The Lorentz Spaces) For u measurable on Ω let

$$u^{**}(t) = \frac{1}{t}\int_0^t u^*(s)\, ds,$$

that is, the average value of u^* over $[0, t]$. Since u^* is nonincreasing, we have $u^*(t) \le u^{**}(t)$.

For $1 \le p \le \infty$ we define the functional

$$\left\|u\,; L^{p,q}(\Omega)\right\| = \begin{cases} \left(\int_0^\infty \left(t^{1/p} u^{**}(t)\right)^q \frac{dt}{t}\right)^{1/q} & \text{if } 1 \le q < \infty \\ \sup\limits_{t>0} t^{1/p} u^{**}(t) & \text{if } q = \infty. \end{cases}$$

The *Lorentz space* $L^{p,q}(\Omega)$ consists of those measurable functions u on Ω for which $\|u\,; L^{p,q}(\Omega)\| < \infty$. Theorem 7.26 below shows that if $1 < p < \infty$, then $L^{p,q}(\Omega)$ is, in fact, identical to the intermediate space $\left(L^1(\Omega), L^\infty(\Omega)\right)_{(p-1)/p,q;K}$ and $\|u\,; L^{p,q}(\Omega)\| = \|u\|_{(p-1)/p,q;K}$. Thus $L^{p,q}(\Omega)$ is a Banach space under the norm $\|u\,; L^{p,q}(\Omega)\|$. It is also a Banach space if $p = 1$ or $p = \infty$.

The second corollary to Theorem 7.26 shows that if $1 < p < \infty$, then $L^{p,q}(\Omega)$ coincides with the set of measurable u for which $\left[u; L^{p,q}(\Omega)\right] < \infty$, where

$$\left[u; L^{p,q}(\Omega)\right] = \begin{cases} \left(\int_0^\infty \left(t^{1/p} u^*(t)\right)^q \frac{dt}{t}\right)^{1/q} & \text{if } 1 \le q < \infty \\ \sup\limits_{t>0} t^{1/p} u^*(t) & \text{if } q = \infty, \end{cases}$$

and that

$$\left[u; L^{p,q}(\Omega)\right] \le \left\|u\,; L^{p,q}(\Omega)\right\| \le \frac{p}{p-1}\left[u; L^{p,q}(\Omega)\right].$$

The index p in $L^{p,q}(\Omega)$ is called the principal index; q is the secondary index. Unless $q = p$, the functional $\left[\,\cdot\,;\, L^{p,q}(\Omega)\right]$ is not a norm since it does not satisfy the triangle inequality; it, however, is a quasi-norm since

$$\left[u + v;\, L^{p,q}(\Omega)\right] \le 2\left(\left[u;\, L^{p,q}(\Omega)\right] + \left[v;\, L^{p,q}(\Omega)\right]\right).$$

For $1 < p < \infty$ it is evident that $\left[\,\cdot\,;\, L^{p,p}(\Omega)\right] = \|\cdot\|_{p,\Omega}$, and therefore $L^{p,p}(\Omega) = L^p(\Omega)$. Moreover, if we recall the definition of the space weak-$L^p(\Omega)$ given in Paragraph 2.55 and having quasi-norm given (for $p < \infty$) by

$$[u]_p = [u]_{p,\Omega} = \left(\sup_{t>0} t^p \delta_u(t)\right)^{1/p},$$

we can show that $L^{p,\infty}(\Omega) = \text{weak-}L^p(\Omega)$. This is also clear for $p = \infty$. If $1 < p < \infty$ and $K > 0$, then for all $t > 0$ we have, putting $s = K^p t^{-p}$,

$$\delta_u(t) \le K^p t^{-p} = s \qquad \Longleftrightarrow \qquad u^*(s) \le t = K s^{-1/p}.$$

Hence $[u]_p \le K$ if and only if $\left[u;\, L^{p,\infty}(\Omega)\right] \le K$, and these two quasi-norms are, in fact, equal.

For $p = 1$ the situation is a little different. Observe that

$$\left\|u\,;\, L^{1,\infty}(\Omega)\right\| = \sup_{t>0} t u^{**}(t) = \sup_{t>0} \int_0^t u^*(s)\, ds = \int_0^\infty u^*(s)\, ds = \|u\|_1$$

so $L^1(\Omega) = L^{1,\infty}(\Omega)$ (*not* $L^{1,1}(\Omega)$ which contains only the zero function).

For $p = \infty$ we have $L^{\infty,\infty}(\Omega) = L^\infty(\Omega)$ since

$$\left\|u\,;\, L^{\infty,\infty}(\Omega)\right\| = \sup_{t>0} u^{**}(t) = \sup_{t>0} \frac{1}{t} \int_0^t u^*(s)\, ds = u^*(0) = \|u\|_\infty .$$

7.26 THEOREM If $u \in L^1(\Omega) + L^\infty(\Omega)$, then for $t > 0$ we have

$$K(t; u) = \int_0^t u^*(s)\, ds = t u^{**}(t). \tag{14}$$

Therefore, if $1 < p < \infty$, $1 \le q \le \infty$, and $\theta = 1 - (1/p)$,

$$L^{p,q}(\Omega) = \left(L^1(\Omega),\, L^\infty(\Omega)\right)_{\theta,q;K}$$

with equality of norms: $\|u\,;\, L^{p,q}(\Omega)\| = \|u\|_{\theta,q;K}$.

Proof. The second conclusion follows immediately from the representation (14) which we prove as follows.

Since $K(t;u) = K(t;|u|)$ we can assume that u is real-valued and nonnegative. Let $u = v + w$ where $v \in L^1(\Omega)$ and $w \in L^\infty(\Omega)$. In order to calculate

$$K(t;u) = \inf_{u=v+w} \left(\|v\|_1 + t\,\|w\|_\infty \right) \tag{15}$$

we can also assume that v and w are real-valued functions since, in any event, $u = \operatorname{Re} v + \operatorname{Re} w$ and $\|\operatorname{Re} v\|_1 \le \|v\|_1$ and $\|\operatorname{Re} w\|_\infty \le \|w\|_\infty$. We can also assume that v and w are nonnegative, for if

$$v_1(x) = \begin{cases} \min\{v(x), u(x)\} & \text{if } v(x) \ge 0 \\ 0 & \text{if } v(x) < 0 \end{cases} \qquad \text{and} \qquad w_1(x) = u(x) - v_1(x),$$

then $0 \le v_1(x) \le |v(x)|$ and $0 \le w_1(x) \le |w(x)|$. Thus the infimum in (15) does not change if we restrict to nonnegative functions v and w.

Thus we consider $u = v + w$, where $v \ge 0$, $v \in L^1(\Omega)$, $w \ge 0$, and $w \in L^\infty(\Omega)$. Let $\lambda = \|w\|_\infty$ and define $u_\lambda(x) = \min\{\lambda, u(x)\}$. Evidently $w(x) \le u_\lambda(x)$ and $u(x) - u_\lambda(x) \le u(x) - w(x) = v(x)$. Now let

$$g(t,\lambda) = \|u - u_\lambda\|_1 + t\lambda \le \|v\|_1 + t\,\|w\|_\infty .$$

Then $K(t;u) = \inf_{0<\lambda<\infty} g(t,\lambda)$. We want to show that this infimum is, in fact, a minimum and is assumed at $\lambda = \lambda_t = \inf\{\tau : \delta_u(\tau) < t\}$.

If $\lambda > \lambda_t$, then $u_\lambda(x) - u_{\lambda_t}(x) \le \lambda - \lambda_t$ if $u(x) > \lambda_t$, and $u_\lambda(x) - u_{\lambda_t}(x) = 0$ if $u(x) \le \lambda_t$. Since $\delta_u(\lambda_t) \le t$, we have

$$\begin{aligned} g(t,\lambda) - g(t,\lambda_t) &= -\int_\Omega \left(u_\lambda(x) - u_{\lambda_t}(x) \right) dx + t(\lambda - \lambda_t) \\ &\ge (\lambda - \lambda_t)\left(t - \delta_u(\lambda_t)\right) \ge 0. \end{aligned}$$

Thus $K(t;u) \le g(t,\lambda_t)$.

On the other hand, if $g(t,\lambda^*) < \infty$ for some $\lambda^* < \lambda_t$, then $g(t,\lambda)$ is a continuous function of λ for $\lambda \ge \lambda^*$ and so for any $\epsilon > 0$ there exists λ such that $\lambda^* \le \lambda < \lambda_t$ and

$$|g(t,\lambda) - g(t,\lambda_t)| < \epsilon.$$

Now $u_\lambda(x) - u_{\lambda^*}(x) = \lambda - \lambda^*$ if $u(x) > \lambda$, and since $\delta_u(\lambda) \ge t$ we have

$$\begin{aligned} g(t,\lambda^*) - g(t,\lambda) &= \int_\Omega \left(u_\lambda(x) - u_{\lambda^*}(x) \right) dx - t(\lambda - \lambda^*) \\ &\ge (\lambda - \lambda^*)\left(\delta_u(\lambda) - t\right) \ge 0. \end{aligned}$$

Thus

$$g(t,\lambda^*) - g(t,\lambda_t) \ge g(t,\lambda^*) - g(t,\lambda) - |g(t,\lambda) - g(t,\lambda_t)| \ge -\epsilon.$$

Since ϵ is arbitrary, $g(t,\lambda^*) \ge g(t,\lambda_t)$ and $K(t;u) \ge g(t,\lambda_t)$. Thus

$$K(t;u) = g(t,\lambda_t) = \left\|u - u_{\lambda_t}\right\|_1 + t\lambda_t.$$

Now $u(x) - u_{\lambda_t}(x) = 0$ except where $u(x) > \lambda_t$ and $\lambda_t = u^*(s)$ for $\delta_u(\lambda_t) \le s \le t$. Therefore, by (13),

$$\begin{aligned} K(t;u) &= \int_0^{\delta_u(\lambda_t)} \left(u^*(s) - \lambda_t\right) ds + t\lambda_t = \int_0^{\delta_u(\lambda_t)} u^*(s)\, ds - \lambda_t \delta_u(\lambda_t) + t\lambda_t \\ &= \int_0^{\delta_u(\lambda_t)} u^*(s)\, ds + \int_{\delta_u(\lambda_t)}^t u^*(s)\, ds = \int_0^t u^*(s)\, ds \end{aligned}$$

which completes the proof. ∎

7.27 COROLLARY If $1 \le p_1 < p < p_2 \le \infty$ and $1/p = (1-\theta)/p_1 + \theta/p_2$, then by the Reiteration Theorem 7.21, up to equivalence of norms,

$$L^{p,q}(\Omega) = \left(L^{p_1}(\Omega), L^{p_2}(\Omega)\right)_{\theta,q;K}.$$

7.28 COROLLARY For $1 < p < \infty$, $1 \le q \le \infty$, and $\theta = 1 - (1/p)$, we have

$$\left[u; L^{p,q}(\Omega)\right] \le \left\|u\,; L^{p,q}(\Omega)\right\| \le \frac{p}{p-1}\left[u; L^{p,q}(\Omega)\right].$$

Proof. Since u^* is decreasing, (14) implies that $tu^*(t) \le K(t;u)$. Thus

$$\begin{aligned} \left[u; L^{p,q}(\Omega)\right] &= \left(\int_0^\infty \left(t^{1/p} u^*(t)\right)^q \frac{dt}{t}\right)^{1/q} \\ &\le \left(\int_0^\infty \left(t^{-\theta} K(t;u)\right)^q \frac{dt}{t}\right)^{1/q} = \|u\|_{\theta,q;K} = \left\|u\,; L^{p,q}(\Omega)\right\|. \end{aligned}$$

On the other hand,

$$\begin{aligned} t^{-\theta} K(t;u) &= \int_0^t t^{-\theta} u^*(s)\, ds \\ &= \int_1^\infty \sigma^{-\theta} \left(\frac{t}{\sigma}\right)^{1-\theta} u^*\left(\frac{t}{\sigma}\right) \frac{d\sigma}{\sigma} = f * g(t), \end{aligned}$$

where

$$f(t) = t^{1-\theta}u^*(t) = t^{1/p}u^*(t), \qquad \text{and} \qquad g(t) = \begin{cases} t^{-\theta} & \text{if } t \geq 1 \\ 0 & \text{if } 0 \leq 1, \end{cases}$$

and the convolution is with respect to the measure dt/t. Since we have $\|f ; L_*^q\| = [u; L^{p,q}(\Omega)]$ and $\|g ; L_*^1\| = 1/\theta = p/(p-1)$, Young's inequality (see Paragraph 7.5) gives

$$\|u ; L^{p,q}(\Omega)\| = \|u\|_{\theta,q;K} = \|f * g ; L_*^q\| \leq \frac{p}{p-1}[u; L^{p,q}(\Omega)]. \quad \blacksquare$$

7.29 REMARK Working with Lorentz spaces and using the real interpolation method allows us to sharpen the cases of the Sobolev imbedding theorem where $p > 1$ and $mp < n$. In those cases, the proof in Chapter IV used Lemma 4.18, where convolution with the kernel ω_m was first shown to be of weak type (p, p^*) (where $p* = np/(n - mp)$) for all such indices p. Then other such indices p_1 and p_2 were chosen with $p_1 < p < p_2$, and Marcinkiewicz interpolation implied that this linear convolution operator must be of strong type (p, p^*).

We can instead apply the the Exact Interpolation Theorem 7.23 and Lorentz interpolation as in Corollary 7.27, to deduce, from the weak-type estimates above, that convolution with ω_m maps $L^p(\Omega)$ into $L^{p^*,p}(\Omega)$; this target space is strictly smaller than $L^{p^*}(\Omega)$, since $p < p^*$. It follows that $W^{m,p}(\Omega)$ imbeds in the smaller spaces $L^{p^*,p}(\Omega)$ when $p > 1$ and $mp < n$.

Recall too that convolution with ω_m is *not* of strong type $(1, 1^*)$ when $m < n$, but an averaging argument, in Lemma 4.24, showed that $W^{m,1}(\Omega) \subset L^{1^*}(\Omega)$ in that case. That argument can be refined as in Fournier [F] to show that in fact $W^{m,1}(\Omega) \subset L^{1^*,1}(\Omega)$ in these cases. This sharper endpoint imbedding had been proved earlier by Poornima [Po] using another method, and also in a dual form in Faris [Fa].

An ideal context for applying interpolation is one where there are apt endpoint estimates from which everything else follows. We illustrate that idea for convolution with ω_m. It is easy, via Fubini's theorem, to verify that if $f \in L^1(\Omega)$ then $\|f * g_0\|_\infty \leq \|f\|_1 \|g_0\|_\infty$ and $\|f * g_1\|_1 \leq \|f\|_1 \|g_1\|_1$ for all functions g_0 in $L^\infty(\Omega)$ and g_1 in $L^1(\Omega)$. Fixing f and interpolating between thc endpoint conditions on the functions g gives that $\|f * g ; L^{p,q}(\Omega)\| \leq C_p \|f\|_1 \|g ; L^{p,q}(\Omega)\|$ for all indices p and q in the intervals $(1, \infty)$ and $[1, \infty]$ respectively. Apply this with $g = \omega_m$, which belongs to $L^{n/(n-m),\infty}(\Omega) = L^{1^*,\infty}(\Omega) = \text{weak-}L^{1^*}(\Omega)$ to deduce that convolution with ω_m maps $L^1(\Omega)$ into $L^{1^*,\infty}(\Omega)$. On the other hand, if $f \in L^{(1^*)',1}(\Omega) = L^{n/m,1}(\Omega)$, then

$$\begin{aligned}
|\omega_m * f(x)| &\le \int_{R^n} |\omega_m(x-y) f(y)|\, dy \\
&\le \int_0^\infty (\omega_m)^*(t) f^*(t)\, dt = \int_0^\infty [t^{1/1^*}(\omega_m)^*(t)][t^{1/(1^*)'} f^*(t)] \frac{dt}{t} \\
&\le \left\| \omega_m \,;\, L^{1^*,\infty}(\Omega) \right\| \int_0^\infty [t^{1/(1^*)'} f^*(t)] \frac{dt}{t} \le C_m \left\| f \,;\, L^{(1^*)',1}(\Omega) \right\|.
\end{aligned}$$

That is, convolution with ω_m maps $L^1(\Omega)$ into $L^{1^*,\infty}(\Omega)$ and $L^{(1^*)',1}(\Omega)$ into $L^\infty(\Omega)$. Real interpolation then makes this convolution a bounded mapping of $L^{p,q}(\Omega)$ into $L^{p^*,q}(\Omega)$ for all indices p in the interval $(1, (1^*)') = (1, n/m)$ and all indices q in $[1, \infty]$.

These conclusions are sharper than those coming from Marcinkiewicz interpolation. On the other hand, the latter applies to mappings of weak-type $(1, 1)$, a case not covered by the K and J methods for Banach spaces, since weak L^1 is not a Banach space. The statement of the Marcinkiewicz Theorem 2.58 also applies to sublinear operators of weak-type (p, q) rather than just linear operators. It is easy, however, to extend the J and K machinery to cover sublinear operators between L^p spaces and Lorentz spaces. As above, this gives target spaces $L^{q,p}$ that are strictly smaller than L^q when $p < q$. Marcinkiewicz does not apply when $p > q$, but the J and K methods still apply, with target spaces $L^{q,p}$ that are larger than L^q in these cases.

Besov Spaces

7.30 The real interpolation method also applies to scales of spaces based on smoothness. For Sobolev spaces on sufficiently smooth domains the resulting intermediate spaces are called Besov spaces. Before defining them, we first establish the following theorem which shows that if $0 < k < m$, then $W^{k,p}(\Omega)$ is suitably intermediate between $L^p(\Omega)$ and $W^{m,p}(\Omega)$ provided Ω is sufficiently regular. Since the proof requires both Theorem 5.2, for which the cone condition suffices, and the approximation property of Paragraph 5.31 which we know holds for $\mathbb{R}^n$ and by extension for any domain satisfying the strong local Lipschitz condition, which implies the cone condition, we state the theorem for domains satisfying the strong local Lipschitz condition even though it holds for some domains which do not satisfy this condition. (See Paragraph 5.31.)

7.31 THEOREM If $\Omega \subset \mathbb{R}^n$ satisfies the strong local Lipschitz condition and if $0 < k < m$ and $1 \le p < \infty$, then

$$W^{k,p}(\Omega) \in \mathscr{H}\big(k/m;\, L^p(\Omega),\, W^{m,p}(\Omega)\big).$$

Proof. In this context we deal with the function norms

$$J(t;u) = \max\{\|u\|_p, t\|u\|_{m,p}\}$$
$$K(t;u) = \inf\{\|u_0\|_p + t\|u_1\|_{m,p} : u = u_0 + u_1, u \in L^p(\Omega), u_1 \in W^{m,p}(\Omega)\}.$$

We must show that

$$\|u\|_{k,p} \le C\, t^{-(k/m)} J(t;u) \tag{16}$$
$$K(t;u) \le C\, t^{k/m} \|u\|_{k,p}. \tag{17}$$

Now Theorem 5.2 asserts that for some constant C and all $u \in W^{m,p}(\Omega)$

$$\|u\|_{k,p} \le C\, \|u\|_p^{1-(k/m)} \|u\|_{m,p}^{k/m}.$$

The expression on the right side is C times the minimum value of

$$t^{-k/m} J(t;u) = \max\{t^{-k/m}\|u\|_p, t^{1-(k/m)}\|u\|_{m,p}\},$$

which occurs for $t = \|u\|_p / \|u\|_{m,p}$, the value of t making both terms in the maximum equal. This proves (16).

We show that (17) is equivalent to the approximation property. If $u \in W^{k,p}(\Omega)$, then

$$K(t;u) \le \|u\|_p + t\|0\|_{m,p} = \|u\|_p \le \|u\|_{k,p}.$$

Thus $t^{-k/m}K(t,u) \le \|u\|_p$ when $t \ge 1$, and inequality (17) holds in that case. If $t^{-(k/m)}K(t;u) \le C\|u\|_{k,p}$ also holds for $0 < t \le 1$, then since we can choose $u_0 \in L^p(\Omega)$ and $u_1 \in W^{m,p}(\Omega)$ with $u = u_0 + u_1$ and $\|u_0\|_p + t\|u_1\|_{m,p} \le 2K(t;u)$, we must have

$$\|u - u_1\|_p = \|u_0\|_p \le 2Ct^{k/m}\|u\|_{k,p} \qquad \text{and} \qquad \|u_1\|_{m,p} \le 2Ct^{(k/m)-1}\|u\|_{k,p},$$

so that with $t = \epsilon^m$, $u_\epsilon = u_1$ is a solution of the approximation problem of Paragraph 5.31. Conversely, if the approximation problem has a solution, that is, if for each $\epsilon \le 1$ there exists $u_\epsilon \in W^{m,p}(\Omega)$ satisfying

$$\|u - u_\epsilon\|_p \le C\epsilon^k\|u\|_{k,p} \qquad \text{and} \qquad \|u_\epsilon\|_{m,p} \le C\epsilon^{k-m}\|u\|_{k,p},$$

then, with $\epsilon = t^{1/m}$, we will have

$$t^{-(k/m)}K(t;u) \le t^{-(k/m)}\left(\|u - u_\epsilon\|_p + t\|u_\epsilon\|_{m,p}\right) \le C\|u\|_{k,p}$$

and (17) holds. This completes the proof. ∎

7.32 (The Besov Spaces) We begin with a definition of Besov spaces on general domains by interpolation.

Let $0 < s < \infty$, $1 \le p < \infty$, and $1 \le q \le \infty$. Also let m be the smallest integer larger than s. We define the *Besov space* $B^{s;p,q}(\Omega)$ to be the intermediate space between $L^p(\Omega)$ and $W^{m,p}(\Omega)$ corresponding to $\theta = s/m$, specifically:

$$B^{s;p,q}(\Omega) = \left(L^p(\Omega), W^{m,p}(\Omega)\right)_{s/m,q;J}.$$

It is a Banach space with norm $\|u\,; B^{s;p,q}(\Omega)\| = \left\|u\,; \left(L^p(\Omega), W^{m,p}(\Omega)\right)_{s/m,q;J}\right\|$ and enjoys many other properties inherited from $L^p(\Omega)$ and $W^{m,p}(\Omega)$, for example the density of the subspace $\{\phi \in C^\infty(\Omega) : \|u\|_{m,p} < \infty\}$. Also, imposing the strong local Lipschitz property on Ω guarantees the existence of an extension operator from $W^{m,p}(\Omega)$ to $W^{m,p}(\mathbb{R}^n)$ and so from $B^{s;p,q}(\Omega)$ to $B^{s;p,q}(\mathbb{R}^n)$. On $\mathbb{R}^n$, there are many equivalent definitions $B^{s;p,q}$ (see [J]), each leading to a definition of $B^{s;,p,q}(\Omega)$ by restriction. For domains with good enough extension properties, these definitions by restriction are equivalent to the definition by real interpolation. Although somewhat indirect, that definition is intrinsic. As in Remark 6.47(1), the definitions by restriction can give smaller spaces for domains without extension properties.

For domains for which the conclusion of Theorem 7.31 holds, that theorem and the Reiteration Theorem 7.21 show that, up to equivalence of norms, we get the same space $B^{s;p,q}(\Omega)$ if we use any integer $m > s$ in the definition above. In fact, if $s_1 > s$ and $1 \le q_1 \le \infty$, then

$$B^{s;p,q}(\Omega) = \left(L^p(\Omega), B^{s_1;p,q_1}(\Omega)\right)_{s/s_1,q;J}.$$

More generally, if $0 \le k < s < m$ and $s = (1-\theta)k + \theta m$, then

$$B^{s;p,q}(\Omega) = \left(W^{k,p}(\Omega), W^{m,p}(\Omega)\right)_{\theta,q;J},$$

and if $0 < s_1 < s < s_2$, $s = (1-\theta)s_1 + \theta s_2$, and $1 \le q_1, q_2 \le \infty$, then

$$B^{s;p,q}(\Omega) = \left(B^{s_1;p,q_1}(\Omega), B^{s_2;p,q_2}(\Omega)\right)_{\theta,q;J}.$$

7.33 Theorem 7.31 also implies that for integer m,

$$B^{m;p,1}(\Omega) \to W^{m,p}(\Omega) \to B^{m;p,\infty}(\Omega).$$

In Paragraph 7.67 we will see that

$$\begin{aligned} &B^{m;p,p}(\Omega) \to W^{m,p}(\Omega) \to B^{m;p,2}(\Omega) \qquad \text{for} \quad 1 < p \le 2, \\ &B^{m;p,2}(\Omega) \to W^{m,p}(\Omega) \to B^{m;p,p}(\Omega) \qquad \text{for} \quad 2 \le p < \infty. \end{aligned}$$

The indices here are best possible; even in the case $\Omega = \mathbb{R}^n$ it is not true that $B^{m;p,q}(\Omega) = W^{m,p}(\Omega)$ for any q unless $p = q = 2$.

The following imbedding theorem for Besov spaces requires only that Ω satisfy the cone condition (or even the weak cone condition) since it makes no use of Theorem 7.31.

7.34 THEOREM (An Imbedding Theorem for Besov Spaces) Let Ω be a domain in $\mathbb{R}^n$ satisfying the cone condition, and let $1 \le p < \infty$ and $1 \le q \le \infty$.

(a) If $sp < n$, then $B^{s;p,q}(\Omega) \to L^{r,q}(\Omega)$ for $r = np/(n - sp)$.

(b) If $sp = n$, then $B^{s;p,1}(\Omega) \to C_B^0(\Omega) \to L^\infty(\Omega)$.

(c) If $sp > n$, then $B^{s;p,q}(\Omega) \to C_B^0(\Omega)$.

Proof. Observe that part (a) follows from part (b) and the Exact Interpolation Theorem 7.23 since if $0 < s < s_1$ and $s_1 p = n$, then (b) implies

$$B^{s;p,q}(\Omega) = \big(L^p(\Omega), B^{s_1;p,1}(\Omega)\big)_{s/s_1,q;J} \to \big(L^p(\Omega), L^\infty(\Omega)\big)_{s/s_1,q;J} = L^{r,q}(\Omega),$$

where $r = [1 - (s/s_1)]/p = np/(n - sp)$.

To prove (b) let m be the smallest integer greater than $s = n/p$. Let $u \in B^{n/p;p,1}(\Omega) = \big(L^p(\Omega), W^{m,p}(\Omega)\big)_{n/(mp),1;J}$. By the discrete version of the J-method, there exist functions u_i in $W^{m,p}(\Omega)$ such that the series $\sum_{i=-\infty}^{\infty} u_i$ converges to u in $B^{n/p;p,1}(\Omega)$ and such that the sequence $\{2^{-in/mp} J(2^i; u_i)\}_{i=-\infty}^{\infty}$ belongs to ℓ^1 and has ℓ^1 norm no larger than $C\,\|u\,; B^{n/p;p,1}(\Omega)\|$. Since $mp > n$ and Ω satisfies the cone condition, Theorem 5.8 shows that

$$\|v\|_\infty \le C_1 \|v\|_p^{1-(n/mp)} \|v\|_{m,p}^{n/mp}$$

for all $v \in W^{m,p}(\Omega)$. Thus

$$\begin{aligned}\|u\|_\infty &\le \sum_{i=-\infty}^{\infty} \|u_i\|_\infty \\ &\le C_1 \sum_{i=-\infty}^{\infty} \|u_i\|_p^{1-(n/mp)} \|u_i\|_{m,p}^{n/mp} \\ &\le C_1 \sum_{i=-\infty}^{\infty} 2^{-in/mp} J\big(2^i; u_i\big) \le C_2 \,\big\|u\,; B^{n/p;p,1}(\Omega)\big\|.\end{aligned}$$

Thus $B^{n/p;p,1}(\Omega) \to L^\infty(\Omega)$. The continuity of u follows as in the proof of Part I, Case A of Theorem 4.12 given in Paragraph 4.16.

Part (c) follows from part (b) since $B^{s;p,q}(\Omega) \to B^{s_1;p,1}(\Omega)$ if $s > s_1$. This imbedding holds because $W^{m,p}(\Omega) \to L^p(\Omega)$. ∎

Generalized Spaces of Hölder Continuous Functions

7.35 (The Spaces $C^{j,\lambda,q}(\overline{\Omega})$) If Ω satisfies the strong local Lipschitz condition and $sp > n$, the Besov space $B^{s;p,q}(\Omega)$ also imbeds into an appropriate space of Hölder continuous functions. To formulate that imbedding we begin by generalizing the Hölder space $C^{j,\lambda}(\overline{\Omega})$ to allow for a third parameter. For this purpose we consider the *modulus of continuity* of a function u defined on Ω given by

$$\omega(u;t) = \sup\bigl\{|u(x)-u(y)| \,:\, x,y\in\Omega,\ |x-y|\le t\bigr\}, \qquad (t>0).$$

Observe that $\omega(u;t) = \omega^*_\infty(u;t)$ in the notation of Paragraph 7.46. Also observe that if $0<\lambda\le 1$ and $t^{-\lambda}\omega(t,u)\le k<\infty$ for all $t>0$, then u is uniformly continuous on Ω. Since $C^j(\overline{\Omega})$ is a subspace of $W^{j,\infty}(\Omega)$ with the same norm, $C^{j,\lambda}(\overline{\Omega})$ consists of those $u\in W^{j,\infty}(\Omega)$ for which $t^{-\lambda}\omega(t,D^\alpha u)$ is bounded for all $0<t<\infty$ and all α with $|\alpha|=j$.

We now define the generalized spaces $C^{j,\lambda,q}(\overline{\Omega})$ as follows. If $j\ge 0$, $0<\lambda\le 1$, and $q=\infty$, then $C^{j,\lambda,\infty}(\overline{\Omega}) = C^{j,\lambda}(\overline{\Omega})$ with norm

$$\left\|u\,;C^{j,\lambda,\infty}(\overline{\Omega})\right\| = \left\|u\,;C^{j,\lambda}(\overline{\Omega})\right\| = \|u\|_{j,\infty} + \max_{|\alpha|=j}\sup_{t>0}\frac{\omega(D^\alpha u;t)}{t^\lambda}.$$

For $j\ge 0$, $0<\lambda\le 1$, and $1\le q<\infty$, the space $C^{j,\lambda,q}(\overline{\Omega})$ consists of those functions $u\in W^{j,\infty}(\Omega)$ for which $\left\|u\,;C^{j,\lambda,q}(\overline{\Omega})\right\|<\infty$, where

$$\left\|u\,;C^{j,\lambda,q}(\overline{\Omega})\right\| = \left\|u\,;C^{j}(\overline{\Omega})\right\| + \max_{|\alpha|=j}\left(\int_0^\infty \left(t^{-\lambda}\omega(D^\alpha u;t)\right)^q\frac{dt}{t}\right)^{1/q}.$$

$C^{j,\lambda,q}(\overline{\Omega})$ is a Banach space under the norm $\left\|\cdot\,;C^{j,\lambda,q}(\overline{\Omega})\right\|$.

7.36 LEMMA If $0<\lambda\le 1$ and $0<\theta<1$, then

$$\left(L^\infty(\Omega), C^{0,\lambda}(\overline{\Omega})\right)_{\theta,q;K} \to C^{0,\theta\lambda,q}(\overline{\Omega}).$$

Proof. Let $u\in C^{0,\lambda}(\overline{\Omega})_{\theta,q;K}$. Then there exists $v\in L^\infty(\Omega)$ and $w\in C^{0,\lambda}(\overline{\Omega})$ such that $u=v+w$ and

$$\|v\|_\infty + t^\lambda\left\|w\,;C^{0,\lambda}(\overline{\Omega})\right\| \le 2K(t^\lambda;u) \qquad \text{for } t>0.$$

If $|h|\le t$, then

$$\begin{aligned}|u(x+h)-u(x)| &\le |v(x+h)|+|v(x)| + \frac{|w(x+h)-w(x)|}{|h|^\lambda}|h|^\lambda \\ &\le 2\|v\|_\infty + \left\|w\,;C^{0,\lambda}(\overline{\Omega})\right\| t^\lambda \le 4K(t^\lambda;u).\end{aligned}$$

Thus $\omega(u;t) \le 4K(t^\lambda; u)$.

Since $\|u\|_\infty \le \|u\,;C^{0,\lambda}(\overline{\Omega})\|$, we have $\|u\|_\infty \le \|u\|_{\theta,q;K}$. Thus, if $1 \le q < \infty$,

$$\begin{aligned}\|u\,;C^{0,\lambda\theta,q}(\overline{\Omega})\| &= \|u\|_\infty + \left(\int_0^\infty \left(t^{-\lambda\theta}\omega(u;t)\right)^q \frac{dt}{t}\right)^{1/q} \\ &\le \|u\|_{\theta,q;K} + 4\left(\int_0^\infty \left(t^{-\lambda\theta}K(t^\lambda;u)\right)^q \frac{dt}{t}\right)^{1/q} \\ &= \|u\|_{\theta,q;K} + 4\lambda^{-1/q}\left(\int_0^\infty \left(\tau^{-\theta}K(\tau;u)\right)^q \frac{d\tau}{\tau}\right)^{1/q} \\ &\le (1+4\lambda^{-1/q})\,\|u\|_{\theta,q;K}\,.\end{aligned}$$

Similarly, for $q=\infty$, we obtain

$$\|u\,;C^{0,\lambda\theta,\infty}(\overline{\Omega})\| \le \|u\|_{\theta,\infty;K} + 4\sup_t t^{-\lambda\theta}K(t^\lambda;u) \le 5\,\|u\|_{\theta,\infty;K}\,.$$

This completes the proof. ∎

7.37 THEOREM Let Ω be a domain in $\mathbb{R}^n$ satisfying the strong local Lipschitz condition. Let $m-1-j \le n/p < s \le m-j$ and $1 \le q \le \infty$. If $\mu = s - n/p$, then

$$B^{s;p,q}(\Omega) \to C^{j,\mu,q}(\overline{\Omega}).$$

Proof. It is sufficient to prove this for $j=0$. By Theorem 7.34(b),

$$B^{n/p;p,1}(\Omega) \to C_B^0(\Omega) \to L^\infty(\Omega).$$

By Part II of Theorem 4.12,

$$W^{m,p}(\Omega) \to C^{0,\lambda}(\overline{\Omega}), \qquad \text{where} \quad \lambda = m - \frac{n}{p}.$$

Now $B^{s;p,q}(\Omega) = \left(B^{n/p;p,1}(\Omega), W^{m,p}(\Omega)\right)_{\theta,q;K}$, where

$$(1-\theta)\frac{n}{p} + \theta m = s.$$

Since $\lambda\theta = \mu$, we have by the Exact Interpolation Theorem and the previous Lemma,

$$B^{s;p,q}(\Omega) \to \left(L^\infty(\Omega), C^{0,\lambda}(\overline{\Omega})\right)_{\theta,q;K} \to C^{j,\mu,q}(\overline{\Omega}). \quad ∎$$

Characterization of Traces

7.38 As shown in the Sobolev imbedding theorem (Theorem 4.12) functions in $W^{m,p}(\mathbb{R}^{n+1})$ (where $mp < n+1$) have traces on $\mathbb{R}^n$ that belong to $L^q(\mathbb{R}^n)$ for $p \le q \le np/(n+1-mp)$. The following theorem asserts that these traces are exactly the functions that belong to $B^{m-(1/p);p,p}(\mathbb{R}^n)$. This is an instance of the phenomenon that passing from functions in $W^{m,p}(\Omega)$ to their traces on surfaces of codimension 1 results in a loss of smoothness corresponding to $1/p$ of a derivative.

In the following we denote points in $\mathbb{R}^{n+1}$ by (x,t) where $x \in \mathbb{R}^n$ and $t \in \mathbb{R}$. The trace $u(x)$ of a smooth function $U(x,t)$ defined on $\mathbb{R}^{n+1}$ is therefore given by $u(x) = U(x,0)$.

7.39 THEOREM (The Trace Theorem) If $1 < p < \infty$, the following conditions on a measurable function u on $\mathbb{R}^n$ are equivalent.

(a) There is a function U in $W^{m,p}(\mathbb{R}^{n+1})$ so that u is the trace of U.

(b) $u \in B^{m-(1/p);p,p}(\mathbb{R}^n)$. ▮

As the proof of this theorem is rather lengthy, we split it into two lemmas; (a) implies (b) and (b) implies (a).

7.40 LEMMA Let $1 < p < \infty$. If $U \in W^{m,p}(\mathbb{R}^{n+1})$, then its trace u belongs to the space $B = B^{m-(1/p);p,p}(\mathbb{R}^n)$ and

$$\|u\|_B \le K\,\|U\|_{m,p,\mathbb{R}^{n+1}}, \tag{18}$$

for some constant K independent of U.

Proof. We represent

$$B \equiv B^{m-(1/p);p,p}(\mathbb{R}^n) = \left(W^{m-1,p}(\mathbb{R}^n),\, W^{m,p}(\mathbb{R}^n)\right)_{\theta,p;J},$$

where

$$\theta = 1 - \frac{1}{p} = \frac{1}{p'}$$

and use the discrete version of the J-method; we have $u \in B^{m-(1/p);p,p}(\mathbb{R}^n)$ if and only if there exist functions u_i in $W^{m-1,p}(\mathbb{R}^n) \cap W^{m,p}(\mathbb{R}^n) = W^{m,p}(\mathbb{R}^n)$ for $-\infty < i < \infty$ such that the series $\sum_{i=-\infty}^{\infty} u_i$ converges to u in norm in the space $W^{m-1,p}(\mathbb{R}^n) + W^{m,p}(\mathbb{R}^n) = W^{m-1,p}(\mathbb{R}^n)$, and such that the sequences $\{2^{-i/p'}\|u_i\|_{m-1,p}\}$ and $\{2^{i/p}\|u_i\|_{m,p}\}$ both belong to ℓ^p. We verify (18) by splitting U into pieces U_i with traces u_i that satisfy these conditions.

Let Φ be an even function on the real line satisfying the following conditions:

(i) $\Phi(t) = 1$ if $-1 \le t \le 1$,

(ii) $\Phi(t) = 0$ if $|t| \ge 2$,

(iii) $|\Phi(t)| \le 1$ for all t,

(iv) $|\Phi^{(j)}(t)| \le C_j < \infty$ for all $j \ge 1$ and all t.

For each integer i let $\Phi_i(t) = \Phi(t/2^i)$; then Φ_i takes the value 1 on the interval $[-2^i, 2^i]$ and takes the value 0 on the intervals $[2^{i+1}, \infty)$ and $(-\infty, -2^{i+1}]$. Also, $|\Phi(t)| \le 1$ and $|\Phi_i'(t)| \le 2^{-i}C_1$ for all t.

Let $\phi_i = \Phi_{i+1} - \Phi_i$. Then $\phi_i(\tau)$ vanishes outside the open intervals $(2^i, 2^{i+2})$ and $(-2^{i+2}, -2^i)$; in particular it vanishes at the endpoints of these intervals. Also $\|\phi_i\|_\infty = 1$ and $\|\phi_i'\|_\infty \le 2^{-i}C_1$.

Now suppose that $U \in C_0^\infty(\mathbb{R}^{n+1})$. Then for each t we have

$$U(x,t) = -\int_t^\infty \frac{\partial U}{\partial \tau}(x,\tau)\,d\tau = -\int_t^\infty D^{(0,1)}U(x,\tau)\,d\tau.$$

Let

$$U_i(x,t) = -\int_t^\infty \phi_i(\tau)D^{(0,1)}U(x,\tau)\,d\tau.$$

Let $u(x) = U(x,0)$ be the trace of U on $\mathbb{R}^n$, and let u_i be the corresponding trace of U_i. Since U has compact support, the functions U_i and u_i vanish when i is sufficiently large. Moreover, $U_i(x,t) = 0$ for all i when $|x|$ is sufficiently large. Therefore the trace u vanishes except on a compact set, on which the series $\sum_{i=\infty}^{\infty} u_i(x)$ converges uniformly to $u(x)$. The terms in this series also vanish off that compact set and taking any partial derivative term-by-term gives a series that converges uniformly on that compact set to the corresponding partial derivative of u.

We use two representations of $u_i(x) = U_i(x,0)$, namely

$$u_i(x) = -\int_{2^i}^{2^{i+2}} \phi_i(\tau)D^{(0,1)}U(x,\tau)\,d\tau = \int_{2^i}^{2^{i+2}} \phi_i'(\tau)U(x,\tau)\,d\tau, \tag{19}$$

where the second expression follows from the first by integration by parts. If $|\alpha| \le m-1$ we obtain from the first representation a corresponding representations of $D^\alpha u_i(x)$:

$$D^\alpha u_i(x) = -\int_{2^i}^{2^{i+2}} \phi_i(\tau)D^{(\alpha,1)}U(x,\tau)\,d\tau,$$

so that, by Hölder's inequality,

$$|D^\alpha u_i(x)| \le (2^{i+2})^{1/p'}\left(\int_{2^i}^{2^{i+2}} \left|D^{(\alpha,1)}U(x,\tau)\right|^p\,d\tau\right)^{1/p}.$$

Each positive number τ lies in exactly two of the intervals $[2^i, 2^{i+1})$ over which the integrals above run. Multiplying by $2^{-i/p'}$, taking p-th powers on both sides, summing with respect to i, and integrating x over $\mathbb{R}^n$ shows that the p-th power of the ℓ^p norm of the sequence $\left\{2^{-i/p'}\|D^\alpha u_i\|_p\right\}_{i=-\infty}^{\infty}$ is no larger than

$$2^{1+2p/p'}\int_{R_+^{n+1}} \left|D^{(\alpha,1)}U(x,\tau)\right|^p \, d\tau\, dx.$$

Thus that ℓ^p norm is bounded by a constant times $\|U\|_{m,p,\mathbb{R}^{n+1}}$.

Using the second representation of u_i in (19), our bound on $\|\phi_i'\|_\infty$, and Hölder's inequality gives us a second estimate

$$|D^\alpha u_i(x)| \le 2^{-i}C_1(2^{i+2})^{1/p'}\left(\int_{2^i}^{2^{i+2}} |D^{(\alpha,0)}U(x,\tau)|^p \, d\tau\right)^{1/p},$$

this one valid for any α with $|\alpha| \le m$. Multiplying by $2^{i/p}$, taking p-th powers on both sides, and summing with respect to i shows that the p-th power of the ℓ^p norm of the sequence $\left\{2^{i/p}\|D^\alpha u_i\|_p\right\}_{i=-\infty}^{\infty}$ is no larger than

$$2^{1+2p/p'}C_1^p\int_{R_+^{n+1}} |D^{(\alpha,0)}U(x,\tau)|^p \, d\tau\, dx.$$

Thus that ℓ^p norm is also bounded by a constant times $\|U\|_{m,p,\mathbb{R}^{n+1}}$.

Together, these estimates show that the norm of u in $B^{m-(1/p);p,p}(\mathbb{R}^n)$ is bounded by a constant times the norm of U in $W^{m,p}(\mathbb{R}^{n+1})$ whenever $U \in C_0^\infty(\mathbb{R}^{n+1})$. Since the latter space is dense in $W^{m,p}(\mathbb{R}^{n+1})$, the proof is complete. ∎

7.41 LEMMA Let $1 < p < \infty$ and $B = B^{m-(1/p);p,p}(\mathbb{R}^n)$. If $u \in B$, then u is the trace of a function $U \in W^{m,p}(\mathbb{R}^{n+1})$ satisfying

$$\|U\|_{m,p,\mathbb{R}^{n+1}} \le K\,\|u\|_B \tag{20}$$

for some constant K independent of u.

Proof. In this proof it is convenient to use a characterization of B different (if $m > 1$) from the one used in the previous lemma, namely

$$B = B^{m-(1/p);p,p}(\mathbb{R}^n) = \left(L^p(\mathbb{R}^n), W^{m,p}(\mathbb{R}^n)\right)_{\theta,p;J},$$

where $\theta = 1 - (1/mp)$. Again we use the discrete version of the J-method. For $u \in B$ we can find $u_i \in L^p(\mathbb{R}^n) \cap W^{m,p}(\mathbb{R}^n) = W^{m,p}(\mathbb{R}^n)$ (for $-\infty < i < \infty$)

such that $\sum_{i=-\infty}^{\infty} u_i$ converges to u in $L^p(\mathbb{R}^n) + W^{m,p}(\mathbb{R}^n) = L^p(\mathbb{R}^n)$, and such that

$$\left\| \{2^{-\theta i} \|u_i\|_p\} ; \ell^p \right\| \le K_1 \|u\|_B ,$$
$$\left\| \{2^{(1-\theta)i} \|u_i\|_{m,p}\} ; \ell^p \right\| \le K_1 \|u\|_B .$$

These estimates imply that $\sum_{i=-\infty}^{\infty} u_i$ converges to u in B. We will construct an extension $U(x,t)$ of $u(x)$ defined on $\mathbb{R}^{n+1}$ such that (20) holds.

It is sufficient to extend the partial sums $s_k = \sum_{i=-k}^{k} u_i$ to S_k on $\mathbb{R}^{n+1}$ with control of the norms:

$$\|S_k\|_{m,p,\mathbb{R}^{n+1}} \le K_1 \left\| s_k ; B^{m-(1/p);p,p}(\mathbb{R}^n) \right\| ,$$

since $\{S_k\}$ will then be a Cauchy sequence in $W^{m,p}(\mathbb{R}^{n+1})$ and so will converge there. Furthermore, we can assume that the functions u and u_i are smooth since the mollifiers $J_\epsilon * u$ and $J_\epsilon * u_i$ (as considered in Paragraphs 2.28 and 3.16) converge to u and u_i in norm in $W^{m,p}(\mathbb{R}^n)$ as $\epsilon \to 0+$. Accordingly, therefore, in the following construction we assume that the functions u and u_i are smooth and that all but finitely many of the u_i vanish identically on $\mathbb{R}^n$.

Let $\Phi(t)$ be as defined in the previous lemma. Here, however, we redefine Φ_i as follows:

$$\Phi_i(t) = \Phi\left(\frac{t}{2^{i/m}}\right), \qquad -\infty < i < \infty.$$

The derivatives of Φ_i then satisfy $|\Phi_i^{(j)}(t)| \le 2^{-ij/m} C_j$. Also, note that for $j \ge 1$, $\Phi_i^{(j)}$ is zero outside the two intervals $(-2^{(i+1)/m}, -2^{i/m})$ and $(2^{i/m}, 2^{(i+1)/m})$, which have total length not exceeding $2^{1+(i/m)}$.

We define the extension of u as follows:

$$U(x,t) = \sum_{i=-\infty}^{\infty} U_i(x,t), \qquad \text{where} \qquad U_i(x,t) = \Phi_i(t) u_i(x).$$

Note that the sum is actually a finite one under the current assumptions. In order to verify (20) it is sufficient to bound by multiples of $\|u\|_B$ the L^p-norms of U and all its mth order derivatives; the Ehrling-Nirenberg-Gagliardo interpolation theorem 5.2 then supplies similar bounds for intermediate derivatives. The mth order derivatives are of three types: $D^{(0,m)}U$, $D^{(\alpha,j)}U$ for $1 \le j \le m-1$ and $|\alpha| + j = m$, and $D^{(\alpha,0)}U$ for $|\alpha| = m$. We examine each in turn.

Since $D^{(0,m)}U_i(x,t) = 2^{-i}\Phi^{(m)}(t/2^{i/m})u_i(x)$, we have

$$\begin{aligned} &\int_{\mathbb{R}^{n+1}} |D^{(0,m)}U_i(x,t)|^p \, dx \, dt \\ &\le \left(\int_{-2^{(i+1)/m}}^{-2^{i/m}} dt + \int_{2^{i/m}}^{2^{(i+1)/m}} dt \right) \int_{\mathbb{R}^n} |D^{(0,m)}U_i(x,t)|^p \, dx \\ &\le 2^{1+(i/m)} 2^{-ip} C_m^p \|u_i\|_p^p = 2C_m^p 2^{-\theta ip} \|u_i\|_p^p . \end{aligned}$$

Since the functions $\Phi_i^{(m)}$ have non-overlapping supports, we can sum the above inequality on i to obtain

$$\left\| D^{(0,m)} U \right\|_{p,\mathbb{R}^{n+1}}^p \le 2C_m^p \sum_{i=-\infty}^{\infty} \left(2^{-\theta i} \|u_i\|_p\right)^p$$
$$= 2C_m^p \left\| \left\{2^{-\theta i} \|u_i\|_p\right\} ; \ell^p \right\|^p \le 2C_m^p \|u\|_B^p$$

and the required estimate for $D^{(0,m)}U$ is proved.

Now consider $D^{(\alpha,j)}U_i(x,t) = 2^{-ij/m}\Phi^{(j)}(t/2^{i/m})D^\alpha u_i(x)$ for which we obtain similarly

$$\int_{\mathbb{R}^{n+1}} |D^{(\alpha,j)}U_i(x,t)|^p \, dx \, dt \le C_j^p 2^{-i(jp-1)/m} \|D^\alpha u_i\|_p^p .$$

Since $|\alpha| = m - j$, we can replace the L^p-norm of $D^\alpha u_i$ with the seminorm $|u_i|_{m-j,p}$, and again using the non-overlapping of the supports of the $\Phi_i^{(j)}$ (since $j \ge 1$) to get

$$\left\| D^{(\alpha,j)} U \right\|_{p,\mathbb{R}^{n+1}}^p \le C_j^p \sum_{i=-\infty}^{\infty} 2^{-i(jp-1)/m} |u_i|_{m-jp}^p .$$

As remarked in Paragraph 5.7, for $1 \le j \le m-1$ Theorem 5.2 assures us that there exists a constant K_2 such that for any $\epsilon > 0$ and any i

$$|u_i|_{m-j,p}^p \le K_2\left(\epsilon^p |u_i|_{m,p}^p + \epsilon^{-(m-j)p/j} \|u\|_p^p\right).$$

Let $\epsilon = 2^{ij/m}$. Then we have

$$\left\| D^{(\alpha,j)} U \right\|_{p,\mathbb{R}^{n+1}}^p \le C_j^p K_2 \sum_{i=-\infty}^{\infty} \left(2^{i/m} |u_i|_{m,p}^p + 2^{-ip(1-(1/mp))} \|u_i\|_p^p\right)$$
$$= C_j^p K_2 \sum_{i=-\infty}^{\infty} \left(2^{(1-\theta)ip} |u_i|_{m,p}^p + 2^{-\theta ip} \|u_i\|_p^p\right)$$
$$\le C_j^p K_2 \left(\left\| \left\{2^{(1-\theta)i} \|u_i\|_{m,p}\right\} ; \ell^p \right\|^p + \left\| \left\{2^{-\theta i} \|u_i\|_p\right\} ; \ell^p \right\| \right)$$
$$\le 2K_1^p C_j^p K_2 \|u\|_B$$

and the bound for $D^{(\alpha,j)}U$ is proved.

Finally, we consider U and $D^{(\alpha,0)}U$ together. (We allow $0 \le |\alpha| \le m$.) Unlike their derivatives, the functions Φ_i have nested rather than non-overlapping supports. We must proceed differently than in the previous cases. Consider

$D^{(\alpha,0)}U(x,t)$ on the strip $2^{j/m} < t \le 2^{(j+1)/m}$ in $\mathbb{R}^{n+1}$. Since $|\Phi_i(t)| \le 1$ and since $U_i(x,t) = 0$ on this strip if $i < j-1$, we have

$$|D^{(\alpha,0)}U(x,t)| \le \sum_{i=j-1}^{\infty} |D^{(\alpha,0)}U_i(x,t)| = \sum_{i=j-1}^{\infty} 2^{-i/mp}a_i,$$

where $a_i = 2^{i/mp}|D^\alpha u_i(x)|$. Thus,

$$\begin{aligned} b_j \equiv \left(\int_{2^{j/m}}^{2^{(j+1)/m}} |D^{(\alpha,0)}U(x,t)|^p\,dt\right)^{1/p} &\le \sum_{i=j-1}^{\infty} 2^{j/mp}\,2^{-i/mp}\,a_i \\ &= \sum_{i=j-1}^{\infty} 2^{(j-i)/mp}a_i = (c*a)_j, \end{aligned}$$

where $c_j = 2^{j/mp}$ when $-\infty < j \le 1$ and $c_j = 0$ otherwise. Observe that $c \in \ell^1$ (say, $\|c\,;\ell^1\| = K_3$), and so by Young's inequality for sequences

$$\|b\,;\ell^p\| \le K_3\,\|a\,;\ell^p\|.$$

Taking pth powers and summing on j now leads to

$$\int_0^\infty \left|D^{(\alpha,0)}U(x,t)\right|^p dt \le K_3^p\,\left\|\{2^{i/mp}|D^\alpha u_i(x)|\}\,;\ell^p\right\|^p.$$

Integrating x over $\mathbb{R}^n$ and taking pth roots then gives

$$\begin{aligned} \left\|D^{(\alpha,0)}U\right\|_{0,p,\mathbb{R}^{n+1}_+} &\le K_3\,\left\|\{2^{i/mp}\,\|D^\alpha u_i\|_p\}\,;\ell^p\right\| \\ &\le K_3\,\left\|\{2^{(1-\theta)i}\,\|u_i\|_{m,p}\}\,;\ell^p\right\| \le K_1K_3\,\|u\|_B\,. \end{aligned}$$

A similar estimate holds for $\left\|D^{(\alpha,0)}U\right\|_{0,p,\mathbb{R}^{n+1}_-}$, so the proof is complete. ∎

7.42 We can now complete the imbedding picture for Besov spaces by proving an analog of the trace imbedding part of the Sobolev Imbedding Theorem 4.12 for Besov spaces. We will show in Lemma 7.44 below that the trace operator T defined for smooth functions U on $\mathbb{R}^{n+1}$ by

$$(TU)(x) = U(x,0)$$

is linear and bounded from $B^{1/p;p,1}(\mathbb{R}^{n+1})$ into $L^p(\mathbb{R}^n)$. Since Theorem 7.39 assures us that T is also bounded from $W^{m,p}(\mathbb{R}^{n+1})$ onto $B^{m-1/p;p;p}(\mathbb{R}^n)$ for every $m \geq 1$, by the exact interpolation theorem (Theorem 7.23), it is bounded from $B^{s;p,q}(\mathbb{R}^{n+1})$ into $B^{s-1/p;p;q}(\mathbb{R}^n)$, that is,

$$B^{s;p,q}(\mathbb{R}^{n+1}) \to B^{s-1/p;p;q}(\mathbb{R}^n),$$

for every $s > 1/p$ and $1 \leq q \leq \infty$. (Although Theorem 7.39 does not apply if $p = 1$, we already know from the Sobolev Theorem 4.12 that traces of functions in $W^{m,1}(\mathbb{R}^{n+1})$ belong to $W^{m-1,1}(\mathbb{R}^n)$.)

We can now take traces of traces. If $n - k < sp < n$ (so that $s - (n-k)/p > 0$), then

$$B^{s;p,q}(\mathbb{R}^n) \to B^{s-(n-k)/p;p;q}(\mathbb{R}^k),$$

We can combine this imbedding with Theorem 7.34 to obtain for $n - k < sp < n$ and $r = kp/(n - sp)$,

$$B^{s;p,p}(\mathbb{R}^n) \to B^{s-(n-k)/p;p;p}(\mathbb{R}^k) \to L^{r,p}(\mathbb{R}^k) \to L^r(\mathbb{R}^k).$$

More generally:

7.43 THEOREM (Trace Imbeddings for Besov Spaces on $\mathbb{R}^n$) If k is an integer satisfying $1 < k < n$, $n - k < sp < n$, and $r = kp/(n - sp)$, then

$$B^{s;p,q}(\mathbb{R}^n) \to B^{s-(n-k)/p;p;q}(\mathbb{R}^k) \to L^{r,q}(\mathbb{R}^k), \quad \text{and}$$
$$B^{s;p,q}(\mathbb{R}^n) \to L^r(\mathbb{R}^k) \quad \text{for } q \leq r.$$

To establish this theorem, we need only prove the following lemma.

7.44 LEMMA The trace operator T defined by $(TU)(x) = U(x, 0)$ imbeds $B^{1/p;p,1}(\mathbb{R}^{n+1})$ into $L^p(\mathbb{R}^n)$.

Proof. Suppose that U belongs to $B \equiv B^{1/p;p,1}(\mathbb{R}^{n+1})$ and, without loss of generality, that $\|U\|_B \leq 1$. Then there exist functions U_i for $-\infty < i < \infty$ such that $U = \sum_i U_i$ and

$$\sum_i 2^{-i/p} \|U_i\|_{p,\mathbb{R}^{n+1}} \leq C \qquad \text{and} \qquad \sum_i 2^{i/p'} \|U_i\|_{1,p,\mathbb{R}^{n+1}} \leq C$$

for some constant C. As in the proof of Lemma 7.40, we can assume that only finitely many of the functions U_i have nonzero values and that they are smooth functions. For any of these functions we have, for $2^i \leq h \leq 2^{i+1}$,

$$\begin{aligned} |U_i(x,0)| &\leq \int_0^h \left|D^{(0,1)}U_i(x,t)\right| dt + |U_i(x,h)| \\ &\leq \int_0^{2^{i+1}} \left|D^{(0,1)}U_i(x,t)\right| dt + |U_i(x,h)|. \end{aligned}$$

Averaging h over $[2^i, 2^{i+1}]$ then gives the estimate

$$|U_i(x,0)| \le \int_0^{2^{i+1}} \left|D^{(0,1)}U_i(x,t)\right| dt + \frac{1}{2^i}\int_{2^i}^{2^{i+1}} |U_i(x,t)|\, dt.$$

By Hölder's inequality,

$$\begin{aligned} |U_i(x,0)| &\le 2^{(i+1)/p'}\left(\int_0^{2^{i+1}} \left|D^{(0,1)}U_i(x,t)\right|^p dt\right)^{1/p} \\ &\quad + \frac{2^{i/p'}}{2^i}\left(\int_{2^i}^{2^{i+1}} |U_i(x,t)|^p\, dt\right)^{1/p} \\ &= a_i(x) + b_i(x), \quad \text{say.} \end{aligned}$$

Then $\|a_i\|_{p,\mathbb{R}^n} \le 2(2^{i/p'})\left\|U_j\right\|_{1,p,\mathbb{R}^{n+1}}$ and $\|b_i\|_{p,\mathbb{R}^n} \le 2^{-i/p}\left\|U_j\right\|_{p,\mathbb{R}^{n+1}}$. We now have

$$\begin{aligned} \|U(\cdot,0)\|_{p,\mathbb{R}^n} &\le \sum_i \|U_i(\cdot,0)\|_{p,\mathbb{R}^n} \\ &\le 2\left(\sum_i 2^{i/p'}\left\|U_j\right\|_{1,p,\mathbb{R}^{n+1}} + \sum_i 2^{-i/p}\left\|U_j\right\|_{p,\mathbb{R}^{n+1}}\right) \le 4C. \end{aligned}$$

This completes the proof. ∎

7.45 REMARKS

1. Theorems 7.39 and 7.43 extend to traces on arbitrary planes of sufficiently high dimension, and, as a consequence of Theorem 3.41, to traces on sufficiently smooth surfaces of sufficiently high dimension.
2. Both theorems also extend to traces of functions in $B^{s;p,q}(\Omega)$ on the intersection of the domain Ω in $\mathbb{R}^n$ with planes or smooth surfaces of dimension k satisfying $k > n - sp$, provided there exists a suitable extension operator for Ω. This will be the case if, for example, Ω satisfies a strong local Lipschitz condition. (See Theorem 5.21.)
3. Before Besov spaces were fully developed, Gagliardo [Ga3] identified the trace space as a space defined by a version of the intrinsic condition (c) in the characterization of Besov spaces in Theorem 7.47 below, where $q = p$ and $s = m - (1/p)$.

Direct Characterizations of Besov Spaces

7.46 The K functional for the pair $\left(L^p(\Omega), W^{m,p}(\Omega)\right)$ measures how closely a given function u can be approximated in L^p norm by functions whose $W^{m,p}$ norm are not too large. For instance, a splitting $u = u_0 + u_1$ with $\|u_0\|_p + t\,\|u_1\|_{m,p} \le 2K(t; u)$ provides such an approximation u_1 to u; then the error $u - u_1 = u_0$ has $L^p(\Omega)$ norm at most $2K(t; u)$ and the approximation u_1 has $W^{m,p}(\Omega)$ norm at most $(2/t)K(t; u)$. So, in principle, the definition of $B^{s;p,q}(\Omega)$ by real interpolation characterizes functions in $B^{s;p,q}(\Omega)$ by the way in which they can be approximated in $L^p(\Omega)$ norm by functions in $W^{m,p}(\Omega)$.

Like many other descriptions of Besov spaces, the one above seems indirect, but it can yield useful upper bounds for Besov norms. On $\mathbb{R}^n$, more direct characterizations come from considering the L^p*-modulus of continuity* and higher-order versions of that modulus. Given a point h in $\mathbb{R}^n$ and a function u in $L^p(\mathbb{R}^n)$, let u_h be the function mapping x to $u(x - h)$, let $\Delta_h u = u - u_h$, let $\omega_p(u; h) = \|\Delta_h u\|_p$, and for positive integers m, let $\omega_p^{(m)}(u; h) = \|(\Delta_h)^m u\|_p$. When $1 \le p < \infty$, mollification shows that $\omega_p(u; h)$ tends to 0 as $h \to 0$, and the same is true for $\omega_p^{(m)}(u; h)$; as stated below, when $m > s$, the rate of the latter convergence to 0 determines whether $u \in B^{s;p,q}(\mathbb{R}^n)$. We also define functions on $\mathbb{R}_+$ by letting $\omega_p^*(u; t) = \sup\{\omega_p(u; h); |h| \le t\}$ and letting $\omega_p^{(m)*}(u; t) = \sup\{\omega_p^{(m)}(u; h); |h| \le t\}$.

7.47 THEOREM (Intrinsic Characterization of $B^{s;p,q}(\mathbb{R}^n)$) Whenever $m > s > 0$, $1 < p < \infty$, and $1 \le q < \infty$, the following conditions on a function u in $L^p(\mathbb{R}^n)$ are equivalent. If $q = \infty$ condition (a) is equivalent to the versions of conditions (b) and (c) with the integrals replaced by the suprema of the quantities inside the square brackets.

(a) $u \in B^{s;p,q}(\mathbb{R}^n)$.

(b) $\displaystyle\int_0^\infty \left[t^{-s}\omega_p^{(m)*}(u; t)\right]^q \frac{dt}{t} < \infty.$

(c) $\displaystyle\int_{R^n} \left[|h|^{-s}\omega_p^{m}(u; h)\right]^q \frac{dh}{|h|^n} < \infty.$ ∎

Before proving this theorem, we observe a few things. First, the moduli of continuity in parts (b) and (c) are never larger than $2^m\,\|u\|_p$; so we get conditions equivalent to (b) and (c) respectively if we use integrals with $t \le 1$ and $|h| \le 1$. Next, the equivalence of conditions (b) and (c) with condition (a), where m does not appear, means that if (b) or (c) holds for some $m > s$, then both conditions hold for all $m > s$.

It follows from our later discussion of Fourier decompositions that if $1 < p < \infty$, then these conditions are equivalent to requiring that the derivatives of u of order k, where k is the largest integer less than s, belong to $L^p(\mathbb{R}^n)$ and satisfy the versions of condition (b) or (c) with $m = 1$ and s replaced with $s - k$.

While we assumed $1 < p < \infty$ in the statement of the theorem, the only part of the proof that requires this is the part showing that (c) ⇒ (a) when $m > 1$. The rest of the proof is valid for $1 \le p \le \infty$.

7.48 (The Proof of Theorem 7.47 for $m = 1$) We assume, for the moment, that $m = 1$ and $s < 1$; in the next Paragraph we will outline with rather less detail how to modify the argument for the case $m > 1$. We show that (a) $\Rightarrow$ (b) $\Rightarrow$ (c) $\Rightarrow$ (a).

The first part is similar to the proof of Lemma 7.36. Suppose first that condition (a) holds and consider condition (b) with $m = 1$. Fix a positive value of the parameter t and split a nontrivial function u as $v + w$ with $\|v\|_p + t\|w\|_{1,p} \le 2K(t; u)$. Then $\Delta_h u = \Delta_h v + \Delta_h w$, and it suffices to control the L^p norms of the these two differences. For the first term, just use the fact that $\|\Delta_h v\|_p \le 2\|u\|_p$.

For the second term, we use mollification to replace v and w with smooth functions satisfying the same estimate on their L^p and $W^{1,p}$ norms respectively. We majorize $|w(x-h) - w(x)|$ by the integral of $|\mathrm{grad}\, w|$ along the line segment joining $x - h$ to x, and use Hölder's inequality to majorize that by $|h|^{1/p'}$ times the one-dimensional L^p norm of the restriction of $|\mathrm{grad}\, w|$ to that segment. Finally, we take p-th powers, integrate with respect to x, and take a p-th root to get that $\|\Delta_h w\|_p \le |h||w|_{1,p}$. When $|h| \le t$ we then obtain

$$\|\Delta_h u\|_p \le \|\Delta_h v\|_p + \|\Delta_h w\|_p \le 2\|v\|_p + t|w|_{1,p} \le 4K(t; u),$$

so condition (a) implies condition (b).

Since t^{-s} decreases and $\omega_p(u; t)$ increases as t increases, condition (b) holds with $m = 1$ if and only if the sequence $\{2^{-is}\omega_p^*(u; 2^i)\}_{i=-\infty}^{\infty}$ belongs to ℓ^q. To deduce condition (c) with $m = 1$, we split the integral in (c) into dyadic pieces with $2^i < |h| \le 2^{i+1}$. The integral of the measure $dh/|h|^n$ over each such piece is the same. In the i-th piece, $\omega_p(u; h) \le \omega_p^*(u; 2^{i+1})$ by the definition of the latter quantity. And in that piece, $|h|^{-s} \le 2^s 2^{-s(i+1)}$. So the integral in (c) is majorized by a constant time the q-th power of the ℓ^q norm of the sequence $\{2^{-(i+1)s}\omega_p^*(f; 2^{(i+1)})\}_{i=-\infty}^{\infty}$, and (c) follows from (b).

We now show that (c) $\Rightarrow$ (a) when $m = 1 > s > 0$. Choose a nonnegative smooth function Φ vanishing outside the ball of radius 2 centred at 0 and inside the ball of radius 1, and satisfying

$$\int_{R^n} \Phi(x)\, dx = 1.$$

For fixed $t > 0$ let $\Phi_t(x) = t^{-n}\Phi(x/t)$; this nonnegative function also integrates to 1, and it vanishes outside the ball of radius $2t$ centred at 0 and inside the ball of radius t.

For u satisfying condition (c), split $u = v + w$ where $w = u * \Phi_t$ and $v = u - w$.

The fact that the density Φ_t has mass 1 ensures that

$$v(x) = \int_{\mathbb{R}^n} \Phi_t(h)[u(x) - u(x-h)]\, dh = \int_{R^n} \Phi_t(h)\Delta_h u(x)\, dh$$
$$= \int_{t<|h|<2t} \Phi_t(h)\Delta_h u(x)\, dh.$$

The function v belongs to $L^p(\mathbb{R}^n)$, being the difference of two functions in that space. To estimate its norm, we use the converse of Hölder's inequality to linearize that norm as the supremum of $\int_{\mathbb{R}^n} |v(x)|g(x)\, dx$ over all nonnegative functions g in the unit ball of $L^{p'}(\mathbb{R}^n)$. For each such function g, we find that

$$\int_{\mathbb{R}^n} |v(x)|g(x)\, dx \le \int_{t<|h|<2t} \Phi_t(h)\left[\int_{\mathbb{R}^n} g(x)|\Delta_h u(x)|\, dx\right] dh$$
$$\le \int_{t<|h|<2t} \Phi_t(h)\|g\|_{p'}\|\Delta_h u\|_p\, dh$$
$$= \int_{t<|h|<2t} \Phi_t(h)\|\Delta_h u\|_p\, dh.$$

Since $\|\Phi_t\|_\infty \le C/t^n$, the last integral above is in turn bounded above by

$$\frac{C}{t^n}\int_{t<|h|<2t} \|\Delta_h u\|_p\, dh \le C\int_{t<|h|<2t} \|\Delta_h u\|_p \frac{dh}{|h|^n}$$
$$\le C_q\left(\int_{t<|h|<2t} [\|\Delta_h u\|_p]^q \frac{dh}{|h|^n}\right)^{1/q},$$

where the last step uses Hölder's inequality and the fact that the coronas $\{h \in R^n : t < |h| < 2t\}$ all have the same measure. Thus we have shown that

$$\|v\|_p \le C_q\left(\int_{t<|h|<2t} [\|\Delta_h u\|_p]^q \frac{dh}{|h|^n}\right)^{1/q}. \tag{21}$$

To bound $K(t; u)$ for the interpolation pair $(L^p(\mathbb{R}^n), W^{1,p}(\mathbb{R}^n))$, we also require a bound for $\|w\|_{1,p} = \|u * \Phi_t\|_{1,p}$. Note that $\|w\|_p \le \|u\|_p\|\Phi_t\|_1 = \|u\|_p$. Moreover,

$$\operatorname{grad} w(x) = [u * \operatorname{grad}(\Phi_t)](x) = \int_{t<|h|<2t} u(x-h)\operatorname{grad}(\Phi_t)(h)\, dh$$
$$= \int_{t<|h|<2t} \left[u(x-h) - u(x)\right]\operatorname{grad}(\Phi_t)(h)\, dh$$
$$= \int_{t<|h|<2t} \Delta_h u(x)\operatorname{grad}(\Phi_t)(h)\, dh,$$

where we used the fact that the average value of $\nabla(\Phi_t)(h)$ is $\mathbf{0}$ to pass from the first line above to the second line. Linearizing as we did for v leads to an upper bound like (21) for $\|\operatorname{grad} w\|_p$, except that $\|\Phi_t\|_\infty$ is replaced by $\|\operatorname{grad}\Phi_t\|_\infty$, which is bounded by $\check{C}/t^{n+1}$ rather than by C/t^n. This division by an extra factor of t leads to the estimate

$$|w|_{1,p} \le \frac{C_q^*}{t}\left(\int_{t<|h|<2t}[\|(\|\Delta_h u\|_p]^q \frac{dh}{|h|^n}\right)^{1/q}.$$

Therefore

$$\begin{aligned} K(t;\, u) &\le \|v\|_p + t\|w\|_{1,p} \\ &\le \text{const.}\left(\int_{t<|h|<2t}[\|\Delta_h u\|_p]^q \frac{dh}{|h|^n}\right)^{1/q} + t\|u\|_p. \end{aligned} \tag{22}$$

We also have the cheap estimate $K(t;\, u) \le \|u\|_p$ from the splitting $u = u + 0$. We use the discrete version of the K method to describe $B^{s;p,q}(\mathbb{R}^n)$. The cheap estimate suffices to make $\sum_{i=1}^{\infty}[2^{-is}K(2^i;\, u)]^q$ finite. When $i \le 0$ we use inequality (22) with $t = 2^i$, and we find that distinct indices i lead to disjoint coronas for the integral appearing in (22). It follows that the part of the ℓ^q norm with $i \le 0$ is bounded above by a constant times $\|u\|_p$ plus a constant times the quantity

$$\left(\int_{|h|\le 2}[|h|^{-s}\omega_p(u;\, h)]^q \frac{dh}{|h|^n}\right)^{1/q}.$$

This completes the proof when $m = 1$ and $1 \le q < \infty$. The proof when $m = 1$ and $q = \infty$ is similar. ∎

7.49 (The Proof of Theorem 7.47 for $m > 1$) We can easily modify some parts of the above proof for the case where $m = 1$ to work when $m > 1$. In particular, to prove that condition (b) implies condition (c) when $m > 1$, simply take the argument for $m = 1$ and replace ω_p^* by $\omega_p^{(m)*}$ and ω_p by $\omega_p^{(m)}$.

To get from (a) to (b) when $m > 1$, consider $B^{s;p,q}(\mathbb{R}^n)$ as a real interpolation space between $X_0 = L^p(\mathbb{R}^n)$ and $X_1 = W^{m,p}(\mathbb{R}^n)$ with $\theta = s/m$; since $m > s$, we have $\theta < 1$. Given a value of t, split u as $v + w$ with $\|v\|_p + t^m\|w\|_{1,p} \le 2K(t^m;\, u)$. Then $\|\Delta_h^m v\|_p \le 2^m\|v\|_p \le 2^{m+1}\|v\|_p$.

Again we can mollify w and then write differences of w as integrals of derivatives of w. When $m = 1$ we found that $\Delta_h w$ was an integral of a first directional derivative of w with respect to path length along the line segment from $x - h$ to x. Denote that directional derivative by $D_h w$. Then Δ_h^2 is equal to the integral along the same line segment of $\Delta_h(D_h w)$. That integrand is itself equal to an integral along a line segment of length $|h|$ with integrand $D_h^2 w$. This represents $\Delta_h^2 w(x)$ as an iterated double integral of $D_h^2 w$, with both integrations running over intervals of length $|h|$. Iteration then represents $\Delta_h^m w$ as an m-fold iterated integral of $D_h^m w$ over intervals of length $|h|$. Applying Hölder's inequality to that integral and then integrating p-th powers over $\mathbb{R}^n$ yields the estimate $\|\Delta_h^m w\|_p \le C|h|^m |w|_{m,p}$. It follows that $\omega_p^{(m)*}(u;t) \le \hat{C} K(t^m; u)$. Thus

$$\begin{aligned}\int_{-\infty}^{\infty} \left[t^{-s}\omega_p^{(m)*}(u;t)\right]^q \frac{dt}{t} &\le \hat{C}^q \int_{-\infty}^{\infty} \left[t^{-s}K(t^m;u)\right]^q \frac{dt}{t} \\ &= \hat{C}^q \int_{-\infty}^{\infty} \left[(t^m)^{-s/m}K(t^m;u)\right]^q \frac{dt}{t} \\ &= \check{C} \int_{-\infty}^{\infty} [\tau^{-\theta}K(\tau;u)\,\frac{d\tau}{\tau},\end{aligned}$$

after the change of variable $\tau = t^m$. So condition (a) still implies condition (b) when $m > 1$.

We now give an outline of the proof that (c) implies (a). See [BB, pp. 192–194] for more details on some of what we do. Since condition (c) for any value of m implies the corresponding condition for larger values of m, we free to assume that m is even, and we do so.

Given a function u satisfying condition (c) for an even index $m > \max\{1, s\}$, and given an integer $i \le 0$, we can split $u = v_i + w_i$, where v_i is an averaged m-fold integral of $\Delta_h^m u$; each single integral in this nest runs over an interval of length comparable to $t = 2^i$, and the averaging involves dividing by a multiple of t^m. The outcome is that we can estimate $\|v_i\|_p$ by the average of $\|\Delta_h^m u\|_p$ over a suitable h-corona. As in the case where $m = 1$, this leads to an estimate for the ℓ^q norm of the sequence $\left\{2^{-is}\|v_i\|_p\right\}_{i=-\infty}^{\infty}$ in terms of the integral in condition (c).

There is still a cheap estimate to guarantee for the pair $X_0 = L^p(\mathbb{R}^n)$ and $X_1 = W^{m,p}(\mathbb{R}^n)$ that the half-sequence $\left\{2^{-is}K(2^{im};u)\right\}_{i=1}^{\infty}$ belongs to ℓ^q. This leaves the problem of suitably controlling the ℓ^q norm of the half-sequence $\left\{2^{i(m-s)}\|w_i\|_{1,p}\right\}_{i=-\infty}^{0}$. We can represent w_i as a sum of m terms, each involving an average, with an m-fold iterated integral, of translates of u in a fixed direction. We can use this representation to estimate the norms in $L^p(\mathbb{R}^n)$ of m-fold directional derivatives of w_i in any fixed direction. In particular, we can do this for the unmixed partial derivatives $D_j^m w_i$, in each case getting an L^p norm that we can control with the part of (c) corresponding to a suitable corona. It is known that L^p estimates for all unmixed derivatives of even order m imply similar estimate for all mixed mth-order derivatives derivatives, and thus for $|w_i|_{m,p}$. (See [St, p. 77]; this is the place where we need m to be even and $1 < p < \infty$.)

Finally, for $K(2^{im};u)$ we also need estimates for $\|w_i\|_p$. Since w_i comes from averages of translates of u, these estimates take the form $\|w_i\|_p \le C\|u\|_p$. For the half sequence $\{2^{-is}K(2^{im}; u)\}_{i=-\infty}^{0}$ we then need to multiply by 2^{im} and 2^{-is}; again the outcome is a finite ℓ^q norm, since $i \le 0$ and $m > s$. ∎

Other Scales of Intermediate Spaces

7.50 The Besov spaces are not the only scale of intermediate spaces that can fill the gap between Sobolev spaces of integer order. Several other such scales have been constructed, each slightly different from the others and each having properties making it useful in certain contexts. As we have seen, the Besov spaces are particularly useful for characterizing traces of functions in Sobolev spaces. However, except when $p = 2$, the Sobolev spaces do not actually belong to the scale of Besov spaces.

Two other scales we will introduce below are:

(a) the scale of fractional order Sobolev spaces (also called spaces of Bessel potentials), denoted $W^{s,p}(\Omega)$, which we will define for positive, real s by a complex interpolation method introduced below. It will turn out that if $s = m$, a positive integer and Ω is reasonable, then the space obtained coincides with the usual Sobolev space $W^{m,p}(\Omega)$.

(b) the scale of Triebel-Lizorkin spaces, $F^{s;p,q}(\mathbb{R}^n)$, which we will define only on $\mathbb{R}^n$ but which will provide a link between the Sobolev, Bessel potential, and Besov spaces, containing members of each of those scales for appropriate choices of the parameters s, p, and q.

We will use Fourier transforms to characterize both of the scales listed above, and will therefore normally work on the whole of $\mathbb{R}^n$. Some results can be extended to more general domains for which suitable extension operators exist.

For the rest of this chapter we will present only descriptive introductions to the topics considered and will eschew formal proofs, choosing to refer the reader to the available literature, e.g., [Tr1, Tr2, Tr3, Tr4], for more information. We particularly recommend the first chapter of [Tr4].

We begin by describing another interpolation method for Banach spaces; this one is based on properties of analytic functions in the complex plane.

7.51 (The Complex Interpolation Method) Let $\{X_0, X_1\}$ be an interpolation pair of complex Banach spaces defined as in Paragraph 7.7 so that $X_0 + X_1$ is a Banach space with norm

$$\|u\|_{X_0+X_1} = \inf\{\|u_0\|_{X_0} + \|u_1\|_{X_1} : u = u_0 + u_1, u_0 \in X_0, u_1 \in X_1\}.$$

Let $\mathscr{F} = \mathscr{F}(X_0, X_1)$ be the space of all functions f of the complex variable $\zeta = \theta + i\tau$ with values in $X_0 + X_1$ that satisfy the following conditions:

(a) f is continuous and bounded on the strip $0 \le \theta \le 1$ into $X_0 + X_1$.

(b) f is analytic from $0 < \theta < 1$ into $X_0 + X_1$ (i.e., the derivative $f'(\zeta)$ exists in $X_0 + X_1$ if $0 < \theta = \operatorname{Re} \zeta < 1$).

(c) f is continuous on the line $\theta = 0$ into X_0 and

$$\|f(i\tau)\|_{X_0} \to 0 \quad \text{as} \quad |\tau| \to \infty.$$

(d) f is continuous on the line $\theta = 1$ into X_1 and

$$\|f(1+i\tau)\|_{X_1} \to 0 \quad \text{as} \quad |\tau| \to \infty.$$

7.52 $\mathscr{F}$ is a Banach space with norm

$$\|f\,;\mathscr{F}\| = \max\left\{\sup_\tau \|f(i\tau)\|_{X_0}, \sup_\tau \|f(1+i\tau)\|_{X_1}\right\}.$$

Given a real number θ in the interval (0, 1), we define

$$X_\theta = [X_0, X_1]_\theta = \left\{u \in X_0 + X_1 \,:\, u = f(\theta) \quad \text{for some} \quad f \in \mathscr{F}\right\}.$$

X_θ is called a *complex interpolation space* between X_0 and X_1; it is a Banach space with norm

$$\|u\|_{X_\theta} = \|u\|_{[X_0, X_1]_\theta} = \inf\left\{\|f\,;\mathscr{F}\| \,:\, f(\theta) = u\right\}.$$

It follows from the above definitions that an analog of the Exact Interpolation Theorem (Theorem 7.23) holds for the complex interpolation method too. (See Calderón [Ca2, p. 115] and [BL, chapter 4].) If $\{X_0, X_1\}$ and $\{Y_0, Y_1\}$ are two interpolation pairs and T is a bounded linear operator from $X_0 + X_1$ into $Y_0 + Y_1$ such that T is bounded from X_0 into Y_0 with norm M_0 and from X_1 into Y_1 with norm M_1, then T is also bounded from X_θ into Y_θ with norm $M \le M_0^{1-\theta} M_1^\theta$ for each θ in the interval [0, 1].

There is also a version of the Reiteration Theorem 7.21 for complex interpolation; if $0 < \theta_0 < \theta_1 < 1$, $0 < \lambda < 1$, and $\theta = (1-\lambda)\theta_0 + \lambda\theta_1$, then

$$\left[[X_0, X_1]_{\theta_0}, [X_0, X_1]_{\theta_1}\right]_\lambda = [X_0, X_1]_\theta$$

with equivalent norms. This was originally proved under the assumption that $X_0 \cap X_1$ is dense in $[X_0, X_1]_{\theta_0} \cap [X_0, X_1]_{\theta_1}$, but this restriction was removed by Cwikel [Cw].

7.53 (Banach Lattices on Ω) Most of the Banach spaces considered in this book are spaces of (equivalence classes of almost everywhere equal) real-valued or complex-valued functions defined in a domain Ω in $\mathbb{R}^n$. Such a Banach space

B is called a *Banach lattice on* Ω if, whenever $u \in B$ and v is a measurable, real- or complex-valued function on Ω satisfying $|v(x)| \le |u(x)|$ a.e. on Ω, then $v \in B$ and $\|v\|_B \le \|u\|_B$. Evidently only function spaces whose norms depend only on the size of the function involved can be Banach lattices. The Lebesgue spaces $L^p(\Omega)$ and Lorentz spaces $L^{p,q}(\Omega)$ are Banach lattices on Ω, but Sobolev spaces $W^{m,p}(\Omega)$ (where $m \ge 1$) are not, since their norms also depend on the size of derivatives of their member functions.

We say that a Banach lattice B on Ω has the *dominated convergence property* if, whenever $u \in B$, $u_j \in B$ for $1 \le j < \infty$, and $|u_j(x)| \le |u(x)|$ a.e. in Ω, then

$$\lim_{j\to\infty} u_j(x) = 0 \text{ a.e.} \quad \Longrightarrow \quad \lim_{j\to\infty} \left\|u_j\right\|_B = 0.$$

The Lebesgue spaces $L^p(\Omega)$ and Lorentz spaces $L^{p,q}(\Omega)$ have this property provided both p and q are finite, but $L^\infty(\Omega)$, $L^{p,\infty}(\Omega)$, and $L^{\infty,q}(\Omega)$ do not. (As a counterexample for L^∞, consider a sequence of translates with non-overlapping supports of dilates of a nontrivial bounded function with bounded support.)

7.54 (The spaces $X_0^{1-\theta}X_1^\theta$) Now suppose that X_0 and X_1 are two Banach lattices on Ω and let $0 < \theta < 1$. We denote by $X_0^{1-\theta}X_1^\theta$ the collection of measurable functions u on Ω for each of which there exists a positive number λ and non-negative real-valued functions $u_0 \in X_0$ and $u_1 \in X_1$ such that $\|u_0\|_{X_0} = 1$, $\|u_1\|_{X_1} = 1$ and

$$|u(x)| \le \lambda u_0(x)^{1-\theta} u_1(x)^\theta. \tag{23}$$

Then $X_0^{1-\theta}X_1^\theta$ is a Banach lattice on Ω with respect to the norm

$$\left\|u\,;\,X_0^{1-\theta}X_1^\theta\right\| = \inf\{\lambda : \text{inequality (23) holds}\}.$$

The key result concerning the complex interpolation of Banach lattices is the following theorem of Calderón [Ca2, p.125] which characterizes the intermediate spaces.

7.55 THEOREM Let X_0 and X_1 be Banach lattices at least one of which has the dominated convergence property. If $0 < \theta < 1$, then

$$[X_0, X_1]_\theta = X_0^{1-\theta}X_1^\theta$$

with equality of norms. ∎

7.56 EXAMPLE It follows by factorization and Hölder's inequality that if $1 \le p_i \le \infty$ for $i = 0, 1$, $p_1 \ne p_2$, and $0 < \theta < 1$, then

$$\left[L^{p_0}(\Omega), L^{p_1}(\Omega)\right]_\theta = L^p(\Omega),$$

with equality of norms, where

$$\frac{1}{p} = \frac{1-\theta}{p_0} + \frac{\theta}{p_1}.$$

Moreover, if also $1 \le q_i \le \infty$ and at least one of the pairs (p_0, q_0) and (p_1, q_1) has finite components, then

$$\left[L^{p_0,q_0}(\Omega), L^{p_1,q_1}(\Omega)\right]_\theta = L^{p,q}(\Omega),$$

with equivalence of norms, where

$$\frac{1}{p} = \frac{1-\theta}{p_0} + \frac{\theta}{p_1} \quad \text{and} \quad \frac{1}{q} = \frac{1-\theta}{q_0} + \frac{\theta}{q_1}.$$

7.57 (Fractional Order Sobolev Spaces) We can define a scale of fractional order spaces by complex interpolation between L^p and Sobolev spaces. Specifically, if $s > 0$ and m is the smallest integer greater than s and Ω is a domain in $\mathbb{R}^n$, we define the space $W^{s,p}(\Omega)$ as

$$W^{s,p}(\Omega) = \left[L^p(\Omega), W^{m,p}(\Omega)\right]_{s/m}.$$

Again, as for Besov spaces, we can use the Reiteration Theorem to replace m with a larger integer and also observe that $W^{s,p}(\Omega)$ is an appropriate complex interpolation space between $W^{s_0,p}(\Omega)$ and $W^{s_1,p}(\Omega)$ if $0 < s_0 < s < s_1$. We will see later that if s is a positive integer and Ω has a suitable extension property, then $W^{s,p}(\Omega)$ coincides with the usual Sobolev space with the same name.

Because $W^{m,p}(\Omega)$ is not a Banach lattice on Ω we cannot use Theorem 7.55 to characterize $W^{s,p}(\Omega)$. Instead we will use properties of the Fourier transform on $\mathbb{R}^n$ for this purpose. Therefore, as we did for Besov spaces, we will normally work only with $W^{s,p}(\mathbb{R}^n)$, and rely on extension theorems to supply results for domains $\Omega \subset \mathbb{R}^n$.

We begin by reviewing some basic aspects of the Fourier transform.

7.58 (The Fourier Transform) The *Fourier transform* of a function u belonging to $L^1(\mathbb{R}^n)$ is the function $\hat{u}$ defined on $\mathbb{R}^n$ by

$$\hat{u}(y) = \frac{1}{(2\pi)^{n/2}} \int_{\mathbb{R}^n} e^{-ix\cdot y} u(x)\, dx.$$

By dominated convergence the function $\hat{u}$ is continuous; moreover, we have $\|\hat{u}\|_\infty \le (2\pi)^{-n/2} \|u\|_1$. If $u \in C^1(\mathbb{R}^n)$ and both u and $D_j u$ belong to $L^1(\mathbb{R}^n)$, then $\widehat{D_j u}(y) = iy_j\hat{u}(y)$ by integration by parts. Similarly, if both u and the function mapping x to $|x|u(x)$ belong to $L^1(\mathbb{R}^n)$, then $\hat{u} \in C^1(\mathbb{R}^n)$; in this case $D_j\hat{u}(y)$ is the value at y of the Fourier transform of the function mapping x to $-ix_j u(x)$.

7.59 (The Space of Rapidly Decreasing Functions) Let $\mathscr{S} = \mathscr{S}(\mathbb{R}^n)$ denote the space of all functions u in $C^\infty(\mathbb{R}^n)$ such that for all multi-indices $\alpha \geq 0$ and $\beta \geq 0$ the function mapping x to $x^\alpha D^\beta u(x)$ is bounded on $\mathbb{R}^n$. Unlike functions in $\mathscr{D}(\mathbb{R}^n)$, functions in $\mathscr{S}$ need not have compact support; nevertheless, they must approach 0 at infinity faster than any rational function of x. For this reason the elements of $\mathscr{S}$ are usually called *rapidly decreasing functions.*

The properties of the Fourier transform mentioned above extend to verify the assertion that the Fourier transform of an element of $\mathscr{S}$ also belongs to $\mathscr{S}$.

The *inverse Fourier transform* $\check{u}$ of an element u of $L^1(\mathbb{R}^n)$ is defined for $x \in \mathbb{R}^n$ by

$$\check{u}(x) = \frac{1}{(2\pi)^{n/2}} \int_{\mathbb{R}^n} e^{ix\cdot y} u(y)\, dy.$$

The *Fourier inversion theorem* [RS, chapter 9] asserts that if $u \in \mathscr{S}$, then the inverse Fourier transform of $\hat{u}$ is u ($\check{\hat{u}} = u$), and, moreover, that the same conclusion holds under the weaker assumptions that $u \in L^1(\mathbb{R}^n) \cap C(\mathbb{R}^n)$ and $\hat{u} \in L^1(\mathbb{R}^n)$. One advantage of considering the Fourier transform on $\mathscr{S}$ is that $u \in \mathscr{S}$ guarantees that $\hat{u} \in L^1(\mathbb{R}^n)$, and the same is true for the function mapping y to $y^\alpha \hat{u}(y)$ for any multi-index $\alpha \geq 0$. In fact, the inverse Fourier transforms of functions in $\mathscr{S}$ also belong to $\mathscr{S}$ and the transform of the inverse transform also returns the original function. Thus the Fourier transform is a one-to-one mapping of $\mathscr{S}$ onto itself.

7.60 (The Space of Tempered Distributions) Given a linear functional F on the space $\mathscr{S}$, we can define another such functional $\hat{F}$ by requiring $\hat{F}(u) = F(\hat{u})$ for all $u \in \mathscr{S}$. Fubini's theorem shows that if F operates by integrating functions in $\mathscr{S}$ against a fixed integrable function f, then $\hat{F}$ operates by integrating against the transform $\hat{f}$:

$$F(u) = \int_{\mathbb{R}^n} f(x)u(x)\, dx, \quad f \in L^1(\mathbb{R}^n),$$

$$\implies \quad \hat{F}(v) = \int_{\mathbb{R}^n} \hat{f}(y)v(y)\, dy. \tag{24}$$

There exists a locally convex topology on $\mathscr{S}$ such that the mapping $F \to \hat{F}$ maps the dual space $\mathscr{S}' = \mathscr{S}'(\mathbb{R}^n)$ in a one-to-one way onto itself. The elements of this dual space $\mathscr{S}'$ are called *tempered distributions*. As was the case for $\mathscr{D}'(\Omega)$, not all tempered distributions can be represented by integration against functions.

7.61 (The Plancherel Theorem) An easy calculation shows that if u and v belong to $L^1(\mathbb{R}^n)$, then $\widehat{u*v} = (2\pi)^{n/2}\hat{u}\hat{v}$; Fourier transformation converts convolution products into pointwise products. If $u \in L^1(\mathbb{R}^n)$, let $\tilde{u}(x) = \overline{u(-x)}$. Then $\widehat{\tilde{u}} = \overline{\hat{u}}$, and $\widehat{u * \tilde{u}} = (2\pi)^{n/2}|\hat{u}|^2$. If $u \in \mathscr{S}$, then both $u * \tilde{u}$ and $|\hat{u}|^2$ also belong to $\mathscr{S}$. Applying the Fourier inversion theorem to $u * \tilde{u}$ at $x = 0$ then gives the following result, known as *Plancherel's Theorem*.

$$u \in \mathscr{S} \quad \Longrightarrow \quad \|\hat{u}\|_2^2 = \|u\|_2^2 .$$

That is, the Fourier transform maps the space $\mathscr{S}$ equipped with the L^2-norm isometrically onto itself. Since $\mathscr{S}$ is dense in $L^2(\mathbb{R}^n)$, the isometry extends to one mapping $L^2(\mathbb{R}^n)$ onto itself. Also, $L^2(\mathbb{R}^n) \subset \mathscr{S}'$ and the distributional Fourier transforms of an L^2 function is the same L^2 function as defined by the above isometry. (That is, the Fourier transform of an element of $\mathscr{S}'$ that operates by integration against L^2 functions as in (24) does itself operate in that way.)

7.62 (Characterization of $W^{s,2}(\mathbb{R}^n)$) Given $u \in L^2(\mathbb{R}^n)$ and any positive integer m, let

$$u_m(y) = (1 + |y|^2)^{m/2}\hat{u}(y). \tag{25}$$

It is easy to verify that $u \in W^{m,2}(\mathbb{R}^n)$ if and only if u_m belongs to $L^2(\mathbb{R}^n)$, and the L^2-norm of u_m is equivalent to the $W^{m,2}$-norm of u. So the Fourier transform identifies $W^{m,2}(\mathbb{R}^n)$ with the Banach lattice of functions w for which $(1 + |\cdot|^2)^{m/2}\hat{w}(\cdot)$ belongs to $L^2(\mathbb{R}^n)$. For each positive integer m that lattice has the dominated convergence property. It follows that $u \in W^{s,2}(\mathbb{R}^n)$ if and only if $(1 + |\cdot|^2)^{s/2}\hat{u}(\cdot)$ belongs to $L^2(\mathbb{R}^n)$.

7.63 (Characterization of $W^{s,p}(\mathbb{R}^n)$) The description of $W^{s,p}(\mathbb{R}^n)$ when $1 < p < 2$ or $2 < p < \infty$ is more complicated. If $u \in L^p(\mathbb{R}^n)$ with $1 < p < 2$, then $u \in L^1(\mathbb{R}^n) + L^2(\mathbb{R}^n)$; this guarantees that $\hat{u} \in L^\infty(\mathbb{R}^n) + L^2(\mathbb{R}^n)$, and in particular that the distribution $\hat{u}$ is a function. Moreover, it follows by complex interpolation that $\hat{u} \in L^{p'}(\mathbb{R}^n)$ and by real interpolation that $\hat{u} \in L^{p',p}(\mathbb{R}^n)$. But the set of such transforms of L^p functions is not a lattice when $1 < p < 2$. This follows from the fact (see [FG]) that every set of positive measure contains a subset E of positive measure so that if the Fourier transform of an L^p function, where $1 < p < 2$, vanishes off E, then the function must be 0. If $u \in L^p(\mathbb{R}^n)$ and E is such a subset on which $\hat{u}(y) \neq 0$, then the function that equals $\hat{u}$ on E and 0 off E is not trivial but would have to be trivial if the set of Fourier transforms of L^p functions were a lattice.

A duality argument shows that the set of (distributional) Fourier transforms of functions in $L^p(\mathbb{R}^n)$ for $p > 2$ cannot be a Banach lattice either. Moreover (see [Sz]), there are functions in $L^p(\mathbb{R}^n)$ whose transforms are not even functions.

Nevertheless, the product of any tempered distribution and any sufficiently smooth function that has at most polynomial growth is always defined. For any distribution $u \in \mathscr{S}'$ we can then define the distribution u_m by analogy with formula (25); we multiply the tempered distribution $\hat{u}$ by the smooth function $(1+|\cdot|^2)^{m/2}$. When $1 < p < 2$ or $2 < p < \infty$, the theory of singular integrals [St, p. 138] then shows that $u \in W^{m,p}(\mathbb{R}^n)$ if and only if the function u_m is the Fourier transform of some function in $L^p(\mathbb{R}^n)$. Again the space $W^{s,p}(\mathbb{R}^n)$ is characterized by the version of this condition with m replaced by s. In particular, if $s = m$ we obtain the usual Sobolev space $W^{m,p}(\mathbb{R}^n)$ up to equivalence of norms, when $1 < p < \infty$. The fractional order Sobolev spaces spaces are natural generalizations of the Sobolev spaces that allow for fractional orders of smoothness.

One can pass between spaces $W^{s,p}(\mathbb{R}^n)$ having the same index p but different orders of smoothness s by multiplying or dividing Fourier transforms by factors of the form $(1+|\cdot|^2)^{-r/2}$. When $r > 0$ these radial factors are constant multiples of Fourier transforms of certain Bessel functions; for this reason the spaces $W^{s,p}(\mathbb{R}^n)$ are often called *spaces of Bessel potentials*. (See [AMS].)

In order to show the relationship between the fractional order Sobolev spaces and the Besov spaces, it is, however, more useful to refine the scale of spaces $W^{s,p}(\mathbb{R}^n)$ using a dyadic splitting of the Fourier transform.

7.64 (An Alternate Characterization of $W^{s,p}(\mathbb{R}^n)$) In proving the Trace Theorem 7.39 we used a splitting of a function in $W^{m,p}(\mathbb{R}^{n+1})$ into dyadic pieces supported in slabs parallel to the subspace $\mathbb{R}^n$ of the traces. Here we are going to use a similar splitting of the Fourier transform of an L^p function into dyadic pieces supported between concentric spheres.

Recall the C^∞ function ϕ_i defined in the proof of Lemma 7.40 and having support in the interval $(2^i, 2^{i+2})$. For each integer i and y in $\mathbb{R}^n$, let $\psi_i(y) = \phi_i(|y|)$. Each of these radially symmetric functions belongs to $\mathscr{S}$ and so has an inverse transform, Ψ_i say, that also belongs to $\mathscr{S}$.

Fix an index p in the interval $(1,\infty)$ and let $u \in L^p(\mathbb{R}^n)$. For each integer i let $T_i u$ be the convolution of u with $(2\pi)^{-n/2}\Psi_i$; thus $\widehat{T_i u}(y) = \psi_i(y)\cdot\hat{u}(y)$. One can regard the functions $T_i u$ as dyadic parts of u with nearly disjoint frequencies. Littlewood-Paley theory [FJW] shows that the L^p-norm of u is equivalent to the L^p-norm of the function mapping x to $[\sum_{i=-\infty}^{\infty}|T_i u(x)|^2]^{1/2}$. That is

$$\|u\|_p \approx \left(\int_{R^n}\left[\sum_{i=-\infty}^{\infty}|T_i u(x)|^2\right]^{p/2} dx\right)^{1/p}.$$

To estimate the norm of u in $W^{s,p}(\mathbb{R}^n)$ we should replace each term $T_i u$ by the function obtained by not only multiplying $\hat{u}$ by ψ_i, but also multiplying the transform by the function mapping y to $(1+|y|^2)^{s/2}$.

On the support of ψ_i the values of that second Fourier multiplier are all roughly equal to $1+2^{si}$. It turns out that $u \in W^{s,p}(\mathbb{R}^n)$ if and only if

$$\|u\|_{s,p} = \left\|u\,;W^{s,p}(\mathbb{R}^n)\right\| \approx \left(\int_{R^n}\left[\sum_{i=-\infty}^{\infty}(1+2^{si})^2|T_iu(x)|^2\right]^{p/2}dx\right)^{1/p} < \infty.$$

This is a complicated but intrinsic characterization of the space $W^{s,p}(\mathbb{R}^n)$. That is, the following steps provide a recipe for determining whether an L^p function u belongs to $W^{s,p}(\mathbb{R}^n)$:

(a) Split u into the the pieces T_iu by convolving with the functions Ψ_i or by multiplying the distribution $\hat{u}$ by ψ_i and then taking the inverse transform. For each point x in $\mathbb{R}^n$ this gives a sequence $\{T_iu(x)\}$.

(b) Multiply the i-th term in that sequence by $(1+2^{si})$ and compute the ℓ^2-norm of the result. This gives a function of x.

(c) Compute the L^p-norm of that function.

The steps in this recipe can be modified to produce other scales of spaces.

7.65 (The Triebel-Lizorkin Spaces) Define $F^{s;p,q}(\mathbb{R}^n)$ to be the space obtained by using steps (a) to (c) above but taking an ℓ^q-norm rather than an ℓ^2-norm in step (b). This gives the family of Triebel-Lizorkin spaces; if $1 \le q < \infty$,

$$\left\|u\,;F^{s;p,q}(\mathbb{R}^n)\right\| \approx \left(\int_{R^n}\left[\sum_{i=-\infty}^{\infty}(1+2^{si})^q|T_iu(x)|^q\right]^{p/q}dx\right)^{1/p} < \infty.$$

Note that $F^{m;p,2}(\mathbb{R}^n)$ coincides with $W^{m,p}(\mathbb{R}^n)$ when m is a positive integer, and $F^{s;p,2}(\mathbb{R}^n)$ coincides with $W^{s,p}(\mathbb{R}^n)$ when s is positive.

7.66 REMARKS

1. The space $F^{0;p,2}(\mathbb{R}^n)$ coincides with $L^p(\mathbb{R}^n)$ when $1 < p < \infty$.
2. The definitions of $W^{s,p}(\mathbb{R}^n)$ and $F^{s;p,q}(\mathbb{R}^n)$ also make sense if $s < 0$, and even if $0 < p, q < 1$. However they may contain distributions that are not functions if $s < 0$, and they will not be Banach spaces unless $p \ge 1$ and $1 \le q < \infty$.
3. If $s > 0$, the recipes for characterizing $W^{s,p}(\mathbb{R}^n)$ and $F^{s;p,q}(\mathbb{R}^n)$ given above can be modified to replace the multiplier $(1+2^{si})$ by 2^{si} and restricting the summations in the ℓ^2 or ℓ^p norm expressions to $i \ge 0$, provided we also explicitly require $u \in L^p(\mathbb{R}^n)$. Thus, for example, $u \in F^{s;p,q}(\mathbb{R}^n)$ if and only if

$$\|u\|_p + \left(\int_{R^n}\left[\sum_{i=0}^{\infty}2^{siq}|T_iu(x)|^q\right]^{p/q}dx\right)^{1/p} < \infty.$$

4. If $s > 0$ and we modify the recipe for $F^{s;p,q}(\mathbb{R}^n)$ by replacing the multiplier $(1+2^{si})$ by 2^{si} but continuing to take the summation over all integers i, then we obtain the so-called *homogeneous Triebel-Lizorkin space* $\dot{F}^{s;p,q}(\mathbb{R}^n)$ which contain equivalence classes of distributions modulo polynomials of low enough degree. Only smoothness and not size determines whether a function belongs to this homogeneous space.

7.67 (An Alternate Definition of the Besov Spaces) It turns out that the Besov spaces $B^{s;p,q}(\mathbb{R}^n)$ arises from the variant of the recipe given in Paragraph 7.64 where the last two steps are modified as follows.

(b') Multiply the i-th term in the sequence $\{T_i u(x)\}$ by $(1 + 2^{si})$ and compute the L^p-norm of the result. This gives a sequence of nonnegative numbers.

(c') Compute the ℓ^q-norm of that sequence.

$$\left\| u \,;\, B^{s;p,q}(\mathbb{R}^n) \right\| \approx \left[\sum_{i=-\infty}^{\infty} \left(\int_{R^n} (1 + 2^{si})^p |T_i u(x)|^p \, dx \right)^{q/p} \right]^{1/q}.$$

This amounts to reversing the order in which the two norms are computed. That order does not matter when $q = p$; thus $B^{s;p,p}(\mathbb{R}^n) = F^{s;p,p}(\mathbb{R}^n)$ with equivalent norms. When $q \neq p$, Minkowski's inequality for sums and integrals reveals that in comparing the outcomes of steps (b) and (c), the larger norm and the smaller space of functions arises when the larger of the indices p and q is used first. That is,

$$\begin{cases} F^{s;p,q}(\mathbb{R}^n) \subset B^{s;p,q}(\mathbb{R}^n) & \text{if } q < p \\ B^{s;p,q}(\mathbb{R}^n) \subset F^{s;p,q}(\mathbb{R}^n) & \text{if } q > p. \end{cases}$$

For fixed s and p the inclusions between the Besov spaces $B^{s;p,q}(\mathbb{R}^n)$ are the same as those between ℓ^q spaces, and the same is true for the Triebel-Lizorkin spaces $F^{s;p,q}(\mathbb{R}^n)$.

Finally, the only link with the scale of fractional order Sobolev spaces and in particular with the Sobolev spaces occurs through the Triebel-Lizorkin scale with $q = 2$. We have

$$\begin{cases} W^{s,p}(\mathbb{R}^n) = F^{s;p,2}(\mathbb{R}^n) \subset F^{s;p,q}(\mathbb{R}^n) & \text{if } q \geq 2 \\ F^{s;p,q}(\mathbb{R}^n) \subset F^{s;p,2}(\mathbb{R}^n) = W^{s,p}(\mathbb{R}^n) & \text{if } q \leq 2. \end{cases}$$

As another example, the trace of $W^{m,p}(\mathbb{R}^{n+1})$ on $\mathbb{R}^n$ is exactly the space $B^{m-1/p;p,p}(\mathbb{R}^n) = F^{m-1/p;p,p}(\mathbb{R}^n)$. When $p \leq 2$, this trace space is included in the space $F^{m-1/p;p,2}(\mathbb{R}^n)$ and thus in the space $W^{m-1/p,p}(\mathbb{R}^n)$. When $p \geq 2$, this inclusion is reversed.

7.68 REMARKS

1. Appropriate versions of Remarks 7.66 for the Triebel-Lizorkin spaces apply to the above characterization of the Besov spaces too. In particular, modifying recipe item (b') to use the multiplier 2^{si} instead of $1 + 2^{si}$ results in a *homogeneous Besov space* $\dot{B}^{s;p,q}(\mathbb{R}^n)$ of equivalence classes of distributions modulo certain polynomials. Again membership in this space depends only on smoothness and not on size.
2. The K-version of the definition of $B^{s;p,q}(\mathbb{R}^n)$ as an intermediate space obtained by the real method is a condition on how well $u \in L^p(\mathbb{R}^n)$ can be approximated by functions in $W^{m,p}(\mathbb{R}^n)$ for some integer $m > s$. But the J-form of the definition requires a splitting of u into pieces u_i with suitable control on the norms of the functions u_i in the spaces $L^p(\mathbb{R}^n)$ and $W^{m,p}(\mathbb{R}^n)$. The Fourier splitting also gives us pieces $T_i u$ for which we can control those two norms, and these can serve as the pieces u_i. Conversely, if we have pieces u_i with suitable control on appropriate norms, and it we apply Fourier decomposition to each piece, we would find that the norms $\|T_j u_i\|_p$ are negligible when $|j-i|$ is large, leading to appropriate estimates for the norms $\|T_j u\|_p$.

7.69 (Extensions for General Domains) Many of the properties of the scales of Besov spaces, spaces of Bessel potentials, and Triebel-Lizorkin spaces on $\mathbb{R}^n$ can be extended to more general domains Ω via the use of an extension operator. Rychkov [Ry] has constructed a linear total extension operator $\mathscr{E}$ that simultaneously and boundedly extends functions in $F^{s;p,q}(\Omega)$ to $F^{s;p,q}(\mathbb{R}^n)$ and functions in $B^{s;p,q}(\Omega)$ to $B^{s;p,q}(\mathbb{R}^n)$ provided Ω satisfies a strong local Lipschitz condition. The same operator $\mathscr{E}$ works for both scales, all real s, and all $p > 0$, $q > 0$; it is an extension operator in the sense that $\mathscr{E}u|_\Omega = u$ in $\mathscr{D}'(\Omega)$ for every u in any of the Besov or Triebel-Lizorkin spaces defined on Ω as restrictions in the sense of $\mathscr{D}'(\Omega)$ of functions in the corresponding spaces on $\mathbb{R}^n$.

The existence of this operator provides, for example, an intrinsic characterization of $B^{s;p,q}(\Omega)$ in terms of that for $B^{s;p,q}(\mathbb{R}^n)$ obtained in Theorem 7.47.

Wavelet Characterizations

7.70 We have seen above how membership of a function u in a space $B^{s;p,q}(\mathbb{R}^n)$ can be determined by the size of the sequence of norms $\|T_i u\|_p$, while its membership in the space $F^{s;p,q}(\mathbb{R}^n)$ requires pointwise information about the sizes of the functions $T_i u$ on $\mathbb{R}^n$. Both characterizations use the functions $T_i u$ of a dyadic decomposition of u defined as inverse Fourier transforms of products of $\hat{u}$ with dilates of a suitable smooth function ϕ. We conclude this chapter by describing how further refining these decompositions to the level of wavelets reduces questions about membership of u in these smoothness classes to questions about the sizes of the (scalar) coefficients of u in such decompositions. These coefficients do form a Banach lattice.

This contrasts dramatically with the situation for Fourier transforms of L^p functions with $1 < p < 2$, where these transforms fail to form a lattice.

7.71 (Wavelet Analysis) An *analyzing wavelet* is a nontrivial function on $\mathbb{R}^n$ satisfying some decay conditions, some cancellation conditions, and some smoothness conditions. Different versions of these conditions are appropriate in different contexts. Two classical examples of wavelets on $\mathbb{R}$ are the following:

(a) The basic *Haar* function h given by

$$h(x) = \begin{cases} 1 & \text{if } \leq x < 1/2 \\ -1 & \text{if } 1/2 \leq x < 1 \\ 0 & \text{otherwise.} \end{cases}$$

(b) A basic Shannon wavelet S defined as the inverse Fourier transform of the function $\hat{S}$ satisfying

$$\hat{S}(y) = \begin{cases} 1 & \text{if } \pi \leq |y| < 2\pi \\ 0 & \text{otherwise.} \end{cases}$$

The Haar wavelet has compact support, and *a fortiori* decays extremely rapidly. The only cancellation condition it satisfies is that $\int_{\mathbb{R}} ch(x)\,dx = 0$ for all constants c. It fails to be smooth, but compensates for that by taking only two nonzero values and thus being simple to use numerically.

The Shannon wavelet does *not* have compact support; instead it decays like $1/|x|$, that is, at a fairly slow rate. However, it has very good cancellation properties, since

$$\int_{\mathbb{R}} x^m S(x)\,dx = 0 \quad \text{for all nonnegative integers } m.$$

(These integrals are equal to constants times the values at $y = 0$ of derivatives of $\hat{S}(y)$. Since $\hat{S}$ vanishes in a neighbourhood of 0, those derivatives all vanish at 0.) Also, $S \in C^\infty(\mathbb{R})$ and even extends to an entire function on the complex plane.

To get a better balance between these conditions, we will invert the roles of function and Fourier transform from the previous section, and use below a wavelet ϕ defined on $\mathbb{R}^n$ as the inverse Fourier transform of a nontrivial smooth function that vanishes outside the annulus where $1/2 < |y| < 2$. Then ϕ has all the cancellation properties of the Shannon wavelet, for the same reasons. Also ϕ decays very rapidly because $\hat{\phi}$ is smooth, and ϕ is smooth because $\hat{\phi}$ decays rapidly. Again the compact support of $\hat{\phi}$ makes ϕ the restriction of an entire function.

Given an analyzing wavelet, w say, we consider some or all of its translates mapping x to $w(x - h)$ and some or all of its (translated) dilates, mapping x to

$w(2^r x - h)$. These too are often called wavelets. Translation preserves L^p norms; dilation does *not* do so, except when $p = \infty$; however, we will use the multiple $2^{rn/2} w(2^r x - h)$ to preserve L^2 norms.

If we apply the same operations to the complex exponential that maps x to e^{ixy} on $\mathbb{R}$, we find that dilation produces other such exponentials, but that translation just multiplies the exponential by a complex constant and so does not produce anything really new. In contrast, the translates of the basic Haar wavelet by integer amounts have disjoint supports and so are orthogonal in $L^2(\mathbb{R})$. A less obvious fact is that translating the Shannon wavelet by integers yields orthogonal functions, this time without disjoint supports.

In both cases, dilating by factors 2^i, where i is an integer, yields other wavelets that are orthogonal to their translates by 2^i times integers, and these wavelets are orthogonal to those in the same family at other dyadic scales. Moreover, in both examples, this gives an orthogonal basis for $L^2(\mathbb{R})$.

Less of this orthogonality persists for wavelets like the one we called ϕ above. But it can still pay to consider the *wavelet transform* of a given function u which maps positive numbers a and vectors h in $\mathbb{R}^n$ to

$$\frac{1}{\sqrt{a^n}} \int_{\mathbb{R}^n} u(x) \phi\left(\frac{x-h}{a}\right) dx.$$

For our purposes it will suffice to consider only those dilations and translates mapping x to $\phi_{i,k}(x) = 2^{in/2}\phi(2^i x - k)$, where i runs through the set of integers, and k runs through the integer lattice in $\mathbb{R}^n$. Integrating u against such wavelets yields *wavelet coefficients* that we can index by the pairs (i, k) and use to characterize membership of u in various spaces.

For much more on wavelets, see [Db].

7.72 (Wavelet Characterization of Besov Spaces) Let ϕ be a function in $\mathscr{S}$ whose Fourier transform $\hat{\phi}$ satisfies the following two conditions:

(i) $\hat{\phi}(y) = 0$ if $|y| < 1/2$ or $|y| > 2$.

(ii) $|\hat{\phi}(y)| > c > 0$ if $3/5 < |y| < 5/3$.

Note that the conditions on $\hat{\phi}$ imply that

$$\int_{\mathbb{R}^n} P(x)\phi(x)\, dx = 0$$

for any polynomial P.

Also, it can be shown (see [FJW, p. 54]) that there exists a dual function $\psi \in \mathscr{S}$ satisfying the same conditions (i) and (ii) and such that

$$\sum_{i=-\infty}^{\infty} \overline{\hat{\phi}(2^{-i}y)}\, \hat{\psi}(2^{-i}y)\, dy = 1 \quad \text{for all } y \neq 0.$$

Let $\mathbb{Z}$ denote the set of all integers. For each $i \in \mathbb{Z}$ and each n-tuple $k = (k_1, \ldots, k_n) \in \mathbb{Z}^n$ we define two wavelet families by using dyadic dilates and translates of ϕ and ψ:

$$\phi_{i,k}(x) = 2^{-in/2}\phi(2^{-i}x - k) \qquad \text{and} \qquad \psi_{i,k}(x) = 2^{in/2}\psi(2^i x - k).$$

Note that the dilations in these two families are in opposite directions and that $\phi_{i,k}$ and $\psi_{i,k}$ have the same L^2 norms as do ϕ and ψ respectively. Moreover, for any polynomial P,

$$\int_{\mathbb{R}^n} P(x)\phi_{i,k}(x)\,dx = 0.$$

Let I denote the set of all indices (i, k) such that $i \in \mathbb{Z}$ and $k \in \mathbb{Z}^n$, and let $\mathscr{F}$ denote the wavelet family $\{\phi_{i,k} : (i, k) \in I\}$. Given a locally integrable function u, we define its wavelet coefficients $c_{i,k}(u)$ with respect to the family $\mathscr{F}$ by

$$c_{i,k}(u) = \int_{\mathbb{R}^n} u(x)\overline{\phi_{i,k}(x)}\,dx,$$

and consider the wavelet series representation

$$u = \sum_{(i,k)\in I} c_{i,k}(u)\psi_{i,k}. \tag{26}$$

The series represents u modulo polynomials as all its terms vanish if u is a polynomial.

It turns out that u belongs to the homogeneous Besov space $\dot{B}^{s;p,q}(\mathbb{R}^n)$ if and only if its coefficients $\{c_{i,k}(u) : (i, k) \in I\}$ belong to the Banach lattice on I having norm

$$\left(\sum_{i=-\infty}^{\infty}\left[2^{i(s+n[1/2-1/p])}\sum_{k\in\mathbb{Z}^n}|c_{i,k}|^p\right]^{q/p}\right)^{1/q}. \tag{27}$$

The condition for membership in the ordinary Besov space $B^{s;p,q}(\mathbb{R}^n)$ is a bit more complicated. We use only the part of the wavelet series (26) with $i \geq 0$ and replace the rest with a new series

$$\sum_{k\in\mathbb{Z}^n} c_k(u)\Psi_k,$$

where $\Psi_k(x) = \Psi(x - k)$ and Ψ is a function in $\mathscr{S}$ satisfying the conditions $\hat{\Psi}(y) = 0$ if $|y| \geq 1$ and $|\hat{\Psi}(y)| > c > 0$ if $|y| \leq 5/6$. Again there is a dual such function Φ with the same properties such that the coefficients $c_k(u)$ are given by

$$c_k(u) = \int_{\mathbb{R}^n} u(x)\overline{\Phi_k(x)}\,dx.$$

We have $u \in B^{s;p,q}(\mathbb{R}^n)$ if and only if the expression

$$\left(\sum_{k\in\mathbb{Z}^n} |c_k|^p\right)^{1/p} + \left(\sum_{i=0}^{\infty}\left[2^{i(s+n[1/2-1/p])}\sum_{k\in\mathbb{Z}^n}|c_{i,k}|^p\right]^{q/p}\right)^{1/q} \tag{28}$$

is finite, and this expression provides an equivalent norm for $B^{s;p,q}(\mathbb{R}^n)$.

Note that, in expressions (27) and (28) the part of the recipe in item 7.66 involving the computation of an L^p-norm seems to have disappeared. In fact, however, for any fixed value of the index i, the wavelet "coefficients" $c_{i,k}$ are actually values of the convolution $u * \phi_{i,0}$ taken at points in the discrete lattice $\{2^i k\}$, where the index i is fixed but k varies. This lattice turns out to be fine enough that the L^p-norm of $u * \phi_{i,0}$ is equivalent to the ℓ^p-norm over this lattice of the values of $u * \phi_{i,0}$.

7.73 (Wavelet Characterization of Triebel-Lizorkin Spaces) Membership in the homogeneous Triebel-Lizorkin space $\dot{F}^{s;p,q}(\mathbb{R}^n)$ is also characterized by a condition where only the sizes of the coefficients $c_{i,k}$ matter, namely the finiteness of

$$\left\|\left(\sum_{i=-\infty}^{\infty}\left[2^{i(s+n/2)}\sum_{k\in\mathbb{Z}}|c_{i,k}|\chi_{i,k}\right]^q\right)^{1/q}\right\|_{p,\mathbb{R}^n}.$$

where $\chi_{i,k}$ is the characteristic function of the cube $2^i k_j \le x_j < 2^i(k_j+)$, $(1 \le j \le n)$. At any point x in $\mathbb{R}^n$ the inner sum above collapses as follows. For each value of the index i the point x belongs to the cube corresponding to i and k for only one value of k, say $k_i(x)$. This reduces matters to the finiteness of

$$\left\|\left(\sum_{i=-\infty}^{\infty}\left[2^{i(s+n/2)}\,|c_{i,k_i(\cdot)}|\right]^q\right)^{1/q}\right\|_{p,\mathbb{R}^n}.$$

We refer to section 12 in [FJ] for information on how to deal in a similar way with the inhomogeneous space $F^{s;p,q}(\mathbb{R}^n)$.

Recall that in the discrete version of the J-method, the pieces u_i in suitable splittings of u are not unique. This flexibility sometimes simplified our analysis, for instance in the proofs of (trace) imbeddings for Besov spaces. The same is true for the related idea of *atomic decomposition*, for which we refer to [FJW] and [FJ] for sharper results and much more information.

8

ORLICZ SPACES AND ORLICZ-SOBOLEV SPACES

Introduction

8.1 In this final chapter we present results on generalizations of Lebesgue and Sobolev spaces in which the role usually played by the convex function t^p is assumed by a more general convex function $A(t)$. The spaces $L_A(\Omega)$, called *Orlicz spaces*, are studied in depth in the monograph by Krasnosel'skii and Rutickii [KR] and also in the doctoral thesis by Luxemburg [Lu], to either of which the reader is referred for a more complete development of the material outlined below. The former also contains examples of applications of Orlicz spaces to certain problems in nonlinear analysis.

It is of some interest to note that a gap in the Sobolev imbedding theorem (Theorem 4.12) can be filled by an Orlicz space. Specifically, if $mp = n$ and $p > 1$, then for suitably regular Ω we have

$$W^{m,p}(\Omega) \to L^q(\Omega), \quad p \le q < \infty, \quad \text{but} \quad W^{m,p}(\Omega) \not\to L^\infty(\Omega);$$

there is no *best*, (i.e., smallest) target L^p-space for the imbedding. In Theorem 8.27 below we will provide an optimal imbedding of $W^{m,p}(\Omega)$ into a certain Orlicz space. This result is due to Trudinger [Td], with precedents in [Ju] and [Pz]. There has been much further work, for instance [Ms] and [Ad1].

Following [KR], we use the class of "N-functions" as defining functions A for Orlicz spaces. This class is not as wide as the class of Young's functions used in

[Lu]; for instance, it excludes $L^1(\Omega)$ and $L^\infty(\Omega)$ from the class of Orlicz spaces. However, N-functions are simpler to deal with, and are adequate for our purposes. Only once, in Theorem 8.39 below, do we need to refer to a more general Young's function.

If the role played by $L^p(\Omega)$ in the definition of the Sobolev space $W^{m,p}(\Omega)$ is assigned instead to the Orlicz space $L_A(\Omega)$, the resulting space is denoted by $W^m L_A(\Omega)$ and is called an *Orlicz-Sobolev space.* Many properties of Sobolev spaces have been extended to Orlicz-Sobolev spaces by Donaldson and Trudinger [DT]. We present some of these results in this chapter.

N-Functions

8.2 (Definition of an N-Function) Let a be a real-valued function defined on $[0, \infty)$ and having the following properties:

(a) $a(0) = 0$, $a(t) > 0$ if $t > 0$, $\lim_{t\to\infty} a(t) = \infty$;

(b) a is nondecreasing, that is, $s > t$ implies $a(s) \ge a(t)$;

(c) a is right continuous, that is, if $t \ge 0$, then $\lim_{s\to t+} a(s) = a(t)$.

Then the real-valued function A defined on $[0, \infty)$ by

$$A(t) = \int_0^t a(\tau)\, d\tau \tag{1}$$

is called an *N-function.*

It is not difficult to verify that any such N-function A has the following properties:

(i) A is continuous on $[0, \infty)$;

(ii) A is strictly increasing that is, $s > t \ge 0$ implies $A(s) > A(t)$;

(iii) A is convex, that is, if $s, t \ge 0$ and $o < \lambda < 1$, then

$$A\big(\lambda s + (1-\lambda)t\big) \le \lambda A(s) + (1-\lambda)A(t);$$

(iv) $\lim_{t\to 0} A(t)/t = 0$, and $\lim_{t\to\infty} A(t)/t = \infty$;

(v) if $s > t > 0$, then $A(s)/s > A(t)/t$.

Properties (i), (iii), and (iv) could have been used to define an N-function since they imply the existence of a representation of A in the form (1) with a having the properties (a)–(c).

The following are examples of N-functions:

$$\begin{aligned}
A(t) &= t^p, && 1 < p < \infty,\\
A(t) &= e^t - t - 1,\\
A(t) &= e^{(t^p)} - 1, && 1 < p < \infty,\\
A(t) &= (1+t)\log(1+t) - t.
\end{aligned}$$

Evidently, $A(t)$ is represented by the area under the graph $\sigma = a(\tau)$ from $\tau = 0$ to $\tau = t$ as shown in Figure 9. Rectilinear segments in the graph of A correspond to intervals on which a is constant, and angular points on the graph of A correspond to discontinuities (i.e., vertical jumps) in the graph of a.

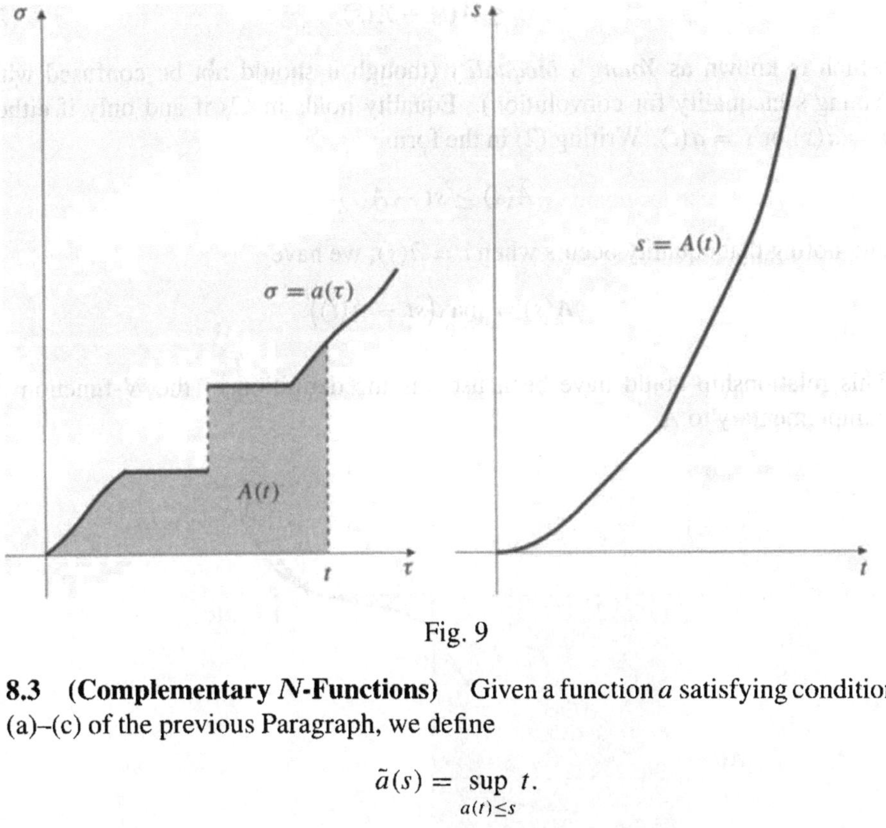

Fig. 9

8.3 (Complementary *N*-Functions) Given a function a satisfying conditions (a)–(c) of the previous Paragraph, we define

$$\tilde{a}(s) = \sup_{a(t) \le s} t.$$

It is readily checked that the function $\tilde{a}$ so defined also satisfies (a)–(c) and that a can be recovered from $\tilde{a}$ via

$$a(t) = \sup_{\tilde{a}(s) \le t} s.$$

If a is strictly increasing then $\tilde{a} = a^{-1}$. The N-functions A and $\tilde{A}$ given by

$$A(t) = \int_0^t a(\tau)\, d\tau, \qquad \tilde{A}(s) = \int_0^s \tilde{a}(\sigma)\, d\sigma$$

are said to be *complementary*; each is the *complement* of the other. Examples of such complementary pairs are:

$$A(t) = \frac{t^p}{p}, \qquad \tilde{A}(s) = \frac{s^{p'}}{p'}, \qquad 1 < p < \infty, \quad \frac{1}{p} + \frac{1}{p'} = 1,$$

and

$$A(t) = e^t - t - 1, \qquad \tilde{A}(s) = (1+s)\log(1+s) - s.$$

$\tilde{A}(s)$ is represented by the area to the left of the graph $\sigma = a(\tau)$ (or, more correctly, $\tau = \tilde{a}(\sigma)$) from $\sigma = 0$ to $\sigma = s$ as shown in Figure 10. Evidently, we have

$$st \le A(t) + \tilde{A}(s), \tag{2}$$

which is known as *Young's inequality* (though it should not be confused with Young's inequality for convolution). Equality holds in (2) if and only if either $t = \tilde{a}(s)$ or $s = a(t)$. Writing (2) in the form

$$\tilde{A}(s) \ge st - A(t)$$

and noting that equality occurs when $t = \tilde{a}(s)$, we have

$$\tilde{A}(s) = \max_{t \ge 0}\big(st - A(t)\big).$$

This relationship could have been used as the definition of the N-function $\tilde{A}$ complementary to A.

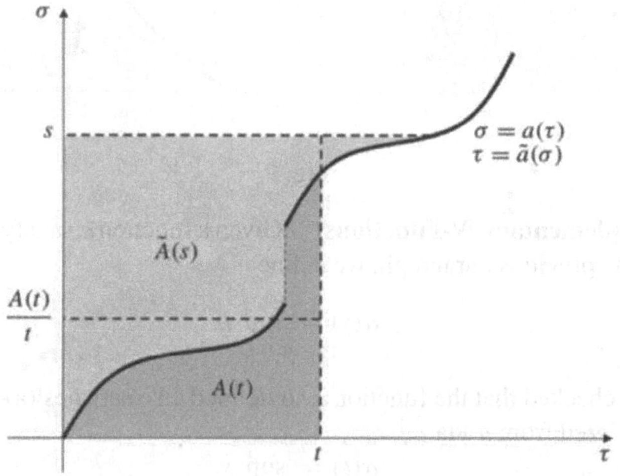

Fig. 10

Since A and $\tilde{A}$ are strictly increasing, they have inverses and (2) implies that for every $t \ge 0$

$$A^{-1}(t)\tilde{A}^{-1}(t) \le A\big(A^{-1}(t)\big) + \tilde{A}\big(\tilde{A}^{-1}(t)\big) = 2t.$$

On the other hand, $A(t) \le ta(t)$, so that, considering Figure 10 again, we have for every $t > 0$,

$$\tilde{A}\left(\frac{A(t)}{t}\right) < \frac{A(t)}{t}t = A(t). \tag{3}$$

Replacing $A(t)$ by t in inequality (3), we obtain

$$\tilde{A}\left(\frac{t}{A^{-1}(t)}\right) < t.$$

Therefore, for any $t > 0$,

$$t < A^{-1}(t)\tilde{A}^{-1}(t) \le 2t. \tag{4}$$

8.4 (Dominance and Equivalence of N-Functions) We shall require certain partial ordering relationships among N-functions. If A and B are two N-functions, we say that B *dominates* A *globally* if there exists a positive constant k such that

$$A(t) \le B(kt) \tag{5}$$

holds for all $t \ge 0$. Similarly, B *dominates* A *near infinity* if there exist positive constants t_0 and k such that (5) holds for all $t \ge t_0$. The two N-functions A and B are *equivalent globally* (or *near infinity*) if each dominates the other globally (or near infinity). Thus A and B are equivalent near infinity if there exist positive constants t_0, k_1, and k_2, such that if $t \ge t_0$, then $B(k_1 t) \le A(t) \le B(k_2 t)$. Such will certainly be the case if

$$0 < \lim_{t\to\infty} \frac{B(t)}{A(t)} < \infty.$$

If A and B have respective complementary N-functions $\tilde{A}$ and $\tilde{B}$, then B dominates A globally (or near infinity) if and only if $\tilde{A}$ dominates $\tilde{B}$ globally (or near infinity). Similarly, A and B are equivalent if and only if $\tilde{A}$ and $\tilde{B}$ are.

8.5 If B dominates A near infinity and A and B are not equivalent near infinity, then we say that A *increases essentially more slowly than* B near infinity. This is the case if and only if for every positive constant k

$$\lim_{t\to\infty} \frac{A(kt)}{B(t)} = 0.$$

The reader may verify that this limit is equivalent to

$$\lim_{t\to\infty} \frac{B^{-1}(t)}{A^{-1}(t)} = 0.$$

Let $1 < p < \infty$ and let A_p denote the N-function

$$A_p(t) = \frac{t^p}{p}, \qquad 0 \le t < \infty.$$

If $1 < p < q < \infty$, then A_p increases essentially more slowly than A_q near infinity. However, A_q does not dominate A_p globally.

8.6 (The Δ_2 Condition) An N-function is said to satisfy a *global* Δ_2-*condition* if there exists a positive constant k such that for every $t \ge 0$,

$$A(2t) \le kA(t). \tag{6}$$

This is the case if and only if for every $r > 1$ there exists a positive constant $k = k(r)$ such that for all $t \ge 0$,

$$A(rt) \le kA(t). \tag{7}$$

Similarly, A satisfies a Δ_2 *condition near infinity* if there exists $t_0 > 0$ such that (6) (or equivalently (7) with $r > 1$) holds for all $t \ge t_0$. Evidently, t_0 may be replaced with any smaller positive number t_1, for if $t_1 \le t \le t_0$, then

$$A(rt) \le \frac{A(rt_0)}{A(t_1)} A(t).$$

If A satisfies a Δ_2-condition globally (or near infinity) and if B is equivalent to A globally (or near infinity), then B also satisfies such a Δ_2-condition. Clearly the N-function $A_p(t) = t^p/p$, $(1 < p < \infty)$, satisfies a global Δ_2-condition. It can be verified that A satisfies a Δ_2-condition globally (or near infinity) if and only if there exists a positive, finite constant c such that

$$\frac{1}{c}\, t\, a(t) \le A(t) \le t\, a(t)$$

holds for all $t \ge 0$ (or for all $t \ge t_0 > 0$), where A is given by (1).

Orlicz Spaces

8.7 (The Orlicz Class $K_A(\Omega)$) Let Ω be a domain in $\mathbb{R}^n$ and let A be an N-function. The *Orlicz class* $K_A(\Omega)$ is the set of all (equivalence classes modulo equality a.e. in Ω of) measurable functions u defined on Ω that satisfy

$$\int_\Omega A(|u(x)|)\, dx < \infty.$$

Since A is convex $K_A(\Omega)$ is always a convex set of functions but it may not be a vector space; for instance, there may exist $u \in K_A(\Omega)$ and $\lambda > 0$ such that $\lambda u \notin K_A(\Omega)$.

We say that the pair (A, Ω) is Δ-*regular* if *either*

(a) A satisfies a global Δ_2-condition, *or*

(b) A satisfies a Δ_2-condition near infinity and Ω has finite volume.

8.8 LEMMA $K_A(\Omega)$ is a vector space (under pointwise addition and scalar multiplication) if and only if (A, Ω) is Δ-regular.

Proof. Since A is convex we have:

(i) $\lambda u \in K_A(\Omega)$ provided $u \in K_A(\Omega)$ and $|\lambda| \le 1$, and

(ii) if $u \in K_A(\Omega)$ implies that $\lambda u \in K_A(\Omega)$ for every complex λ, then $u, v \in K_A(\Omega)$ implies $u + v \in K_A(\Omega)$.

It follows that $K_A(\Omega)$ is a vector space if and only if $\lambda u \in K_A(\Omega)$ whenever $u \in K_A(\Omega)$ and $|\lambda| > 1$.

If A satisfies a global Δ_2-condition and $|\lambda| > 1$, then we have by (7) for $u \in K_A(\Omega)$

$$\int_\Omega A(|\lambda u(x)|)\, dx \le k(|\lambda|) \int_\Omega A(|u(x)|)\, dx < \infty.$$

Similarly, if A satisfies a Δ_2-condition near infinity and $\text{vol}(\Omega) < \infty$, we have for $u \in K_A(\Omega)$, $|\lambda| > 1$, and some $t_0 > 0$,

$$\begin{aligned}\int_\Omega A(|\lambda u(x)|)\, dx &= \left(\int_{\{x\in\Omega:|u(x)|\ge t_0\}} + \int_{\{x\in\Omega:|u(x)|<t_0\}}\right) A(|\lambda u(x)|)\, dx \\ &\le k(|\lambda|) \int_\Omega A(|\lambda u(x)|)\, dx + A(|\lambda| t_0)\text{vol}(\Omega) < \infty.\end{aligned}$$

In either case $K_A(\Omega)$ is seen to be a vector space.

Now suppose that (A, Ω) is not Δ-regular and, if $\text{vol}(\Omega) < \infty$, that $t_0 > 0$ is given. There exists a sequence $\{t_j\}$ of positive numbers such that

(i) $A(2t_j) \ge 2^j A(t_j)$, and

(ii) $t_j \ge t_0 > 0$ if $\text{vol}(\Omega) < \infty$.

Let $\{\Omega_j\}$ be a sequence of mutually disjoint, measurable subsets of Ω such that

$$\text{vol}(\Omega)_j = \begin{cases} 1/[2^j A(t_j)] & \text{if } \text{vol}(\Omega) = \infty \\ A(t_0)\text{vol}(\Omega)/[2^j A(t_j)] & \text{if } \text{vol}(\Omega) < \infty. \end{cases}$$

Let

$$u(x) = \begin{cases} t_j & \text{if } x \in \Omega_j \\ 0 & \text{if } x \in \Omega - \left(\bigcup_{j=1}^\infty \Omega_j\right). \end{cases}$$

Then

$$\int_\Omega A(|u(x)|)\,dx = \sum_{j=1}^{\infty} A(t_j)\mathrm{vol}(\Omega)_j$$

$$= \begin{cases} 1 & \text{if } \mathrm{vol}(\Omega) = \infty \\ A(t_0)\mathrm{vol}(\Omega) & \text{if } \mathrm{vol}(\Omega) < \infty. \end{cases}$$

But

$$\int_\Omega A(|2u(x)|)\,dx \geq \sum_{j=1}^{\infty} 2^j A(t_j)\mathrm{vol}(\Omega)_j = \infty.$$

Thus $K_A(\Omega)$ is not a vector space. ∎

8.9 (The Orlicz Space $L_A(\Omega)$) The *Orlicz space* $L_A(\Omega)$ is the linear hull of the Orlicz class $K_A(\Omega)$, that is, the smallest vector space (under pointwise addition and scalar multiplication) that contains $K_A(\Omega)$. Evidently, $L_A(\Omega)$ contains all scalar multiples λu of elements $u \in K_A(\Omega)$. Thus $K_A(\Omega) \subset L_A(\Omega)$, these sets being equal if and only if (A, Ω) is Δ-regular.

The reader may verify that the functional

$$\|u\|_A = \|u\|_{A,\Omega} = \inf\left\{k > 0 : \int_\Omega A\left(\frac{|u(x)|}{k}\right) dx \leq 1\right\}$$

is a norm on $L_A(\Omega)$. It is called the Luxemburg norm. The infimum is attained. In fact, if k decreases towards $\|u\|_A$ in the inequality

$$\int_\Omega A\left(\frac{|u(x)|}{k}\right) dx \leq 1, \tag{8}$$

we obtain by monotone convergence

$$\int_\Omega A\left(\frac{|u(x)|}{\|u\|_A}\right) dx \leq 1. \tag{9}$$

Equality may fail to hold in (9) but if equality holds in (8), then $k = \|u\|_A$.

8.10 THEOREM $L_A(\Omega)$ is a Banach space with respect to the Luxemburg norm.

The completeness proof is similar to that for the L^p spaces given in Theorem 2.16. The details are left to the reader. We remark that if $1 < p < \infty$ and $A_p(t) = t^p/p$, then

$$L^p(\Omega) = L_{A_p}(\Omega) = K_{A_p}(\Omega).$$

Moreover, $\|u\|_{A_p,\Omega} = p^{-1/p}\,\|u\|_{p,\Omega}$.

8.11 (A Generalized Hölder Inequality) If A and $\tilde{A}$ are complementary N-functions, a generalized version of Hölder's inequality

$$\left| \int_\Omega u(x)v(x)\,dx \right| \le 2\,\|u\|_{A,\Omega}\,\|v\|_{\tilde{A},\Omega} \tag{10}$$

can be obtained by applying Young's inequality (2) to $|u(x)|/\,\|u\|_A$ and $|v(x)|/\,\|v\|_{\tilde{A}}$ and integrating over Ω.

The following elementary imbedding theorem is an analog for Orlicz spaces of Theorem 2.14 for L^p spaces.

8.12 THEOREM (An Imbedding Theorem for Orlicz Spaces) The imbedding

$$L_B(\Omega) \to L_A(\Omega)$$

holds if and only if *either*

(a) B dominates A globally, *or*

(b) B dominates A near infinity and $\mathrm{vol}(\Omega) < \infty$.

Proof. If $A(t) \le B(kt)$ for all $t \ge 0$, and if $u \in L_B(\Omega)$, then

$$\int_\Omega A\left(\frac{|u(x)|}{k\,\|u\|_B}\right) dx \le \int_\Omega B\left(\frac{|u(x)|}{\|u\|_B}\right) dx \le 1.$$

Thus $u \in L_A(\Omega)$ and $\|u\|_A \le k\,\|u\|_B$.

If $\mathrm{vol}(\Omega) < \infty$, let $t_1 = A^{-1}\big((2\mathrm{vol}(\Omega))^{-1}\big)$. If B dominates A near infinity, then there exists positive numbers t_0 and k such that $A(t) \le B(kt)$ for $t \ge t_0$. Evidently, for $t \ge t_1$ we have

$$A(t) \le \max\left\{1, \frac{A(t_0)}{B(kt_1)}\right\} B(kt) = k_1 B(kt).$$

If $u \in L_B(\Omega)$ is given, let $\Omega'(u) = \{x \in \Omega : |u(x)|/[2k_1k\,\|u\|_B] < t_1\}$ and $\Omega''(u) = \Omega - \Omega'(u)$. Then

$$\begin{aligned}
\int_\Omega A\left(\frac{|u(x)|}{2k_1k\,\|u\|_B}\right) dx &= \left(\int_{\Omega'(u)} + \int_{\Omega''(u)}\right) A\left(\frac{|u(x)|}{2k_1k\,\|u\|_B}\right) dx \\
&\le \frac{1}{2\mathrm{vol}(\Omega)} \int_{\Omega'(u)} dx + k_1 \int_{\Omega''(u)} B\left(\frac{|u(x)|}{2k_1\,\|u\|_B}\right) dx \\
&\le \frac{1}{2} + \frac{1}{2}\int_\Omega B\left(\frac{|u(x)|}{\|u\|_B}\right) dx \le 1.
\end{aligned}$$

Thus $u \in L_A(\Omega)$ and $\|u\|_A \le 2k_1k\,\|u\|_B$.

Conversely, suppose that neither of the hypotheses (a) and (b) holds. Then there exist numbers $t_j > 0$ such that

$$A(t_j) \geq B(jt_j), \qquad j = 1, 2, \ldots .$$

If $\mathrm{vol}(\Omega) < \infty$, we may assume, in addition, that

$$t_j \geq \frac{1}{j} B^{-1}\left(\frac{1}{\mathrm{vol}(\Omega)}\right).$$

Let Ω_j be a subdomain of Ω having volume $1/B(jt_j)$, and let

$$u_j(x) = \begin{cases} jt_j & \text{if } x \in \Omega_j \\ 0 & \text{if } x \in \Omega - \Omega_j. \end{cases}$$

Then

$$\int_\Omega A\left(\frac{|u_j(x)|}{j}\right) dx \geq \int_\Omega B(|u_j(x)|)\, dx = 1$$

so that $\|u_j\|_B = 1$ but $\|u_j\|_A \geq j$. Thus $L_B(\Omega)$ is not imbedded in $L_A(\Omega)$. ∎

8.13 (Convergence in Mean) A sequence $\{u_j\}$ of functions in $L_A(\Omega)$ is said to *converge in mean* to $u \in L_A(\Omega)$ if

$$\lim_{j\to\infty} \int_\Omega A\big(|u_j(x) - u(x)|\big)\, dx = 0.$$

The convexity of A implies that for $0 < \epsilon \leq 1$ we have

$$\int_\Omega A\big(|u_j(x) - u(x)|\big)\, dx \leq \epsilon \int_\Omega A\left(\frac{|u_j(x) - u(x)|}{\epsilon}\right) dx$$

from which it follows that norm convergence in $L_A(\Omega)$ implies mean convergence. The converse holds, that is, mean convergence implies norm convergence, if and only if (A, Ω) is Δ-regular. The proof is similar to that of Lemma 8.8 and is left to the reader.

8.14 (The Space $E_A(\Omega)$) Let $E_A(\Omega)$ denote the closure in $L_A(\Omega)$ of the space of functions u which are bounded on Ω and have bounded support in $\overline{\Omega}$. If $u \in K_A(\Omega)$, the sequence $\{u_j\}$ defined by

$$u_j(x) = \begin{cases} u(x) & \text{if } |u(x)| \leq j \text{ and } |x| \leq j, \quad x \in \Omega \\ 0 & \text{otherwise} \end{cases} \tag{11}$$

converges a.e. on Ω to u. Since $A(|u(x) - u_j(x)|) \leq A(|u(x)|)$, we have by dominated convergence that u_j converges to u in mean in $L_A(\Omega)$. Therefore, if

(A, Ω) is Δ-regular, then $E_A(\Omega) = K_A(\Omega) = L_A(\Omega)$. If (A, Ω) is not Δ-regular, then we have

$$E_A(\Omega) \subset K_A(\Omega) \subsetneqq L_A(\Omega)$$

so that $E_A(\Omega)$ is a proper closed subspace of $L_A(\Omega)$ in this case. To verify the first inclusion above let $u \in E_A(\Omega)$ be given. Let v be a bounded function with bounded support such that $0 < \|u - v\|_A < 1/2$. Using the convexity of A and (9), we obtain

$$\frac{1}{\|2u - 2v\|_A} \int_\Omega A\big(|2u(x) - 2v(x)|\big)\, dx \le \int_\Omega A\left(\frac{|2u(x) - 2v(x)|}{\|2u - 2v\|_A}\right) dx \le 1,$$

whence $2u - 2v \in K_A(\Omega)$. Since $2v$ clearly belongs to $K_A(\Omega)$ and $K_A(\Omega)$ is convex, we have $u = (1/2)(2u - 2v) + (1/2)(2v) \in K_A(\Omega)$.

8.15 LEMMA $E_A(\Omega)$ is the maximal linear subspace of $K_A(\Omega)$.

Proof. Let S be a linear subspace of $K_A(\Omega)$ and let $u \in S$. Then $\lambda u \in K_A(\Omega)$ for every scalar λ. If $\epsilon > 0$ and u_j is given by (11), then u_j/ϵ converges to u/ϵ in mean in $L_A(\Omega)$ as noted in Paragraph 8.14. Hence, for sufficiently large values of j we have

$$\int_\Omega A\left(\frac{|u_j(x) - u(x)|}{\epsilon}\right) dx \le 1$$

and therefore u_j converges to u in norm in $L_A(\Omega)$. Thus $S \subset E_A(\Omega)$. ∎

8.16 THEOREM Let Ω have finite volume, and suppose that the N-function A increases essentially more slowly than the N-function B near infinity. Then

$$L_B(\Omega) \to E_A(\Omega).$$

Proof. Since $L_B(\Omega) \to L_A(\Omega)$ is already established we need only show that $L_B(\Omega) \subset E_A(\Omega)$. Since $L_B(\Omega)$ is the linear hull of $K_B(\Omega)$ and $E_A(\Omega)$ is the maximal linear subspace of $K_A(\Omega)$, it is sufficient to show that $\lambda u \in K_A(\Omega)$ whenever $u \in K_B(\Omega)$ and λ is a scalar. But there exists a positive number t_0 such that $A(|\lambda| t) \le B(t)$ for all $t \ge t_0$. Thus

$$\begin{aligned}\int_\Omega A\big(|\lambda u(x)|\big)\, dx &= \left(\int_{\{x\in\Omega:|u(x)|\le t_0\}} + \int_{\{x\in\Omega:|u(x)|>t_0\}}\right) A\big(|\lambda u(x)|\big)\, dx \\ &\le A\big(|\lambda| t_0\big)\mathrm{vol}(\Omega) + \int_\Omega B\big(|u(x)|\big)\, dx < \infty\end{aligned}$$

whence the theorem follows. ∎

Duality in Orlicz Spaces

8.17 LEMMA Given $v \in L_{\tilde{A}}(\Omega)$, the linear functional F_v defined by

$$F_v(u) = \int_\Omega u(x)v(x)\,dx \tag{12}$$

belongs to the dual space $[L_A(\Omega)]'$ and its norm $\|F_v\|$ in that space satisfies

$$\|v\|_{\tilde{A}} \le \|F_v\| \le 2\,\|v\|_{\tilde{A}}\,. \tag{13}$$

Proof. It follows by Hölder's inequality (10) that

$$|F_v(u)| \le 2\,\|u\|_A\,\|v\|_{\tilde{A}}$$

holds for all $u \in L_A(\Omega)$, confirming the second inequality in (13).
To establish the other half of (13) we may assume that $v \ne 0$ in $L_{\tilde{A}}(\Omega)$ so that $\|F_v\| = K > 0$. Let

$$u(x) = \begin{cases} \tilde{A}\left(\dfrac{|v(x)|}{K}\right) \Big/ \dfrac{|v(x)|}{K} & \text{if } v(x) \ne 0 \\ 0 & \text{if } v(x) = 0. \end{cases}$$

If $\|u\|_A > 1$, then for $0 < \epsilon \le \|u\|_A - 1$ we have

$$\frac{1}{\|u\|_A - \epsilon}\int_\Omega A\big(|u(x)|\big)\,dx \ge \int_\Omega A\left(\frac{|u(x)|}{\|u\|_A - \epsilon}\right)dx > 1.$$

Letting $\epsilon \to 0+$ we obtain, using (3),

$$\begin{aligned} \|u\|_A &\le \int_\Omega A\big(|u(x)|\big)\,dx = \int_\Omega A\left(\tilde{A}\left(\frac{|v(x)|}{K}\right) \Big/ \frac{|v(x)|}{K}\right)dx \\ &< \int_\Omega \tilde{A}\left(\frac{|v(x)|}{K}\right)dx = \frac{1}{\|F_v\|}\int_\Omega u(x)v(x)\,dx \le \|u\|_A\,. \end{aligned}$$

This contradiction shows that $\|u\|_A \le 1$. Now

$$\|F_v\| = \sup_{\|u\|_A \le 1} |F_v(u)| \ge \|F_v\| \left| \int_\Omega \tilde{A}\left(\frac{|v(x)|}{\|F_v\|}\right)dx \right|$$

so that

$$\int_\Omega \tilde{A}\left(\frac{|v(x)|}{\|F_v\|}\right)dx \le 1. \tag{14}$$

Thus, $\|v\|_{\tilde{A}} \le \|F_v\|$. ∎

8.18 REMARK The above lemma also holds when F_v is restricted to act on $E_A(\Omega)$. To obtain the first inequality of (13) in this case take $\|F_u\|$ to be the norm of F_v in $[E_A(\Omega)]'$ and replace u in the above proof by $\chi_n u$ where χ_n is the characteristic function of $\Omega_n = \{x \in \Omega : |x| \le n \text{ and } |u(x)| \le n\}$. Evidently, $\chi_n u$ belongs to $E_A(\Omega)$, $\|\chi_n u\|_A \le 1$, and (14) becomes

$$\int_\Omega \chi_n(x)\tilde{A}\left(\frac{|v(x)|}{\|F_v\|}\right) dx \le 1.$$

Since $\chi_n(x)$ increases to unity a.e. on Ω as $n \to \infty$, we obtain (14) again, and $\|v\|_{\tilde{A}} \le \|F_v\|$ as before.

8.19 THEOREM (The Dual of $E_A(\Omega)$) The dual space of $E_A(\Omega)$ is isomorphic and homeomorphic to $L_{\tilde{A}}(\Omega)$.

Proof. We have already shown that any element $v \in L_{\tilde{A}}(\Omega)$ determines a bounded linear functional F_v via (12) on $L_A(\Omega)$ and also on $E_A(\Omega)$, and that in either case the norm of this functional differs from $\|v\|_{\tilde{A}}$ by at most a factor of 2. It remains to be shown that every bounded linear functional on $E_A(\Omega)$ is of the form F_v for some such v.

Let $F \in [E_A(\Omega)]'$ be given. We define a complex measure λ on the measurable subsets S of Ω having finite volume by setting

$$\lambda(S) = F(\chi_S),$$

χ_S being the characteristic function of S. Since

$$\int_\Omega A\left(|\chi_S(x)|A^{-1}\left[\frac{1}{\mathrm{vol}(S)}\right]\right) dx = \int_S \frac{1}{\mathrm{vol}(S)}\, dx = 1 \tag{15}$$

we have

$$|\lambda(S)| \le \|F\|\,\|\chi_S\|_A = \frac{\|F\|}{A^{-1}(1/\mathrm{vol}(S))}.$$

Since the right side tends to zero with $\mathrm{vol}(S)$, the measure λ is absolutely continuous with respect to Lebesgue measure, and so by the Radon-Nikodym Theorem 1.52, λ can be expressed in the form

$$\lambda(S) = \int_S v(x)\, dx,$$

for some v that is integrable on Ω. Thus

$$F(u) = \int_\Omega u(x)v(x)\, dx$$

holds for measurable, simple functions u.

If $u \in E_A(\Omega)$, a sequence of measurable, simple functions u_j can be found that converges a.e. to u and satisfies $|u_j(x)| \le |u(x)|$ on Ω. Since $|u_j(x)v(x)|$ converges a.e. to $|u(x)v(x)|$, Fatou's Lemma 1.49 yields

$$\left|\int_\Omega u(x)v(x)\,dx\right| \le \sup_j \int_\Omega |u_j(x)v(x)|\,dx = \sup_j \left|F\big(|u_j|\operatorname{sgn} v\big)\right|$$
$$\le \|F\| \sup_j \|u_j\|_A \le \|F\|\,\|u\|_A .$$

It follows that the linear functional

$$F_v(u) = \int_\Omega u(x)v(x)\,dx$$

is bounded on $E_A(\Omega)$ whence $v \in L_{\tilde{A}}(\Omega)$ by Remark 8.18. Since F_v and F assume the same values on the measurable, simple functions, a set that is dense in $E_A(\Omega)$ (see Theorem 8.21 below), they agree on $E_A(\Omega)$ and the theorem is proved. ∎

A simple application of the Hahn-Banach Theorem shows that if $E_A(\Omega)$ is a proper subspace of $L_A(\Omega)$ (that is, if (A, Ω) is *not* Δ-regular), then there exists a bounded linear functional F on $L_A(\Omega)$ that is not given by (12) for any $v \in L_{\tilde{A}}(\Omega)$. As an immediate consequence of this fact we have the following theorem.

8.20 THEOREM (Reflexivity of Orlicz Spaces) $L_A(\Omega)$ is reflexive if and only if both (A, Ω) and $(\tilde{A}, \Omega)$ are Δ-regular.

We omit any discussion of uniform convexity of Orlicz spaces. This subject is treated in Luxemburg's thesis [Lu].

Separability and Compactness Theorems

We next generalize to Orlicz spaces the L^p approximation Theorems 2.19, 2.21, and 2.30.

8.21 THEOREM (Approximation of Functions in $E_A(\Omega)$)

(a) $C_0(\Omega)$ is dense in $E_A(\Omega)$.

(b) $E_A(\Omega)$ is separable.

(c) If J_ϵ is the mollifier of Paragraph 2.28, then for each $u \in E_A(\Omega)$ we have $\lim_{\epsilon \to 0+} J_\epsilon * u = u$ in norm in $E_A(\Omega)$.

(d) $C_0^\infty(\Omega)$ is dense in $E_A(\Omega)$.

Proof. Part (a) is proved by the same method used in Theorem 2.19. In approximating $u \in E_A(\Omega)$ first by simple functions we can assume that u is bounded on

Ω and has bounded support. Then a dominated convergence argument shows that the simple functions converge in norm to u in $E_A(\Omega)$. (The details are left to the reader.)

Part (b) follows from part (a) by the same proof given for Theorem 2.21.

Consider part (c). If $u \in E_A(\Omega)$, let u be extended to $\mathbb{R}^n$ so as to vanish identically outside Ω. Let $v \in L_{\tilde{A}}(\Omega)$. Then

$$\left|\int_\Omega \big(J_\epsilon * u(x) - u(x)\big)v(x)\,dx\right| \le \int_{\mathbb{R}^n} J(y)\,dy \int_\Omega |u(x-\epsilon y) - u(x)||v(x)|\,dx$$
$$\le 2\,\|v\|_{\tilde{A},\Omega} \int_{|y|\le 1} J(y)\,\left\|u_{\epsilon y} - u\right\|_{A,\Omega}\,dy$$

by Hölder's inequality (10), where $u_{\epsilon y}(x) = u(x-\epsilon y)$. Thus by (13) and Theorem 8.19,

$$\|J_\epsilon * u - u\|_{A,\Omega} = \sup_{\|v\|_{\tilde{A},\Omega}\le 1} \left|\int_\Omega \big(J_\epsilon * u(x) - u(x)\big)v(x)\,dx\right|$$
$$\le 2\int_{|y|\le 1} J(y)\,\left\|u_{\epsilon y} - u\right\|_{A,\Omega}\,dy.$$

Given $\delta > 0$ we can find $\tilde{u} \in C_0(\Omega)$ such that $\|u - \tilde{u}\|_{A,\Omega} < \delta/6$. Clearly, $\left\|u_{\epsilon y} - \tilde{u}_{\epsilon y}\right\|_{A,\Omega} < \delta/6$ and for sufficiently small ϵ, $\left\|\tilde{u}_{\epsilon y} - \tilde{u}\right\|_{A,\Omega} < \delta/6$ for every y with $|y| \le 1$. Thus $\|J_\epsilon * u - u\|_{A,\Omega} < \delta$ and (c) is established.

Part (d) is an immediate consequence of parts (a) and (c). ∎

8.22 REMARK $L_A(\Omega)$ is not separable unless $L_A(\Omega) = E_A(\Omega)$, that is, unless (A, Ω) is Δ-regular. A proof of this fact may be found in [KR] (Chapter II, Theorem 10.2).

8.23 (Convergence in Measure) A sequence $\{u_j\}$ of measurable functions is said to *converge in measure* on Ω to the function u provided that for each $\epsilon > 0$ and $\delta > 0$ there exists an integer M such that if $j > M$, then

$$\mathrm{vol}(\{x \in \Omega : |u_j(x) - u(x)| > \epsilon\}) \le \delta.$$

Clearly, in this case there also exists an integer N such that if $j, k \ge N$, then

$$\mathrm{vol}(\{x \in \Omega : |u_j(x) - u_k(x)| \ge \epsilon\}) \le \delta.$$

8.24 THEOREM Let Ω have finite volume and suppose that the N-function B increases essentially more slowly than A near infinity. If the sequence $\{u_j\}$ is bounded in $L_A(\Omega)$ and convergent in measure on Ω, then it is convergent in norm in $L_B(\Omega)$.

Proof. Fix $\epsilon > 0$ and let $v_{j,k}(x) = \big(u_j(x) - u_k(x)\big)/\epsilon$. Clearly $\{v_{j,k}\}$ is bounded in $L_A(\Omega)$; say $\|v_{j,k}\|_{A,\Omega} \le K$. Now there exists a positive number t_0 such that if $t > t_0$, then

$$B(t) \le \frac{1}{4} A\left(\frac{t}{K}\right).$$

Let $\delta = 1/[4B(t_0)]$ and set

$$\Omega_{j,k} = \left\{x \in \Omega : |v_{j,k}(x)| \ge B^{-1}\left(\frac{1}{2\mathrm{vol}(\Omega)}\right)\right\}.$$

Since $\{u_j\}$ converges in measure, there exists an integer N such that if $j, k \ge N$, then $\mathrm{vol}(\Omega)_{j,k} \le \delta$. Set

$$\Omega'_{j,k} = \{x \in \Omega_{j,k} : |v_{j,k}(x)| \ge t_0\}, \qquad \Omega''_{j,k} = \Omega_{j,k} - \Omega'_{j,k}.$$

For $j, k \ge N$ we have

$$\begin{aligned}\int_\Omega B(|v_{j,k}(x)|)\,dx &= \left(\int_{\Omega-\Omega_{j,k}} + \int_{\Omega'_{j,k}} + \int_{\Omega''_{j,k}}\right) B(|v_{j,k}(x)|)\,dx \\ &\le \frac{\mathrm{vol}(\Omega)}{2\mathrm{vol}(\Omega)} + \frac{1}{4}\int_{\Omega'_{j,k}} A\left(\frac{|v_{j,k}(x)|}{K}\right) dx + \delta B(t_0) \le 1.\end{aligned}$$

Hence $\|u_j - u_k\|_{B,\Omega} \le \epsilon$ and so $\{u_j\}$ converges in $L_B(\Omega)$. ∎

The following theorem will be useful when we wish to extend the Rellich-Kondrachov Theorem 6.3 to imbeddings of Orlicz-Sobolev spaces.

8.25 THEOREM (Precompact Sets in Orlicz Spaces) Let Ω have finite volume and suppose that the N-function B increases essentially more slowly than A near infinity. Then any bounded subset S of $L_A(\Omega)$ which is precompact in $L^1(\Omega)$ is also precompact in $L_B(\Omega)$.

Proof. Evidently $L_A(\Omega) \to L^1(\Omega)$ since Ω has finite volume. If $\{u_j^*\}$ is a sequence in S, then it has a subsequence $\{u_j\}$ that converges in $L^1(\Omega)$; say $u_j \to u$ in $L^1(\Omega)$. Let $\epsilon, \delta > 0$. Then there exists an integer N such that if $j \ge N$, then $\|u_j - u\|_{1,\Omega} \le \epsilon\delta$. If follows that

$$\mathrm{vol}(\{x \in \Omega : |u_j(x) - u(x)| \ge \epsilon\}) \le \delta.$$

Thus $\{u_j\}$ converges to u in measure on Ω and hence also in $L_B(\Omega)$. ∎

A Limiting Case of the Sobolev Imbedding Theorem

8.26 If $mp = n$ and $p > 1$, the Sobolev Imbedding Theorem 4.12 provides no best (i.e., smallest) target space into which $W^{m,p}(\Omega)$ can be imbedded. In this case, for suitably regular Ω,

$$W^{m,p}(\Omega) \to L^q(\Omega), \qquad p \le q < \infty,$$

but (see Example 4.43)

$$W^{m,p}(\Omega) \not\subset L^\infty(\Omega).$$

If the class of target spaces for the imbedding is enlarged to contain Orlicz spaces, then a best such target space can be found.

We first consider the case of bounded Ω and later extend our consideration to unbounded domains. The following theorem was established by Trudinger [Td]. For other proofs see [B+] and [Ta]; for refinements going beyond Orlicz spaces see [BW] and [MP].

8.27 THEOREM (Trudinger's Theorem) Let Ω be a bounded domain in $\mathbb{R}^n$ satisfying the cone condition. Let $mp = n$ and $p > 1$. Set

$$A(t) = \exp\left(t^{n/(n-m)}\right) - 1 = \exp\left(t^{p/(p-1)}\right) - 1. \tag{16}$$

Then there exists the imbedding

$$W^{m,p}(\Omega) \to L_A(\Omega).$$

Proof. If $m > 1$ and $mp = n$, then $W^{m,p}(\Omega) \to W^{1,n}(\Omega)$. Therefore it is sufficient to prove the theorem for $m = 1$, $p = n > 1$. Let $u \in C^1(\Omega) \cap W^{1,n}(\Omega)$ (a set that is dense in $W^{1,n}(\Omega)$) and let $x \in \Omega$. By the special case $m = 1$ of Lemma 4.15 we have, denoting by C a cone contained in Ω, having vertex at x, and congruent to the cone specifying the cone condition for Ω,

$$\begin{aligned} |u(x)| &\le K_1 \left(\|u\|_{1,C} + \sum_{j=1}^{n} \int_C |D_j u(x)||x-y|^{1-n}\, dy \right) \\ &\le K_1 \left(\|u\|_{1,\Omega} + \sum_{j=1}^{n} \int_\Omega |D_j u(y)||x-y|^{1-n}\, dy \right). \end{aligned}$$

We want to estimate the L^s-norm $\|u\|_s$ for arbitrary $s > 1$. If $v \in L^{s'}(\Omega)$ (where

$(1/s)+(1/s')=1)$, then

$$\int_\Omega |u(x)v(x)|\,dx \le K_1\left(\|u\|_1\int_\Omega |v(x)|\,dx+\sum_{j=1}^n\int_\Omega\int_\Omega \frac{|D_ju(y)||v(x)|}{|x-y|^{n-1}}\,dy\,dx\right)$$

$$\le K_1\,\|u\|_1\,\|v\|_{s'}\,(\mathrm{vol}(\Omega))^{1/s}$$

$$+K_1\sum_{j=1}^n\left(\int_\Omega\int_\Omega \frac{|v(x)|}{|x-y|^{n-(1/s)}}\,dy\,dx\right)^{(n-1)/n}$$

$$\times\left(\int_\Omega\int_\Omega \frac{|D_ju(y)|^n|v(x)|}{|x-y|^{(n-1)/s}}\,dy\,dx\right)^{1/n}.$$

By Lemma 4.64, if $0\le \nu<n$,

$$\int_\Omega \frac{1}{|x-y|^\nu}\,dy \le \frac{K_2}{n-\nu}(\mathrm{vol}(\Omega))^{1-(\nu/n)}.$$

Hence

$$\int_\Omega\int_\Omega \frac{|v(x)|}{|x-y|^{n-(1/s)}}\,dy\,dx \le K_2 s\big(\mathrm{vol}(\Omega)\big)^{1/(sn)}\int_\Omega |v(x)|\,dx$$

$$\le K_3 s\big(\mathrm{vol}(\Omega)\big)^{1/(sn)+1/s}\,\|v\|_{s'}.$$

Also,

$$\int_\Omega\int_\Omega \frac{|D_ju(y)|^n|v(x)|}{|x-y|^{(n-1)/s}}\,dy\,dx \le \int_\Omega |D_ju(y)|^n\,dy\;\|v\|_{s'}\left(\int_\Omega \frac{1}{|x-y|^{n-1}}\,dx\right)^{1/s}$$

$$\le \left\|D_ju\right\|_p^p\,\|v\|_{s'}\left(K_2\big(\mathrm{vol}(\Omega)\big)^{1/n}\right)^{1/s}$$

$$= K_4\left\|D_ju\right\|_n^n\,\|v\|_{s'}\,(\mathrm{vol}(\Omega))^{1/(ns)}.$$

It follows from these estimates that

$$\int_\Omega |u(x)v(x)|\,dx \le K_1\,\|u\|_1\,\|v\|_{s'}\,(\mathrm{vol}(\Omega))^{1/s}$$

$$+K_4\sum_{j=1}^n s^{(n-1)/n}\left\|D_ju\right\|_n\,\|v\|_{s'}\,(\mathrm{vol}(\Omega))^{1/s}.$$

Since $s^{(n-1)/n}>1$ and since $W^{1,n}(\Omega)\to L^1(\Omega)$, we now have

$$\|u\|_s=\sup_{v\in L^{s'}(\Omega)}\frac{1}{\|v\|_{s'}}\int_\Omega |u(x)v(x)|\,dx \le K_5 s^{(n-1)/n}\big(\mathrm{vol}(\Omega)\big)^{1/s}\,\|u\|_{1,n}.$$

The constant K_5 depends only on n and the cone determining the cone condition for Ω. Setting $s = nk/(n-1)$, we obtain

$$\begin{aligned}\int_\Omega |u(x)|^{nk/(n-1)}\,dx &\le \operatorname{vol}(\Omega)\left(\frac{nk}{n-1}\right)^k \left(K_5\,\|u\|_{1,n}\right)^{nk/(n-1)}\\ &= \operatorname{vol}(\Omega)\left(\frac{k}{e^{n/(n-1)}}\right)^k \left(eK_5\left[\frac{n}{n-1}\right]^{(n-1)/n}\|u\|_{1,n}\right)^{nk/(n-1)}\end{aligned}$$

Since $e^{n/(n-1)} > e$, the series $\sum_{k=1}^{\infty}(1/k!)\left(k/e^{n/(n-1)}\right)^k$ converges to a finite sum $K6$. Let $K_7 = \max\{1, K_6\operatorname{vol}(\Omega)\}$ and put

$$K_8 = eK_7K_5\left(\frac{n}{n-1}\right)^{(n-1)/n}\|u\|_{1,n} = K_9\,\|u\|_{1,n}\,.$$

Then

$$\int_\Omega\left(\frac{|u(x)|}{K_8}\right)^{nk/(n-1)}dx \le \frac{\operatorname{vol}(\Omega)}{K_7^{nk/(n-1)}}\left(\frac{k}{e^{n/(n-1)}}\right)^k < \frac{\operatorname{vol}(\Omega)}{K_7}\left(\frac{k}{e^{n/(n-1)}}\right)^k$$

since $K_7 \ge 1$ and $nk/(n-1) > 1$. Expanding $A(t)$ in a power series, we now obtain

$$\begin{aligned}\int_\Omega A\left(\frac{|u(x)|}{K_8}\right)dx &= \sum_{k=1}^{\infty}\frac{1}{k!}\int_\Omega\left(\frac{|u(x)|}{K_8}\right)^{nk/(n-1)}dx\\ &< \frac{\operatorname{vol}(\Omega)}{K_7}\sum_{k=1}^{\infty}\frac{1}{k!}\left(\frac{k}{e^{n/(n-1)}}\right)^k \le 1.\end{aligned}$$

Hence $u \in L_A(\Omega)$ and

$$\|u\|_A \le K_8 = K_9\,\|u\|_{m,p},$$

where K_9 depends on n, $\operatorname{vol}(\Omega)$, and the cone C determining the cone condition for Ω. ∎

8.28 REMARK The imbedding established in Theorem 8.27 is "best possible" in the sense that if there exist an imbedding of the form

$$W_0^{m,p}(\Omega) \to L_B(\Omega),$$

then A dominates B near infinity. A proof of this fact for the case $m = 1$, $p = n > 1$ can be found in [HMT]. The general case is left to the reader as an exercise.

Trudinger's theorem can be generalized to fractional-order spaces. For results in this direction the reader is referred to [Gr] and [P].

Recent efforts have identified non-Orlicz function spaces that are smaller than Trudinger's space into which $W^{m,p}(\Omega)$ can be imbedded in the limiting case $mp = n$. See [MP] in this regard.

8.29 (Extension to Unbounded Domains) If Ω is unbounded and so (satisfying the cone condition) has infinite volume, then the N-function A given by (16) may not decrease rapidly enough at zero to to allow membership in $L_A(\Omega)$ of every $u \in W^{m,p}(\Omega)$ (where $mp = n$). Let k_0 be the smallest integer such that $k_0 \geq p-1$ and define a modified N-function A_0 by

$$A_0(t) = \exp\left(t^{p/(p-1)}\right) - \sum_{j=0}^{k_0-1} \frac{1}{j!} t^{jp/(p-1)}.$$

Evidently A_0 is equivalent to A near infinity so for any domain Ω having finite volume, $L_A(\Omega)$ and $L_{A_0}(\Omega)$ coincide and have equivalent norms. However, A_0 enjoys the further property that for $0 < r \leq 1$,

$$A_0(rt) \leq r^{k_0 p/(p-1)} A_0(t) \leq r^p A_0(t). \tag{17}$$

We show that if $mp = n$, $p > 1$, and Ω satisfies the cone condition (but may be unbounded), then

$$W^{m,p}(\Omega) \to L_{A_0}(\Omega).$$

Lemma 4.22 implies that even an unbounded domain Ω satisfying the cone condition can be written as a union of countably many subdomains Ω_j each satisfying the cone condition specified by a cone independent of j, each having volume satisfying

$$0 < K_1 \leq \operatorname{vol}(\Omega_j) \leq K_2$$

with K_1 and K_2 independent of j, and such that any $M+1$ of the subdomains have empty intersection. It follows from Trudinger's theorem that

$$\|u\|_{A_0,\Omega_j} \leq K_3 \|u\|_{m,p,\Omega_j}$$

with K_3 independent of j. Using (17) with $r = M^{1/p} \|u\|_{m,p,\Omega_j}^{-1} \|u\|_{m,p,\Omega}$ and the finite intersection property of the domains Ω_j, we have

$$\begin{aligned} \int_\Omega A_0\left(\frac{|u(x)|}{M^{1/p} K_3 \|u\|_{m,p,\Omega}}\right) dx &\leq \sum_{j=1}^{\infty} \int_{\Omega_j} A_0\left(\frac{|u(x)|}{M^{1/p} K_3 \|u\|_{m,p,\Omega}}\right) dx \\ &\leq \sum_{j=1}^{\infty} \frac{\|u\|_{m,p,\Omega_j}^p}{M \|u\|_{m,p,\Omega}^p} \leq 1. \end{aligned}$$

Hence $\|u\|_{A_0,\Omega} \le M^{1/p} K_3 \|u\|_{m,p,\Omega}$ as required.

We remark that if $k_0 > p - 1$, the above result can be improved slightly by using in place of A_0 the N-function $\max\{t^p, A_0(t)\}$.

Orlicz-Sobolev Spaces

8.30 (Definitions) For a given domain Ω in $\mathbb{R}^n$ and a given N-function A the *Orlicz-Sobolev space* $W^m L_A(\Omega)$ consists of those (equivalence classes of) functions u in $L_A(\Omega)$ whose distributional derivatives $D^\alpha u$ also belong to $L_A(\Omega)$ for all α with $|\alpha| \le m$. The space $W^m E_A(\Omega)$ is defined in an analogous fashion. It may be checked by the same method used for ordinary Sobolev spaces in Chapter 3 that $W^m L_A(\Omega)$ is a Banach space with respect to the norm

$$\|u\|_{m,A} = \|u\|_{m,A,\Omega} = \max_{0\le|\alpha|\le m} \|D^\alpha u\|_{A,\Omega},$$

and that $W^m E_A(\Omega)$ is a closed subspace of $W^m L_A(\Omega)$ and hence also a Banach space with the same norm. It should be kept in mind that $W^m E_A(\Omega)$ coincides with $W^m L_A(\Omega)$ if and only if (A, Ω) is Δ-regular. If $1 < p < \infty$ and $A_p(t) = t^p$, then $W^m L_{A_p}(\Omega) = W^m E_{A_p}(\Omega) = W^{m,p}(\Omega)$, the latter space having norm equivalent to those of the former two spaces.

As in the case of ordinary Sobolev spaces, $W_0^m L_A(\Omega)$ is taken to be the closure of $C_0^\infty(\Omega)$ in $W^m L_A(\Omega)$. (An analogous definition for $W_0^m E_A(\Omega)$ clearly leads to the same spaces in all cases.)

Many properties of Orlicz-Sobolev spaces are obtained by very straightforward generalization of the proofs of the same properties for ordinary Sobolev spaces. We summarize these in the following theorem and refer the reader to the corresponding results in Chapter 3 for the method of proof. The details can also be found in the article by Donaldson and Trudinger [DT].

8.31 THEOREM (Basic Properties of Orlicz-Sobolev Spaces)

(a) $W^m E_A(\Omega)$ is separable (Theorem 3.6).

(b) If (A, Ω) and $(\tilde{A}, \Omega)$ are Δ-regular, then $W^m E_A(\Omega) = W^m L_A(\Omega)$ is reflexive (Theorem 3.6).

(c) Each element F of the dual space $[W^m E_A(\Omega)]'$ is given by

$$F(u) = \sum_{0\le|\alpha|\le m} \int_\Omega D^\alpha u(x)\, v_\alpha(x)\, ds$$

for some functions $v_\alpha \in L_{\tilde{A}}(\Omega)$, $0 \le |\alpha| \le m$ (Theorem 3.9).

(d) $C^\infty(\Omega) \cap W^m E_A(\Omega)$ is dense in $W^m E_A(\Omega)$ (Theorem 3.17).

(e) If Ω satisfies the segment condition, then $C^\infty(\overline{\Omega})$ is dense in $W^m E_A(\Omega)$ (Theorem 3.22).

(f) $C_0^\infty(\mathbb{R}^n)$ is dense in $W^m E_A(\mathbb{R}^n)$. Thus $W_0^m L_A(\mathbb{R}^n) = W^m E_A(\mathbb{R}^n)$ (Theorem 3.22).

Imbedding Theorems for Orlicz-Sobolev Spaces

8.32 Imbedding results analogous to those obtained for the spaces $W^{m,p}(\Omega)$ in Chapters 4 and 6 can be formulated for the Orlicz-Sobolev spaces $W^m L_A(\Omega)$ and $W^m E_A(\Omega)$. The first results in this direction were obtained by Dankert [Da]. A fairly general imbedding theorem along the lines of Theorems 4.12 and 6.3 was presented by Donaldson and Trudinger [DT] and we develop it below.

As was the case with ordinary Sobolev spaces, most of these imbedding results are obtained for domains satisfying the cone condition. Exceptions are those yielding (generalized) Hölder continuity estimates; these require the strong local Lipschitz condition. Some results below are proved only for bounded domains. The method used in extending the analogous results for ordinary Sobolev spaces to unbounded domains does not seem to extend in a straightforward manner when general Orlicz spaces are involved. In this sense the imbedding picture we present here is incomplete. Best possible Orlicz-Sobolev imbeddings, involving a careful study of rearrangements, have been found recently by Cianchi [Ci]. We settle here for results that follow by methods we used earlier for imbeddings of $W^{m,p}(\Omega)$ and for weighted spaces; that is also how we proved Trudinger's theorem.

8.33 (A Sobolev Conjugate) We concern ourselves for the time being with imbeddings of $W^1 L_A(\Omega)$; the imbeddings of $W^m L_A(\Omega)$ are summarized in Theorem 8.43. As usual, Ω is assumed to be a domain in $\mathbb{R}^n$.

Let A be an N-function. We shall always suppose that

$$\int_0^1 \frac{A^{-1}(t)}{t^{(n+1)/n}}\,dt < \infty, \tag{18}$$

replacing, if necessary, A by another N-function equivalent to A near infinity. (If Ω has finite volume, (18) places no restrictions on A from the point of view of imbedding theory since N-functions equivalent near infinity determine identical Orlicz spaces in that case.)

Suppose also that

$$\int_1^\infty \frac{A^{-1}(t)}{t^{(n+1)/n}}\,dt = \infty. \tag{19}$$

For instance, if $A(t) = A_p(t) = t^p$, $p > 1$, then (19) holds precisely when $p \le n$. With (19) satisfied, we define the *Sobolev conjugate N-function* A_* of A by setting

$$A_*^{-1}(t) = \int_0^t \frac{A^{-1}(\tau)}{\tau^{(n+1)/n}}\,d\tau, \qquad t \ge 0. \tag{20}$$

It may readily be checked that A_* is an N-function. If $1 < p < n$, we have, setting $q = np/(n-p)$ (the normal Sobolev conjugate exponent for p),

$$A_{p*}(t) = q^{1-q} p^{-q/p} A_q(t).$$

It is also readily seen for the case $p = n$ that $A_{n*}(t)$ is equivalent near infinity to the N-function $e^t - t - 1$. In [Ci] a different Sobolev conjugate is used; it is equivalent when $p = n$ to the N-function in Trudinger's theorem.

8.34 LEMMA Let $u \in W^{1,1}_{\mathrm{loc}}(\Omega)$ and let f satisfy a Lipschitz condition on $\mathbb{R}$. If $g(x) = f(|u(x)|)$, then $g \in W^{1,1}_{\mathrm{loc}}(\Omega)$ and

$$D_j g(x) = f'(|u(x)|) \operatorname{sgn} u(x) \cdot D_j u(x).$$

Proof. Since $|u| \in W^{1,1}_{\mathrm{loc}}(\Omega)$ and $D_j|u(x)| = \operatorname{sgn} u(x) \cdot D_j u(x)$ it is sufficient to establish the lemma for positive, real-valued functions u so that $g(x) = f\big(u(x)\big)$. Let $\phi \in \mathscr{D}(\Omega)$ and let $\{e_j\}_{j=1}^n$ be the standard basis in $\mathbb{R}^n$. Then

$$\begin{aligned} -\int_\Omega f\big(u(x)\big) D_j \phi(x)\,dx &= -\lim_{h\to 0} \int_\Omega f\big(u(x)\big) \frac{\phi(x) - \phi(x - he_j)}{h}\,dx \\ &= \lim_{h\to 0} \int_\Omega \frac{f\big(u(x+he_j)\big) - f\big(u(x)\big)}{h} \phi(x)\,dx \\ &= \lim_{h\to 0} \int_\Omega Q(x,h) \frac{u(x+he_j) - u(x)}{h} \phi(x)\,dx, \end{aligned}$$

where, since f satisfies a Lipschitz condition, for each h the function $Q(\cdot,h)$ is defined a.e. on Ω by

$$Q(x,h) = \begin{cases} \dfrac{f\big(u(x+he_j)\big) - f\big(u(x)\big)}{u(x+he_j) - u(x)} & \text{if } u(x+he_j) \neq u(x) \\ f'\big(u(x)\big) & \text{otherwise.} \end{cases}$$

Moreover, $\|Q(\cdot,h)\|_{\infty,\Omega} \le K$ for some constant K independent of h. A well-known theorem in functional analysis tells us that for some sequence of values of h tending to zero, $Q(\cdot,h)$ converges to $f'\big(u(\cdot)\big)$ in the weak-star topology of $L^\infty(\Omega)$. On the other hand, since $u \in W^{1,1}(\operatorname{supp}(\phi))$ we have

$$\lim_{h\to 0} \frac{u(x+he_j) - u(x)}{h} \phi(x) = D_j u(x) \cdot \phi(x)$$

in $L^1(\operatorname{supp}(\phi))$. It follows that

$$-\int_\Omega f\big(u(x)\big) D_j \phi(x)\,dx = \int_\Omega f'\big(u(x)\big) D_j u(x) \phi(x)\,dx,$$

which implies the lemma. ∎

8.35 THEOREM (Imbedding Into an Orlicz Space) Let Ω be bounded and satisfying the cone condition in $\mathbb{R}^n$. If (18) and (19) hold, then

$$W^1L_A(\Omega) \to L_{A_*}(\Omega),$$

where A_* is given by (20). Moreover, if B is any N-function increasing essentially more slowly than A_* near infinity, then the imbedding

$$W^1L_A(\Omega) \to L_B(\Omega)$$

(exists and) is compact.

Proof. The function $s = A_*(t)$ satisfies the differential equation

$$A^{-1}(s)\frac{ds}{dt} = s^{(n+1)/n}, \tag{21}$$

and hence, since $s < A^{-1}(s)\tilde{A}^{-1}(s)$ (see (4)),

$$\frac{ds}{dt} \le s^{1/n}\tilde{A}^{-1}(s).$$

Therefore $\sigma(t) = \big(A_*(t)\big)^{(n-1)/n}$ satisfies the differential inequality

$$\frac{d\sigma}{dt} \le \frac{n-1}{n}\tilde{A}^{-1}\Big(\big(\sigma(t)\big)^{n/(n-1)}\Big). \tag{22}$$

Let $u \in W^1L_A(\Omega)$ and suppose, for the moment, that u is bounded on Ω and is not zero in $L_A(\Omega)$. Then $\int_\Omega A_*(|u(x)|/\lambda)\,dx$ decreases continuously from infinity to zero as λ increases from zero to infinity, and, accordingly, assumes the value unity for some positive value K of λ. Thus

$$\int_\Omega A_*\left(\frac{|u(x)|}{K}\right)dx = 1, \qquad K = \|u\|_{A_*}. \tag{23}$$

Let $f(x) = \sigma\big(|u(x)|/K\big)$. Evidently, $u \in W^{1,1}(\Omega)$ and σ is Lipschitz on the range of $|u|/K$ so that, by the previous lemma, f belongs to $W^{1,1}(\Omega)$. By Theorem 4.12 we have $W^{1,1}(\Omega) \to L^{n/(n-1)}(\Omega)$ and so

$$\begin{aligned}\|f\|_{n/(n-1)} &\le K_1\left(\sum_{j=1}^n \|D_jU\|_1 + \|f\|_1\right)\\ &= K_1\left[\sum_{j=1}^n \frac{1}{K}\int_\Omega \sigma'\left(\frac{|u(x)|}{K}\right)|D_ju(x)|\,dx + \int_\Omega \sigma\left(\frac{|u(x)|}{K}\right)dx\right].\end{aligned} \tag{24}$$

By (23) and Hölder's inequality (10), we obtain

$$1 = \left(\int_\Omega A_* \left(\frac{|u(x)|}{K} \right) dx \right)^{(n-1)/n} = \|f\|_{n/(n-1)}$$
$$\leq \frac{2K_1}{K} \sum_{j=1}^{n} \left\| \sigma' \left(\frac{|u|}{K} \right) \right\|_{\tilde{A}} \|D_j u\|_A + K_1 \int_\Omega \sigma' \left(\frac{|u(x)|}{K} \right) dx. \tag{25}$$

Making use of (22), we have

$$\left\| \sigma' \left(\frac{|u|}{K} \right) \right\|_{\tilde{A}} \leq \frac{n-1}{n} \left\| \tilde{A}^{-1} \left(\left(\sigma \left(\frac{|u|}{K} \right) \right)^{n/(n-1)} \right) \right\|_{\tilde{A}}$$
$$= \frac{n-1}{n} \inf \left\{ \lambda > 0 : \int_\Omega \tilde{A} \left(\frac{\tilde{A}^{-1}(A_*(|u(x)|/K))}{\lambda} \right) dx \leq 1 \right\}.$$

Suppose $\lambda > 1$. Then

$$\int_\Omega \tilde{A} \left(\frac{\tilde{A}^{-1}(A_*(|u(x)|/K))}{\lambda} \right) dx \leq \frac{1}{\lambda} \int_\Omega A_* \left(\frac{|u(x)|}{K} \right) dx = \frac{1}{\lambda} < 1.$$

Thus

$$\left\| \sigma' \left(\frac{|u|}{K} \right) \right\|_{\tilde{A}} \leq \frac{n-1}{n}. \tag{26}$$

Let $g(t) = A_*(t)/t$ and $h(t) = \sigma(t)/t$. It is readily checked that h is bounded on finite intervals and $\lim_{t\to\infty} g(t)/h(t) = \infty$. Thus there exists a constant t_0 such that $h(t) \leq g(t)/(2K)$ if $t \geq t_0$. Putting $K_2 = K_2 \sup_{0\leq t\leq t_0} h(t)$, we have, for all $t \geq 0$,

$$\sigma(t) \leq \frac{1}{2K_1} A_*(t) + \frac{K_2}{K_1} t.$$

Hence

$$K_1 \int_\Omega \sigma \left(\frac{|u(x)|}{K} \right) dx \leq \frac{1}{2} \int_\Omega A_* \left(\frac{|u(x)|}{K} \right) dx + \frac{K_2}{K_1} \int_\Omega |u(x)|\, dx$$
$$\leq \frac{1}{2} + \frac{K_3}{K} \|u\|_A, \tag{27}$$

where $K_3 = 2K_2 \|1\|_{\tilde{A}} < \infty$ since Ω has finite volume.
Combining (25)–(27), we obtain

$$1 \leq \frac{2K_1}{K}(n-1) \|u\|_{1,A} + \frac{1}{2} + \frac{K_3}{K} \|u\|_A,$$

so that

$$\|u\|_{A_*} = K \le K_4 \|u\|_{1,A}, \tag{28}$$

where K_4 depends only on n, A, $\mathrm{vol}(\Omega)$, and the cone determining the cone condition for Ω.

To extend (28) to arbitrary $u \in W^1 L_A(\Omega)$ let

$$u_k(x) = \begin{cases} |u(x)| & \text{if } |u(x)| \le k \\ k \operatorname{sgn} u(x) & \text{if } |u(x)| > k. \end{cases} \tag{29}$$

Clearly u_k is bounded and it belongs to $W^1 L_A(\Omega)$ by the previous lemma. Moreover, $\|u_k\|_{A_*}$ increases with k but is bounded by $K_4 \|u\|_A$. Therefore, $\lim_{k\to\infty} \|u_k\|_{A_*} = K$ exists and $K \le K_4 \|u\|_{1,A}$. By Fatou's lemma 1.49

$$\int_\Omega A_*\left(\frac{|u(x)|}{K}\right) dx \le \lim_{k\to\infty} \int_\Omega A_*\left(\frac{|u_k(x)|}{K}\right) dx \le 1$$

whence $u \in L_{A_*}(\Omega)$ and (28) holds.

Since Ω has finite volume we have

$$W^1 L_A(\Omega) \to W^{1,1}(\Omega) \to L^1(\Omega),$$

the latter imbedding being compact by Theorem 6.3. A bounded subset of $W^1 L_A(\Omega)$ is bounded in $L_{A_*}(\Omega)$ and precompact in $L^1(\Omega)$, and hence precompact in $L_B(\Omega)$ by Theorem 8.25 whenever B increases essentially more slowly than A_* near infinity. ∎

Theorem 8.35 extends to arbitrary (even unbounded) domains Ω provided W is replaced by W_0.

8.36 THEOREM Let Ω be an arbitrary domain in $\mathbb{R}^n$. If the N function A satisfies (18) and (19), then

$$W_0^m L_A(\Omega) \to L_{A_*}(\Omega).$$

Moreover, if Ω_0 is a bounded subdomain of Ω, then the imbedding

$$W_0^m L_A(\Omega) \to L_B(\Omega_0)$$

exists and is compact for any N-function B increasing essentially more slowly that A_* near infinity.

Proof. If $u \in W_0^m L_A(\Omega)$, then the function f in the proof of Theorem 8.35 can be approximated in $W^{1,1}(\Omega)$ by elements of $C_0^\infty(\Omega)$. By Sobolev's inequality

(Theorem 4.31), (24) holds with the term $\|f\|_1$ absent from the right side. Therefore (27) is not needed and the proof does not require that Ω have finite volume. The cone condition is not required either, since Sobolev's inequality holds for all $u \in C_0^\infty(\mathbb{R}^n)$. The compactness arguments are similar to those above. ∎

8.37 REMARK Theorem 8.35 is not optimal in the sense that for some A, L_{A_*} is not necessarily the smallest Orlicz space in which $W^1L_A(\Omega)$ can be imbedded. For instance, if $A(t) = A_n(t) = t^n/n$, then, as noted earlier, $A_*(t)$ is equivalent near infinity to $e^t - t - 1$, an N-function that increases essentially more slowly near infinity than does $\exp\left(t^{n/(n-1)}\right) - 1$. Thus Theorem 8.27 gives a sharper result than Theorem 8.35 in this case. In [DT] Donaldson and Trudinger state that Theorem 8.35 can be improved by the methods of Theorem 8.27 provided A dominates near infinity every t^p with $p < n$, but that Theorem 8.35 gives optimal results if for some $p < n$, t^p dominates A near infinity. The former cases are those where [Ci] improves on Theorem 8.35.

There are also some unbounded domains [Ch] for which some Orlicz-Sobolev imbeddings are compact.

The following theorem generalizes (the case $m = 1$ of) the part of Theorem 4.12 dealing with traces on lower dimensional hyperplanes.

8.38 THEOREM (Traces on Planes) Let Ω be a bounded domain in $\mathbb{R}^n$ satisfying the cone condition, and let Ω_k denote the intersection of Ω with a k-dimensional plane in $\mathbb{R}^n$. Let A be an N-function for which (18) and (19) hold, and let A_* be given by (20). Let $1 \le p < n$ where p is such that the function B defined by $B(t) = A(t^{1/p})$ is an N-function. If either $n - p < k \le n$ or $p = 1$ and $n - 1 \le k \le n$, then

$$W^1L_A(\Omega) \to L_{A_*^{k/n}}(\Omega_k),$$

where $A_*^{k/n}(t) = [A_*(t)]^{k/n}$.

Moreover, if $p > 1$ and C is an N-function increasing essentially more slowly than $A_*^{k/n}$ near infinity, then the imbedding

$$W^1L_A(\Omega) \to L_C(\Omega_k) \tag{30}$$

is compact.

Proof. The problem of verifying that $A_*^{k/n}$ is an N-function is left to the reader. Let $u \in W^1L_A(\Omega)$ be a bounded function. Then

$$\int_{\Omega_k} A_*^{k/n}\left(\frac{|u(y)|}{K}\right) dy = 1, \qquad K = \|u\|_{A_*^{k/n},\Omega_k}. \tag{31}$$

We wish to show that

$$K \le K_1 \|u\|_{1,A,\Omega} \tag{32}$$

with K_1 independent of u. Since this inequality is known to hold for the special case $k = n$ (Theorem 8.35) we may assume without loss of generality that

$$K \ge \|u\|_{A_*,\Omega} = \|u\|_{A_*^{n/n},\Omega_n}. \tag{33}$$

Let $\omega(t) = [A_*(t)]^{1/q}$ where $q = np/(n-p)$. By (case $m = 1$ of) Theorem 4.12 we have

$$\begin{aligned}
\left\|\omega\left(\frac{|u|}{K}\right)\right\|^p_{kp/(n-p),\Omega_k} &\le K_2\left[\sum_{j=1}^n \left\|D_j\omega\left(\frac{|u|}{K}\right)\right\|^p_{p,\Omega} + \left\|\omega\left(\frac{|u|}{K}\right)\right\|^p_{p,\Omega}\right] \\
&= K_2\left[\frac{1}{K^p}\sum_{j=1}^n \int_\Omega \left|\omega'\left(\frac{|u(x)|}{K}\right)\right|^p |D_j u(x)|^p\,dx \right. \\
&\qquad \left. + \int_\Omega \left|\omega\left(\frac{|u(x)|}{K}\right)\right|^p dx\right].
\end{aligned}$$

Using (31) and noting that $\||v|^p\|_{B,\Omega} \le \|v\|^p_{A,\Omega}$, we obtain

$$\begin{aligned}
1 = \left[\int_{\Omega_k}\left(A_*\left(\frac{|u(y)|}{K}\right)\right)^{k/n} dy\right]^{(n-p)/k} &= \left\|\omega\left(\frac{|u|}{K}\right)\right\|^p_{kp/(n-p),\Omega_k} \\
\le \frac{2K_2}{K^p}\sum_{j=1}^n \left\|\left(\omega'\left(\frac{|u|}{K}\right)\right)^p\right\|_{\tilde{B},\Omega} \||D_j u|^p\|_{B,\Omega} &+ K_2\left\|\omega\left(\frac{|u|}{K}\right)\right\|^p_{p,\Omega} \\
\le \frac{2nK_2}{K^p}\left\|\left(\omega'\left(\frac{|u|}{K}\right)\right)^p\right\|_{\tilde{B},\Omega} \|u\|^p_{1,A,\Omega} &+ K_2\left\|\omega\left(\frac{|u|}{K}\right)\right\|^p_{p,\Omega}.
\end{aligned} \tag{34}$$

Now $B^{-1}(t) = \left(A^{-1}(t)\right)^p$ and so, using (21) and (4), we have

$$\begin{aligned}
\left(\omega'(t)\right)^p &= \frac{1}{q^p}\left(A_*(t)\right)^{p(1-q)/q}\left(A'_*(t)\right)^p \\
&= \frac{1}{q^p}A_*(t)\frac{1}{B^{-1}\left(A_*(t)\right)} \le \frac{1}{q^p}\tilde{B}^{-1}\left(A_*(t)\right).
\end{aligned}$$

It follows by (33) that

$$\int_\Omega \tilde{B}\left(\left(\frac{\omega'(|u(x)|/K)}{1/q}\right)^p\right) dx \le \int_\Omega A_*\left(\frac{|u(x)|}{K}\right) dx \le 1.$$

So

$$\left\| \left(\omega' \left(\frac{|u|}{K} \right) \right)^p \right\|_{\tilde{B},\Omega} \le \frac{1}{q^p}. \tag{35}$$

Now set $g(t) = A_*(t)/t^p$ and $h(t) = (\omega(t)/t)^p$. It is readily checked that $\lim_{t\to\infty} g(t)/h(t) = \infty$. In order to see that $h(t)$ is bounded near zero let $s = A_*(t)$ and consider

$$(h(t))^{1/p} = \frac{(A_*(t))^{1/q}}{t} = \frac{s^{(1/p)-(1/n)}}{\displaystyle\int_0^s \frac{A^{-1}(\tau)}{\tau^{(n+1)/n}}\,d\tau} \le \frac{s^{1/p}}{\displaystyle\int_0^s \frac{(B^{-1}(\tau))^{1/p}}{\tau}\,d\tau}.$$

Since B is an N-function $\lim_{\tau\to\infty} B^{-1}(\tau)/\tau = \infty$. Hence, for sufficiently small values of t we have

$$(h(t))^{1/p} \le \frac{s^{1/p}}{\displaystyle\int_0^s \tau^{-1+(1/p)}\,d\tau} = \frac{1}{p}.$$

Therefore, there exists a constant K_3 such that for $t \ge 0$

$$(\omega(t))^p \le \frac{1}{2K_2} A_*(t) + K_3 t^p.$$

Using (33) we now obtain

$$\begin{aligned}
\left\| \omega\left(\frac{|u|}{K}\right) \right\|_{p,\Omega}^p &\le \frac{1}{2K_2} \int_\Omega A_*\left(\frac{|u(x)|}{K}\right) dx + \frac{K_3}{K^p} \int_\Omega |u(x)|^p\, dx \\
&\le \frac{1}{2K_2} + \frac{2K_3}{K^p} \left\| |u|^p \right\|_{B,\Omega} \|1\|_{\tilde{B},\Omega} \\
&\le \frac{1}{2K_2} + \frac{K_4}{K^p} \|u\|_{A,\Omega}^p .
\end{aligned} \tag{36}$$

From (34)–(36) there follows the inequality

$$1 \le \frac{2nK_2}{K^p} \cdot \frac{1}{q^p} \|u\|_{1,A,\Omega}^p + \frac{1}{2} + \frac{K_4 K_2}{K^p} \|u\|_{A,\Omega}^p$$

and hence (32). The extension of (32) to arbitrary $u \in W^1 L_A(\Omega)$ now follows as in the proof of Theorem 8.35.

Since $B(t) = A(t^{1/p})$ is an N-function and Ω is bounded, we have

$$W^1 L_A(\Omega) \to W^{1,p}(\Omega) \to L^1(\Omega_k),$$

the latter imbedding being compact by Theorem 6.3 provided $p > 1$. The compactness of (30) now follows from Theorem 8.25. ∎

8.39 THEOREM (Imbedding Into a Space of Continuous Functions) Let Ω satisfy the cone condition in $\mathbb{R}^n$. Let A be an N-function for which

$$\int_1^\infty \frac{A^{-1}(t)}{t^{(n+1)/n}}\,dt < \infty. \tag{37}$$

Then

$$W^1L_A(\Omega) \to C_B^0(\Omega) = C(\Omega) \cap L^\infty(\Omega).$$

Proof. Let C be a finite cone contained in Ω. We shall show that there exists a constant K_1 depending on n, A, and the dimensions of C such that

$$\|u\|_{\infty,C} \le K_1 \|u\|_{1,A,C}. \tag{38}$$

In doing so, we may assume without loss of generality that A satisfies (18), for if not, and if B is an N-function satisfying (18) and equivalent to A near infinity, then $W^1L_A(C) \to W^1L_B(C)$ with imbedding constant depending on A, B, and $\mathrm{vol}(C)$ by Theorem 8.12. Since B satisfies (37) we would have

$$\|u\|_{\infty,C} \le K_2 \|u\|_{1,B,C} \le K_3 \|u\|_{1,A,C}.$$

Now Ω is a union of congruent copies of some such finite cone C so that (38) clearly implies

$$\|u\|_{\infty,\Omega} \le K_1 \|u\|_{1,A,\Omega}. \tag{39}$$

Since A is assumed to satisfy (18) and (37) we have

$$\int_0^\infty \frac{A^{-1}(t)}{t^{(n+1)/n}}\,dt = K_4 < \infty.$$

Let

$$\Lambda^{-1}(t) = \int_0^t \frac{A^{-1}(\tau)}{\tau^{(n+1)/n}}\,d\tau.$$

The Λ^{-1} maps $[0, \infty)$ in a one-to-one manner onto $[0, K_4)$ and has a convex inverse Λ. We extend the domain of definition of Λ to $[0, \infty)$ by defining $\Lambda(t) = \infty$ for $t \ge K_4$. The function Λ is a *Young's function.* (See Luxemburg [Lu] or O'Neill [O].) Although it is not an N-function in the sense defined early in this chapter, nevertheless the Luxemburg norm

$$\|u\|_{\Lambda,C} = \inf\left\{k > 0 : \int_C \Lambda\left(\frac{|u(x)|}{k}\right) dx \le 1\right\}$$

is easily seen to be a norm on $L^\infty(C)$ equivalent to the usual norm; in fact,

$$\frac{1}{K_4}\|u\|_{\infty,C} \le \|u\|_{\Lambda,C} \le \frac{1}{\Lambda^{-1}(1/\mathrm{vol}(C))}\|u\|_{\infty,C}\,.$$

Moreover, $s = \Lambda(t)$ satisfies the differential equation (21), so that the proof of Theorem 8.35 can be carried over in this case to yield, for $u \in W^1L_A(C)$,

$$\|u\|_{\Lambda,C} \le K_5\,\|u\|_{1,A,C}$$

and inequality (38) follows.

By Theorem 8.31(d) an element $u \in W^mE_A(\Omega)$ can be approximated in norm by functions continuous on Ω. It follows from (39) that u must coincide a.e. on Ω with a continuous function. (See Paragraph 4.16.)

Suppose that an N-function B can be constructed such that the following conditions are satisfied:

(a) $B(t) = A(t)$ near zero.

(b) B increases essentially more slowly than A near infinity.

(c) B satisfies

$$\int_1^\infty \frac{B^{-1}(t)}{t^{(n+1)/n}}\,dt \le 2\int_1^\infty \frac{A^{-1}(t)}{t^{(n+1)/n}}\,dt < \infty.$$

Then, by Theorem 8.16, $u \in W^1L_A(C)$ implies $u \in W^1E_B(C)$ so that we have $W^1L_A(\Omega) \subset C(\Omega)$ as required.

It remains, therefore, to construct an N-function B having the properties (a)–(c). Let $1 < t_1 < t_2 < \cdots$ be such that

$$\int_{t_k}^\infty \frac{A^{-1}(t)}{t^{(n+1)/n}}\,dt = \frac{1}{2^{2k}}\int_1^\infty \frac{A^{-1}(t)}{t^{(n+1)/n}}\,dt.$$

We define a sequence $\{s_k\}$ with $s_k \ge t_k$, and the function $B^{-1}(t)$, inductively as follows.

Let $s_1 = t_1$ and $B^{-1}(t) = A^{-1}(t)$ for $0 \le t \le s_1$. Having chosen $s_1, s_2, \ldots, s_k$ and defined $B^{-1}(t)$ for $0 \le t \le s_{k-1}$, we continue $B^{-1}(t)$ to the right of s_{k-1} along a straight line with slope $(A^{-1})'(s_{k-1}-)$ (which always exists since A^{-1} is concave) until a point t_k' is reached where $B^{-1}(t_k') = 2^{k-1}A^{-1}(t_k')$. Such t_k' exists because $\lim_{t\to\infty} A^{-1}(t)/t = 0$. If $t_k' \ge t_k$, let $s_k = t_k'$. Otherwise let $s_k = t_k$ and extend B^{-1} from t_k' to s_k by setting $B^{-1}(t) = 2^{k-1}A^{-1}(t)$. Evidently B^{-1} is concave and B is an N-function. Moreover, $B(t) = A(t)$ near zero and since

$$\lim_{t\to\infty}\frac{B^{-1}(t)}{A^{-1}(t)} = \infty,$$

B increases essentially more slowly than A near infinity. Finally,

$$\int_1^\infty \frac{B^{-1}(t)}{t^{(n+1)/n}}\,dt \le \int_1^{s_1} \frac{A^{-1}(t)}{t^{(n+1)/n}}\,dt + \sum_{k=2}^{\infty} \int_{s_{k-1}}^{s_k} \frac{2^{k-1}A^{-1}(t)}{t^{(n+1)/n}}\,dt$$
$$\le \int_1^{s_1} \frac{A^{-1}(t)}{t^{(n+1)/n}}\,dt + \sum_{k=2}^{\infty} 2^{k-1} \int_{t_{k-1}}^{\infty} \frac{A^{-1}(t)}{t^{(n+1)/n}}\,dt$$
$$= 2\int_1^\infty \frac{A^{-1}(t)}{t^{(n+1)/n}}\,dt,$$

as required. ∎

8.40 THEOREM (Uniform Continuity) Let Ω be a domain in $\mathbb{R}^n$ satisfying the strong local Lipschitz condition. If the N-function A satisfies

$$\int_1^\infty \frac{A^{-1}(t)}{t^{(n+1)/n}}\,dt < \infty, \tag{40}$$

then there exists a constant K such that for any $u \in W^1L_A(\Omega)$ (which may be assumed continuous by the previous theorem) and all $x, y \in \Omega$ we have

$$|u(x) - u(y)| \le K\,\|u\|_{1,A,\Omega} \int_{|x-y|^{-n}}^\infty \frac{A^{-1}(t)}{t^{(n+1)/n}}\,dt. \tag{41}$$

Proof. We establish (41) for the case where Ω is a cube of unit edge; the extension to more general strongly Lipschitz domains can then be carried out just as in the proof of Lemma 4.28. As in that lemma we let Ω_σ denote a parallel subcube of Ω having edge σ and obtain for $x \in \overline{\Omega_\sigma}$

$$\left| u(x) - \frac{1}{\sigma^n}\int_{\Omega_\sigma} u(z)\,dz \right| \le \frac{\sqrt{n}}{\sigma^{n-1}} \int_0^1 t^{-n}\,dt \int_{\Omega_{t\sigma}} |\operatorname{grad} u|\,dz.$$

By (15), $\|1\|_{\tilde{A},\Omega_{t\sigma}} = 1/\tilde{A}^{-1}(t^{-n}\sigma^{-n})$. It follows by Hölder's inequality and (4) that

$$\int_{\Omega_{t\sigma}} |\operatorname{grad} u|\,dz \le 2\,\|\operatorname{grad} u\|_{A,\Omega_{t\sigma}}\,\|1\|_{\tilde{A},\Omega_{t\sigma}}$$
$$\le \frac{2}{\tilde{A}^{-1}(t^{-n}\sigma^{-n})}\,\|u\|_{1,A,\Omega}$$
$$\le 2\sigma^n t^n A^{-1}(t^{-n}\sigma^{-n})\,\|u\|_{1,A,\Omega}.$$

Hence

$$\left| u(x) - \frac{1}{\sigma^n}\int_{\Omega_\sigma} u(z)\,dz \right| \le 2\sqrt{n}\sigma\,\|u\|_{1,A,\Omega} \int_0^1 A^{-1}\left(\frac{1}{t^n\sigma^n}\right) dt$$
$$= \frac{2}{\sqrt{n}}\,\|u\|_{1,A,\Omega} \int_{\sigma^{-n}}^\infty \frac{A^{-1}(\tau)}{\tau^{(n+1)/n}}\,d\tau.$$

If $x, y \in \Omega$ and $\sigma = |x - y| < 1$, then there exists a subcube Ω_σ with $x, y \in \overline{\Omega_\sigma}$, and it follows from the above inequality applied to both x and y that

$$|u(x) - u(y)| \leq \frac{4}{\sqrt{n}} \|u\|_{1,A,\Omega} \int_{|x-y|^{-n}}^{\infty} \frac{A^{-1}(t)}{t^{(n+1)/n}}\, dt.$$

For $|x - y| \geq 1$, (41) follows directly from (39) and (40). ∎

8.41 (Generalization of Hölder Continuity) Let M denote the class of positive, continuous, increasing functions of $t > 0$. If $\mu \in M$, the space $C_\mu(\overline{\Omega})$, consisting of those functions $u \in C(\overline{\Omega})$ for which the norm

$$\left\| u\,; C_\mu(\overline{\Omega}) \right\| = \left\| u\,; C(\overline{\Omega}) \right\| + \sup_{\substack{x,y\in\Omega \\ x\neq y}} \frac{|u(x) - u(y)|}{\mu(|x - y|)}$$

is finite, is a Banach space under that norm. The theorem above asserts that if (40) holds, then

$$W^1 L_A(\Omega) \to C_\mu(\overline{\Omega}), \qquad \text{where} \quad \mu(t) = \int_{|x-y|^{-n}}^{\infty} \frac{A^{-1}(t)}{t^{(n+1)/n}}\, dt. \tag{42}$$

If $\mu, \nu \in M$ are such that $\mu/\nu \in M$, then for bounded Ω we have, as in Theorem 1.34, that the imbedding

$$C_\mu(\overline{\Omega}) \to C_\nu(\overline{\Omega})$$

exists and is compact. Hence the imbedding

$$W^1 L_A(\Omega) \to C_\nu(\overline{\Omega})$$

is also compact if μ is given as in (42).

8.42 (Generalization to Higher Orders of Smoothness) We now prepare to state the general Orlicz-Sobolev imbedding theorem of Donaldson and Trudinger [DT] by generalizing the framework used for imbeddings of $W^1 L_A(\Omega)$ considered above so that we can formulate imbeddings of $W^m L_A(\Omega)$.

For a given N-function A we define a sequence of N-functions $B_0, B_1, B_2, \ldots$ as follows:

$$B_0(t) = A(t)$$

$$(B_k)^{-1}(t) = \int_0^t \frac{(B_{k-1})^{-1}(\tau)}{\tau^{(n+1)/n}}\, d\tau, \qquad k = 1, 2, \ldots.$$

(Observe that $B_1 = A_*$.) At each stage we assume that

$$\int_0^1 \frac{(B_k)^{-1}(t)}{t^{(n+1)/n}}\, dt < \infty, \tag{43}$$

replacing B_k, if necessary, with another N-function equivalent to it near infinity and satisfying (43).

Let $J = J(A)$ be the smallest nonnegative integer such that

$$\int_1^\infty \frac{(B_J)^{-1}(t)}{t^{(n+1)/n}}\, dt < \infty.$$

Evidently, $J(A) \le n$. If μ belongs to the class M defined in the previous Paragraph, we define the space $C^m_\mu(\overline{\Omega})$ to consist of those functions $u \in C(\overline{\Omega})$ for which $D^\alpha u \in C_\mu(\overline{\Omega})$ whenever $|\alpha| \le m$. The space $C^m_\mu(\overline{\Omega})$ is a Banach space with respect to the norm

$$\left\| u\,;\, C^m_\mu(\overline{\Omega})\right\| = \max_{|\alpha|\le m} \left\| D^\alpha u\,;\, C_\mu(\overline{\Omega})\right\|.$$

8.43 THEOREM (A General Orlicz-Sobolev Imbedding Theorem) Let Ω be a bounded domain in $\mathbb{R}^n$ satisfying the cone condition. Let A be an N-function.

(a) If $m \le J(A)$, then $W^m L_A(\Omega) \to L_{B_m}(\Omega)$. Moreover, if B is an N-function increasing essentially more slowly than B_m near infinity, then the imbedding $W^m L_A(\Omega) \to L_B(\Omega)$ exists and is compact.

(b) If $m > J(A)$, then $W^m L_A(\Omega) \to C^0_B(\Omega) = C^0(\Omega) \cap L^\infty(\Omega)$.

(c) If $m > J(A)$ and Ω satisfies the strong local Lipschitz condition, then $W^m L_A(\Omega) \to C^{m-J-1}_\mu(\overline{\Omega})$ where

$$\mu(t) = \int_{t^{-n}}^\infty \frac{(B_J)^{-1}(\tau)}{\tau^{(n+1)/n}}\, d\tau.$$

Moreover, the imbedding $W^m L_A(\Omega) \to C^{m-J-1}(\overline{\Omega})$ is compact and so is $W^m L_A(\Omega) \to C^{m-J-1}_\nu(\overline{\Omega})$ provided $\nu \in M$ and $\mu/\nu \in M$.

8.44 REMARK Theorem 8.43 follows in a straightforward way from the special cases with $m = 1$ provided earlier. Also, if we replace L_A by E_A in part (a) we get $W^m E_A(\Omega) \to E_{B_m}(\Omega)$ since the sequence $\{u_k\}$ defined by (29) converges to u if $u \in W^1 E_A(\Omega)$. Theorem 8.43 holds without any restrictions on Ω if $W^m L_A(\Omega)$ is replaced with $W^m_0 L_A(\Omega)$.

8.45 REMARK Since Theorem 8.43 implies that $W^m L_A(\Omega) \to W^1 L_{B_{m-1}}(\Omega)$, we will also have $W^m L_A(\Omega) \to L_{[(B_m)^{k/n}]}(\Omega_k)$, where Ω_k is the intersection of Ω with a k-dimensional plane in $\mathbb{R}^n$, provided that (using Theorem 8.38) there exists p satisfying $1 \le p < n$ for which $n - p < k \le n$ (or $n - 1 \le k \le n$ if $p = 1$) and $B(t) = B_m(t^{1/p})$ is an N-function.

REFERENCES

[Ad1] Adams, David R. (1988) A sharp inequality of J. Moser for higher order derivatives. *Ann. of Math.* **(2) 128**, 385–398.

[Ad2] Adams, David R. and Hedberg, Lars Inge. (1996) *Function Spaces and Potential Theory.* Springer-Verlag, Berlin.

[A] Adams, Robert A. (1975) *Sobolev Spaces.* Academic Press, New York.

[A1] Adams, R. A. (1973) Some integral inequalities with applications to the imbedding of Sobolev spaces over irregular domains. *Trans. Amer. Math. Soc.* **178**, 401–429.

[A2] Adams, R. A. (1970) Capacity and compact imbeddings. *J. Math. Mech.* **19**, 923–929.

[A3] Adams, R. A. (1988) Anisotropic Sobolev Inequalities. *Časopis Pro Pěstování Matematiky.* **113**, 267–279.

[A4] Adams, R. A. (1988) Reduced Sobolev Inequalities. *Canad. Math. Bull.* **31**, 159–167.

[AF1] Adams, R. A. and Fournier, John. (1977) Cone conditions and properties of Sobolev spaces, *J. Math. Anal. Appl.* **61**, 713–734.

[AF2] Adams, R. A. and Fournier, John. (1971) Compact imbedding theorems for functions without compact support, *Canad. Math. Bull.* **14**, 305–309.

[AF3] Adams, R. A. and Fournier, John. (1971) Some imbedding theorems for Sobolev spaces, *Canad. J. Math.* **23**, 517–530.

[AF4] Adams, R. A. and Fournier, John. (1978) The real interpolation of Sobolev spaces on subdomains of $\mathbb{R}^n$, *Canad. J. Math.* **30**, 190–214.

[Ag] Agmon, S. (1965) *Lectures on Elliptic Boundary Value Problems.* Van Nostrand-Reinhold, Princeton, New Jersey.

[AMS] Aronszajn, N., Mulla, F., and Szeptycki, P. (1963) On spaces of potentials connected with L^p classes, *Ann. Inst. Fourier (Grenoble).* **13**, 211–306.

[Au] Aubin, Thierry. (1998) *Some Nonlinear Problems in Riemannian Geometry.* Springer-Verlag, Berlin.

[B+] Bakry, D., Coulhon, T., Ledoux, M., Saloff-Coste, L. (1995) Sobolev inequalities in disguise, *Indiana Univ. Math. J.* **44**, 1033–1074.

[BKC] Ball, Keith, Carlen, Eric A., Lieb, Elliott H. (1994) Sharp uniform convexity and smoothness inequalities for trace norms, *Invent. Math.* **115**, 463–482.

[BP] Benedek, A. and Panzone, R. (1961) The spaces L^p with mixed norm *Duke Math. J.* **28**, 301–324.

[BSh] Bennett, Colin and Sharpley, Robert. (1988) *Interpolation of Operators.* Academic Press, Boston.

[BSc] Berger, Melvyn S. and Schechter, Martin. (1972) Embedding theorems and quasi-linear elliptic boundary value problems for unbounded domains, *Trans. Amer. Math. Soc.* **172**, 261–278.

[BL] Bergh, Jöran and Löfström, Jörgen. (1976) *Interpolation Spaces, an Introduction.* Springer-Verlag, Berlin.

[BIN1] Besov, O. V., Il'in, V. P., Nikol'skiĭ, S. M. (1978) *Integral representations of functions and imbedding theorems, Vol. I.* John Wiley & Sons, New York.

[BIN2] Besov, O. V., Il'in, V. P., Nikol'skiĭ, S. M. (1979) *Integral representations of functions and imbedding theorems, Vol. II.* John Wiley & Sons, New York.

[BW] Brèzis, Haïm, Wainger, Stephen. (1980) A note on limiting cases of Sobolev imbeddings and convolution inequalities, *Comm. Partial Differential Equations.* **5**, 773–789.

[Br1] Browder, F. E. (1953) On the eigenfunctions and eigenvalues of general elliptic differential operators. *Proc. Nat. Acad. Sci. USA.* **39**, 433–439.

[Br1] Browder, F. E. (1961) On the spectral theory of elliptic differential operators I. *Math. Ann.* **142**, 22–130.

[Bu2] Burenkov, Victor. (1999) Extension theorems for Sobolev spaces. The Maz´ya Anniversary Collection, Vol 1. *Oper. Theory Adv. Appl.* **109**, Birkhäuser, Basel, 187–200.

[Bu1] Burenkov, Victor. (1998) Sobolev Spaces on Domains. *Teubner-Texte zur Mathematik.* **137**, Stuttgart.

[BB] Butzer, Paul L. and Berens, Hubert. (1967) *Semi-Groups of Operators and Approximation.* Springer-Verlag, New York.

[Ch] Cahill, Ian Graham. (1975) *Compactness of Orlicz-Sobolev space imbeddings for unbounded domains.*, (M. Sc. thesis), University of British Columbia .

[Ca1] Calderón, A. P. (1961) Lebesgue spaces of differentiable functions and distributions, *Proc. Sym. Pure Math..* **4**, 33-49.

[Ca2] Calderón, A. P. (1964) Intermediate spaces and interpolation, the complex method, *Studia Math..* **24**, 113-190.

[CZ] Calderón, A. P. and Zygmund, A. (1952) On the existence of certain singular integrals, *Acta Math.* **88**, 85–139.

[Ci] Cianchi, Andrea. (1996) A sharp embedding theorem for Orlicz-Sobolev spaces, *Indiana Univ. Math. J.* **45**, 39–65.

[Ck] Clark, C. W. (1966) The Hilbert-Schmidt property for embedding maps between Sobolev spaces, *Canad. J. Math.* **18**, 1079–1084.

[Clk] Clarkson, J. A. (1936) Uniformly convex spaces, *Trans. Amer. Math. Soc.* **40**, 396–414.

[Cw] Cwikel, Michael. (1978) Complex interpolation spaces, a discrete definition and reiteration, *Indiana Univ. Math. J.* **27**, 1005–1009.

[Da] Dankert, G. (1966) Sobolev imbedding theorems in Orlicz spaces. (Thesis). Univ. of Cologne.

[Db] Daubechies, Ingrid. (1992) *Ten Lectures on Wavelets.* (CBMS-NSF Regional Conference Series in Applied Mathematics, **61**,, SIAM, Philadelphia).

[DL] Deny, J. and Lions, J. L. (1955) Les espace du type de Beppo Levi. *Ann. Inst. Fourier (Grenoble).* **5**, 305–370.

[DaL] Dautray, R. and Lions, J. L. (1990) *Mathematical Analysis and Numerical Methods for Science and Technology, Vol. 3, Spectral Theory and Applications.*, Springer-Verlag, Berlin..

[DT] Donaldson, T. K. and Trudinger, N. S. (1971) Orlics-Sobolev spaces and imbedding theorems. *J. Funct. Anal.* **8**, 52–75.

[EE] Edmunds, D. E. and Evans, W. D. (1987) *Spectral Theory and Differential Operators.* (Oxford University Press, New York).

[ET] Edmunds, D. E. and Triebel, H. (1996) *Function Spaces, Entropy Numbers, Differential Operators.* (Cambridge Tracts in Mathematics) **120**,, Cambridge University Press.

[E] Ehrling, E. (1954) On a type of eigenvalue problem for certain elliptic differential operators. *Math. Scand.* **2**, 267–285.

[Fa] Faris, William G. (1976) Weak Lebesgue spaces and quantum mechanical binding, *Duke Math. J.* **43**, 365–373.

[FG] Figà-Talamanca, Alessandro and Gaudry, Garth I. (1970) Multipliers and sets of uniqueness of L^p, *Michigan Math. J.* **17**, 179–191.

[Fo] Folland, Gerald B. (1999) *Real Analysis, 2nd Edition.* (John Wiley & Sons, New York).

[F] Fournier, John J. F. (1987) Mixed norms and rearrangements: Sobolev's inequality and Littlewood's inequality, *Ann. Mat. Pura Appl.* **(4) 148**, 51–76.

[FJ] Frazier, M. and Jawerth, B. (1990) A discrete transform and decompositions of distribution spaces, *J. Funct. Anal.* **93**, 34–170.

[FJW] Frazier, M., Jawerth, B., and Weiss, G. (1991) Littlewood-Paley theory and the study of function spaces, *CBMS Regional Conference Series.* **79**, Amer. Math. Soc., Providence.

[Ga1] Gagliardo, E. (1958) Properità di alcune classi di funzioni in piu variabili, *Ricerche Mat.* **7**, 102–137.

[Ga2] Gagliardo, E. (1959) Ulteriori properità di alcune classi di funzioni in piu variabili, *Ricerche Mat.* **8**, 24–51.

[Ga3] Gagliardo, E. (1957) Caratterizzazioni della tracce sulla frontiera relative ad alcune classi di funzioni in n variabili, *Rend. Sem. Mat. Univ. Padova.* **27**, 284–305.

[Gr] Grisvard, P. (1966) Commutativité de deux foncteurs d'interpolation et applications, *J. Math. Pures Appl.* **45**, 143–290.

[HK] Hajłasz, Piotr and Koskela, Pekka. (2000) Sobolev met Poincaré, (Mem. Amer. Math. Soc) **145**, Providence.

[Hb] Hebey, Emmanuel. (1996) Sobolev Spaces on Riemannian Manifolds, Lecture Notes in Mathematics (Springer-Verlag) Berlin.

[Hn] Heinonen, Juha. (2001) Lectures on Analysis on Metric Spaces, (Springer-Verlag) New York.

[HMT] Hempel, J. A., Morris, G. R., and Trudinger, N. S. (1970) On the sharpness of a limiting case of the Sobolev imbedding theorem, *Bull. Austral. Math. Soc.* **3**, 369–373.

[He] Hestenes, M. (1941) Extension of the range of a differentiable function, *Duke J. Math.* **8**, 183–192.

[J] Johnson, Raymond. (1985) Review of *Theory of function spaces* by Hans Triebel, *Bull. Amer. Math. Soc.* **13**, 76–80 [Tr3].

[Jn] Jones, Peter W. (1981) Quasiconformal mappings and extendability of functions in Sobolev spaces, *Acta Math.* **147**, 71–88.

[Ju] Judovič, V. I. (1961) Some estimates connected with integral operators and with solutions of elliptic equations, *Dokl. Akad. Nauk SSSR.* **138**, 805–808.

[K] Kondrachov, V. I. (1945) Certain properties of functions in the spaces L^p, *Dokl. Akad. Nauk SSSR.* **48**, 535–538.

[Ko] Koskela, P. (1998) Extensions and Imbeddings, *J. Functional Anal.*. **159**, 369–383.

[KR] Krasnosel'skii , M. A. and Rutickii, Ya. B. (1961) *Convex Functions and Orlicz Spaces.* (Noordhoff, Groningen, The Netherlands).

[Ku] Kufner, Alois. (1980) *Weighted Sobolev Spaces.* (Teubner Verlagsgesellschaft, Leipzig) [english transl. (John Wiley & Sons) 1985].

[LL] Lieb, Elliott H. and Loss, Michael. (1997) *Analysis.* Graduate Studies in Mathematics **14**, Amer. Math. Society.

[Lj] Lions, J. L. (1965) Problèmes aux Limites dans les Équations aux Derivées Partielles, (Seminar Notes), Univ. of Montreal Press.

[Lp] Lions, P. L. (1982) Symétrie et compacité dans les espaces de Sobolev, *j> Funct. Anal.* **49**, 315–334.

[Lu] Luxemburg, W. (1955) Banach Function Spaces, (Thesis), Technische Hogeschool te Delft, The Netherlands.

[MP] Malý, Jan and Pick, Luboš. (2001) An elementary proof of sharp Sobolev embeddings, *Proc. Amer. Math. Soc.* **130**, 555–563.

[Mk] Marcinkiewicz, J. (1939) Sur l'interpolation d'opérateurs, *C. R. Acad. Sci. Paris.* **208**, 1272–1273.

[Mr] Maurin, K. (1961) Abbildungen vom Hilbert-Schmidtschen Typus und ihre Anwendungen, *Math. Scand.* **9**, 187–215.

[Mz1] Maz´ya, Vladimir G. (1985) *Sobolev Spaces.* (Springer-Verlag, Berlin).

[Mz2] Maz´ya, V. G. and Shaposhnikova, T. O. (1985) *Theory of Multipliers in Spaces of Differentiable Fundtions.* (Pitman, Boston).

[MS] Meyers, N. and Serrin, J. (1964) $H = W$, *Proc. Nat. Acad. Sci. USA.* **51**, 1055-1056.

[Mo] Morrey, C. B. (1940) Functions of several variables and absolute continuity, II, *Duke J. Math.* **6**, 187–215.

[Ms] Moser, J. (1970/71) A sharp form of an inequality by N. Trudinger, *Indiana Univ. Math. J.* **20**, 1077-1092.

[Nr1] Nirenberg, L. (1955) Remarks on strongly elliptic partial differential equations. *Comm. Pure Appl. Math.* **8**, 649–675.

[Nr2] Nirenberg, L. (1966) An extended interpolation inequality. *Ann. Scuola Norm. Sup. Pisa.* **20**, 733–737.

[O] O'Neill, R. (1965) Fractional integration in Orlicz spaces. *Trans. Amer. Math. Soc.* **115**, 300-328.

[P] Peetre, J. (1966) Espaces d'interpolation et théorème de Soboleff, *Ann. Inst. Fourier (Grenoble).* **16**, 279–317.

[Pz] Pohožaev, S. I. (1965) On the eigenfunctions of the equation $\Delta u + \lambda f(u) = 0$, *Dokl. Akad. Nauk SSSR.* **165**, 36–39.

[Po] Poornima, S. (1983) An embedding theorem for the Sobolev space $W^{1,1}$, *Bull. Sci. Math.* **(2) 107**, 253–259.

[RS] Reed, Michael and Simon, Barry. (1972) *Methods of Moderm Mathematical Physics. I. Functional Analysis.* (Academic Press, New York - London).

[Re] Rellich, F. (1930) Ein Satz über mittlere Konvergenz. *Göttingen Nachr..* 30–35.

[Ro] Royden, H. (1988) *Real Analysis, 3rd Edition.* (Macmillan, New York).

[Ru1] Rudin, W. (1973) *Functional Analysis.* (McGraw Hill, New York).

[Ru2] Rudin, W. (1987) *Real and Complex Analysis, 3rd Edition.* (McGraw-Hill, New York).

[Ry] Rychkov, Vyacheslav S. (1999) On restrictions and extensions of the Besov and Triebel-Lizorkin spaces with respect to Lipschitz domains, *J. London Math. Soc.* **(2) 60**, 237–257.

[Sx] Saxe, Karen. (2002) *Beginning Functional Analysis.* (Springer-Verlag, New York).

[Sch] Schwartz, L. (1966) *Théorie des Distributions.* (Hermann, Paris).

[Se] Seeley, R. T. (1964) Extensions of C^∞-functions defined in a half space, *Proc. Amer. Math. Soc.* **15**, 625–626.

[So1] Sobolev, S. L. (1938) On a theorem of functional analysis, *Mat. Sb.* **46**, 471–496.

[So2] Sobolev, S. L. (1988) *Some Applications of Functional Analysis in Mathematical Physics.* Moscow [English transl.: (Amer. Math. Soc. Transl., Math Mono. **90**, (1991)]).

[St] Stein, E. M. (1970) *Singular Integrals and Differentiability Properties of Functions.* (Princeton Math Series, Vol. 30) (Princeton Univ. Press, Princeton, New Jersey).

[SW] Stein, E. M. and Weiss, G. (1972) *Introduction to Fourier Analysis on Euclidean Spaces.* (Princeton Math Series, Vol. 32) (Princeton Univ. Press, Princeton, New Jersey).

[Sr] Strichartz, Robert S. (1967) Multipliers on fractional Sobolev spaces, *J. Math. Mech.* **16**, 1031–1060.

[Sz] Szeptycki, P. (1968) On functions and measures whose Fourier transforms are functions, *Math. Ann.* **179**, 31–41.

[T] Talenti, Giorgio. (1976) Best constant in Sobolev inequality, *Ann. Mat. Pura Appl.* **(4) 110**, 353–372 .

[Ta] Tartar, Luc. (1998) Imbedding theorems of Sobolev spaces into Lorentz spaces, *Boll. Unione Mat. Ital. Sez. B. Artic. Ric. Mat.* **8**, 479–500.

[Tr1] Triebel, Hans. (1978) *Interpolation Theory, Function Spaces, Differential Operators.*(North-Holland Mathematical Library, Vol 18) (North-Holland, Amsterdam).

[Tr2] Triebel, Hans. (1978) *Spaces of Besov-Hardy-Sobolev Type.* (Teubner-Texte zur Mathematik) (Teubner Verlagsgesellschaft, Leipzig).

[Tr3] Triebel, Hans. (1983) *Theory of Function Spaces.*(Monographs in Mathematics, Vol 78) (Birkhäuser Verlag, Basel).

[Tr4] Triebel, Hans. (1992) *Theory of Function Spaces II.*(Monographs in Mathematics, Vol 84) (Birkhäuser Verlag, Basel).

[Td] Trudinger, N. S. (1967) On imbeddings into Orlicz spaces and some applications, *J. Math. Mech.* **17**, 473–483.

[W] Whitney, H. (1934) Analytic extensions of differentiable functions defined in closed sets, *Trans. Amer. Math. Soc.* **36**, 63–89.

[Y] Yosida, K. (1965) *Functional Analysis.* (Springer-Verlag, Berlin).

[Zm] Ziemer, William P. (1989) *Weakly Differentiable Functions. Sobolev Spaces and Functions of Bounded Variation.* (Springer-Verlag, New York).

[Z] Zygmund, A. (1956) On a theorem of Marcinkiewicz concerning interpolation of operators *J. Math. Pures Appl. (9).* **35**, 223–248.

INDEX

图书在版编目（CIP）数据

索伯列夫空间　：第2版 ：英文 / (加) 罗伯特·亚当斯 (Robert A. Adams), (加) 约翰·福尼尔 (John J. F. Fournier) 著. -- 影印本. -- 北京 : 世界图书出版有限公司北京分公司, 2025. 6. -- ISBN 978-7-5232-1860-0

Ⅰ. O177.3

中国国家版本馆CIP数据核字第2025HH6419号

中文书名	索伯列夫空间（第 2 版）
英文书名	Sobolev Spaces 2nd Edition
著　　者	[加] 罗伯特·亚当斯（Robert A. Adams） [加] 约翰·福尼尔（John J. F. Fournier）
责任编辑	陈　亮　董心意
出版发行	世界图书出版有限公司北京分公司
地　　址	北京市东城区朝内大街 137 号
邮　　编	100010
电　　话	010-64038355（发行）　64033507（总编室）
网　　址	http://www.wpcbj.com.cn
邮　　箱	wpcbjst@vip.163.com
销　　售	新华书店
印　　刷	北京建宏印刷有限公司
开　　本	710 mm × 1000 mm　1/16
印　　张	20
字　　数	305 千字
版　　次	2025 年 6 月第 1 版
印　　次	2025 年 6 月第 1 次印刷
版权登记	01-2022-3006
国际书号	ISBN 978-7-5232-1860-0
定　　价	99.00 元

版权所有　翻印必究

（如发现印装质量问题，请与所购图书销售部门联系调换）